BUILDING CONSTRUCTION IN WARM CLIMATES

VOLUME THREE

R. L. FULLERTON

Msc. CEng. MIStructE. FIOB. Sometime Professor of Building, Ahmadu Bello University, Nigeria. Formerly Associate Professor in Building Technology, Kumasi University of Science and Technology, Ghana

OXFORD UNIVERSITY PRESS 1977

Oxford University Press, Walton Street, Oxford OX2 6DP

OXFORD LONDON GLASGOW NEW YORK
TORONTO MELBOURNE WELLINGTON CAPE TOWN
IBADAN NAIROBI DAR ES SALAAM LUSAKA ADDIS ABABA
KUALA LUMPUR SINGAPORE JAKARTA HONG KONG TOKYO
DELHI BOMBAY CALCUTTA MADRAS KARACHI

© Oxford University Press 1977

British Library Catalogueing in Publication Data

Fullerton, Richard Lewis
    Building construction in warm climates.
    Vol. 3. – (Oxford tropical handbooks).
    1. Building – Tropics
    I. Title II. Series
    690'.0913   TH153

ISBN   019 859509 3

Set by Hope Services, Wantage
and printed in Great Britain
by Fletcher & Son Ltd., Norwich

Contents

Abbreviations

Reference should also be made to Vols. 1 & 2.

°C	degrees Celsius	I.H.V.E.	Institute of Heating and Ventilation Engineers
A	area; ampere		
B.O.E.	brick on edge	INT	interior
B.R.	boil resistant	k	conductivity
B.R.E.	Building Research Establishment	k.d.	knock down
		kg/m^3	kilogram(me)s per cubic metre
B.S.	British Standard		
m/s	metres per second	kW	kilowatt
MN/m^2	meganewtons per square metre	kN	kilonewton
		mb	millibar
C.P.	Code of Practice		
C.P.A.	Chipboard Promotion Association	mm	millimetre
		MN	meganewton
C_{pe}	external pressure coefficient	M.R.	moisture resistant
		m.s.	mild steel
C_{pi}	internal pressure coefficient		
c.w.	cold water	N/m^2	newtons per square metre
D.O.E.	Department of the Environment		
		O.D.	ordnance datum
d.p.c.	damp proof course	P.V.A.	polyvinyl acetate
d.p.m.	damp proof membrane	q	dynamic wind pressure
D.T.S.	dense tar surfacings	R	thermal resistance
FIDOR	Fibre Building Board Development Association	r.c.	reinforced concrete
		R.H.S.	rectangular hollow section
F.F.L.	finished floor level	R.H.	relative humidity
g/kg	gram(me)s per kilo-gram(me)	S.S.	stainless steel
		s.w.g.	standard wire gauge
g.m.s.	galvanised mild steel	T.D.	temperature drop
G.R.G.	glass-fibre reinforced gypsum	T.V.	television
		U	thermal transmittance
G.R.C.	glass-fibre reinforced cement	U.B.	universal beam
		U.P.V.C.	unplasticised polyvinyl

G.R.P.	glass-fibre reinforced polyester		chloride
		U.V.	ultra violet
H.D.	high density	V.D.	vapour drop
V.P.	vapour pressure	W.B.	water-borne
V.R.	vapour resistance	W.B.P.	weather, boil proof
W	watt	X.P.M.	expanded metal

Preface

This volume, the third in the series, was intended to cover only the Construction & Concrete Technology syllabus of the Higher Technician Diploma of the City & Guilds of London Institute. In view of current technological developments, however, the scope has been expanded to embrace prefabrication and industrial techniques omitted from earlier volumes, as these are now used increasingly in tropical lands. In this connection, emphasis has been placed on standardized housing and community halls of administration and recreation, which are now in some demand.

Construction techniques and materials necessary to both warm humid and hot dry climates have been included, with alternative methods given as far as possible. Attention has also been paid to the appropriate building regulations, including fire and security regulations.

This book, together with previous volumes, should prove a useful aid to undergraduates and students of architecture, building, surveying and structural engineering, and to those taking degrees and professional examinations in these disciplines. The series should also be a helpful reference to those whose formal studies are completed and who are now in practice.

In view of the number of drawings which have had to be included — there are over a hundred pages of diagrams and tables — it has not been possible to continue the practice of placing captions opposite the illustrations. However, as this volume is intended for the senior student and graduate this may not be found a disadvantage.

R.L.F.

Acknowledgements

In addition to those previously mentioned, the writer is indebted to many people, and particularly to the following:

The Vice-Chancellor and staff of Ahmadu Bello University, Nigeria, for their continuing help and support.

Messrs Taylor Woodrow (Nigeria) Ltd. for permission to reproduce the illustration on the front cover.

Dr. J. Kasinski, M.Sc., Ph.D., Head of Structures, Department of Building, Ahmadu Bello University, for his help and co-operation in full-scale experimental work on precast concrete frames.

Mr. W.C. Norfolk, FIOB, AMBIM, Head of Building & Civil Engineer-Southampton College of Technology, for use of the college's facilities.

Mr. J. Towner, MIOB, and the staff of the Southampton College of Technology for their support and help.

The National Building Commodity Centre Ltd., and the Building Centre, Southampton, for information and catalogues of which use has been freely made.

Extracts from the following British Standards are produced by permission of the British Standards Institution, 101 Pentonville Rd., London. N1 9ND, from whom copies of the complete standards may be purchased:

B.S. 747: Part 4, 1970	Fire tests on building materials and structures.
B.S. 747: Part 2, 1970	Roofing felts.
B.S. 882, 1201: Part 2, 1973 AMD 1975	Aggregates from natural sources for concrete including granolithic
B.S. 1722: Parts 1 to 11, AMD 1974	Fences.
B.S. 4737: AMD 1973	Intruder alarm systems in buildings.
C.P. 3: Chapter V: Part 2, 1972	Wind loads.
C.P. 3: Chapter 1: Part 1, 1964	Daylighting
C.P. 3: Chapter 1(C): 1950	Ventilation.
C.P. 3: Chapter III: AMD 1974	Sound insulation and noise reduction

Extracts from the following Building Research Establishment Notes, Digests and DCED standard drawings are produced by permission of the DOE; Head of Overseas Division, BRE; and the Controller, HMSO:
BRE 8, Built-up felt roofs. BRE 27, Rising damp in walls. BRE 41,

Estimating daylight in Buildings. BRE 54, Damp-proofing solid walls. BRE 67, Soils and foundations. BRE 85, Joints between concrete wall panels: open drain joints. BRE 108, Standard U-values. BRE 110, Condensation. BRE 128, Insulation against external noise. BRE 140, Double glazing and double windows. BRE 143, Sound insulation; basic principles. BRE 150, Concrete; materials. BRE 180, Condensation in roofs. BRE 187, Sound insulation of lightweight dwellings.

DCED/G/1; Standard clayware and concrete drain fittings, cast iron covers, gratings and frames.

DCED/G/7; Standard soakaways with covers and frames.

DOE Tech. Memorandum No. BE5; Parapet systems.

Overseas Building Notes.

1 | Climatic elements

1.1 Sunshine, wind, and water

Sunshine, wind, and water play important roles in climatic control, but in varying degrees and not necessarily in that order. Sunshine, including daylight, in warm climates is usually provided in excess of man's needs; wind and water, however, vary greatly according to location, from cooling breeze to violent storm. The presence of the building itself will modify the microclimate of the site in terms of wind flow patterns, shelter, and sunshine exposure.

The control and measurement of daylight has received close attention from building research establishments, particularly in England, and its application to tropical buildings is essential to the study of the architectural and building professions. Briefly, the amount of indoor daylight needed for a specific purpose can now be determined and is normally expressed as a ratio of indoor to outdoor illumination commly known as the Daylight Factor. This depends on sky conditions, size, shape, and position of windows, and reflectivity of indoor surfaces. Since the publication of Volume 2, improved daylight protractors for various types of sky and slopes of glazing have been produced; these now have two semicircular scales, the main one for the sky component (overcast sky), and an auxiliary one to correct for window lengths. A full set of protractors may be obtained from B.R.E., England.

SOLAR RADIATION. A further phenomenon of sunlight is solar radiation, both ultra-violet and heat-producing. Solar radiation raises the surface temperature of the construction above that of the surrounding air. Thus when the effects of solar radiation are considered in design calculations a compromise has to be made to take account of air temperature also. The result is known as the Sol-Air temperature. Section A6 of the I.H.V.E. guide (U.K.) gives full information on solar data. The ultra-violet component of sunlight is active in the breakdown of a number of materials, which at the same time may also absorb infra-red radiation, with a consequent rise in temperature; and these effects in their turn may cause colour changes, thermal movement, cracking, and changes in conductivity. Flat roofs, for instance,

1

particularly those with a high coefficient of expansion, often need solar reflective treatment. The relationship between construction techniques, method of air handling, and the percentage of glazing in buildings has also been studied.

Transmission of light alone through glass can be fairly easily controlled; but the material itself raises heat problems. Thermal transmittance through glass can be regulated, usually by use of single, double, or triple glazing. Protection can be had from an air-space width of only a few millimetres, though for sound-insulation a wider space is needed. Tinted glass or sealed units are effective against both light and heat; and solar control glass is also available to reduce the transmission of solar heat.

1.2 Comfort and health

Latitude plays its part in man's well-being. Hilly or mountainous regions, the neighbourhoods of lakes and rivers, coastal areas, low-lying jungles, and swamp all affect comfort. Warm humid equatorial zones frequently experience heavy seasonal rainfall combined with relatively little change in temperature and humidity throughout the year. High mountainous inland areas, on the other hand, have much more clearly defined seasons, with cold nights in winter and more variable rainfall.

Buildings need to modify the climate to achieve protection from heat and sometimes from cold. Hot dry climates may be affected by a wide range of diurnal temperature; hence heavy building fabrics are used to retard the flow of heat from outside to inside. They also even out the fluctuation in environmental temperature between day and night, and smooth out the high and low peaks by means of the time lag. Solar radiation heat built up during the day can also be cooled by the night wind.

The combined effects of moisture, sunshine, and high temperature can create demands on the human body not experienced in more temperate zones. A hot dry climate dries the skin, which then becomes less sensitive to extremes of heat and cold; whereas a moist atmosphere can have the opposite effect. This is noticeable in humid climates when hot dry winds penetrate the equatorial belt periodically; and the drying effect may be unpleasant.

TROPICAL GROWTHS. Insect life and fungal growth abound in equatorial zones, encouraged by tropical heat and humidity; and this may lead to attacks upon, or the decay of, materials. Fungal growth on any surface capable of retaining dirt, such as asbestos, bitumen, and paint finishes, can cause deterioration and impair appearance, even when it does not

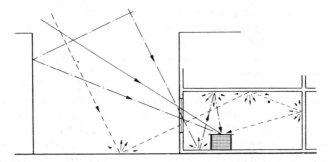

Fig. 1.1 Ways in which daylight reaches indoor position.

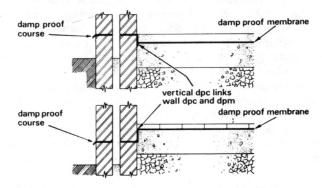

damp proof course

damp proof membrane

vertical dpc links wall dpc and dpm

damp proof course

damp proof membrane

Fig. 1.2 (a) (b)

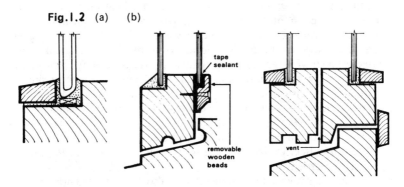

tape sealant

removable wooden beads

vent

Typical factory-sealed double glazing units

Typical glazed in situ double glazing

Typical double windows, coupled type

Fig. 1.3

actually destroy the materials. Algae, lichen, and mosses grow quickly, while fungi can appear as spots or patches of mildew on clothes. Precautions in the form of fly-screens or sprays have to be taken by man against the ravages of insects and micro-organisms, including the ubiquitous mosquito. The effect of termites is well known and has been discussed in Volume 1. Overcrowding and the lack of refuse-disposal facilities, sanitation, and a fresh water supply can lead to plague and disease, particularly when accompanied by damp and poor housing. Life-giving water itself in the form of rivers and lakes can also be responsible for the breeding of much living matter injurious to health. Good building cannot in itself eliminate these problems; but it can help.

NATURAL LIGHT AND TEMPERATURE. The critical factor in the control of light and heat is often the window, where protection from glare and sunshine is usually necessary. Shade devices are numerous – shutters, awnings, canopies, blinds, egg-crate exteriors, *brises-soleil*, perforated screens, metal grilles, and extended floor slabs. Internal light control can be achieved by blinds, curtains, pleated paper, louvres, etc. These are good for shade but less effective in reducing solar gain, which is better accomplished by using highly reflective surfaces or sometimes coloured glass. The latter provides a restful effect, though control is not possible without opening windows. The degree of exposure is also important, and this depends on whether the location is sheltered or not.

AIR MOVEMENT. In areas of high humidity some form of cross-ventilation is necessary; hence buildings are made as open as possible. The introduction of the glass louvre was perhaps the greatest single advance in tropical building in recent years. In inland areas and higher latitudes, however, ventilation needs to be restricted to counter the effect of high wind; and in hot dry climates it is curtailed by the window size necessary to exclude daytime heat and dust. Frequently simple ventilation is insufficient to cope with human needs, for no natural method can make a building cooler than the lowest shade temperature. In a fairly dry climate evaporative coolers may be used; but elsewhere some form of air conditioning is required. The cooling load can be reduced by effective shading devices, by proper insulation, and (rarely) by vapour barriers.

In growing towns air movement can also be restricted where windows are kept closed because of traffic noise. In such cases some form of mechanical cooling is installed – usually fans or unit air conditioners, though the latter can be very noisy if not properly insulated.

SOUND. The effect of sound on human comfort is becoming more of a problem as countries develop. This is due not only to traffic, but also to artificial amplification, mechanical processes, and the passage of aircraft, as well as to natural causes, Fig. 1.5. Insulation against airborne sound and impact-noise is becoming increasingly necessary to reduce noise to acceptable standards. Insulation depends on the type of constructional envelope used, the acoustic demands of the structure, the need for open windows and access, the type of insulation available, the air-cooling devices used, and the extent of vibration- and traffic-noise. Structural vibration in industrial building also needs special consideration.

Hospitals and health centres usually have different requirements, and the combined standards of peace and cleanliness are high. Lecture-, concert-, and conference-halls usually need acoustic treatment to be effective. Direct and multidirectional reverberant sound reflected from walls and ceilings can render speech difficult to understand. Multipurpose halls sometimes have problems with the types of materials used, as these may absorb and reflect sound in varying amounts; and reduction of reverberation-time is usually the first consideration. Fortunately, this can be calculated. Temporary measures involve the use of absorbent materials such as carpets, curtains, and upholstery — all of which tend to increase body temperature and make conditions unpleasant. Reflective sound barriers, baffle boards, loudspeaker systems, double glazing, and other similar means of control could be helpful.

1.3 Effect of environment
The characteristics of different types of material and their effect on buildings may directly affect the mode of construction used. Perhaps the first consideration is the exposure of materials, all of which suffer to some extent, depending on location.

POLLUTION. In growing industrial areas atmospheric pollution is becoming an increasing problem. Some buildings are now being ventilated with a mixture of air, water vapour, grit, dust, smoke and gases, rather than fresh air. In noisy urban situations where solar heat gains are excessive, occupants may be left with a choice of overheated stale air, or noise and polluted air, unless mechanical ventilation can be installed. Pollution can be natural or artificial. Corrosion by salt water or sea spray in coastal areas affects many materials, and rusting of metals is a real problem. Concrete, too, can be eroded, especially when improperly reinforced; while ozone, an allotropic form of oxygen once considered as health-giving, can in fact be harmful when occurring in

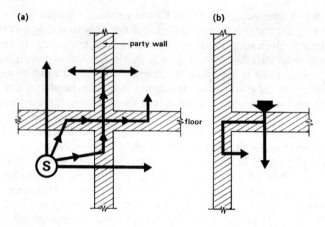

Fig. 1.4 Vertical sections illustrating paths of sound transmission
(a) airborne
(b) impact

high concentrations, as it does in some industrial areas. Sand contaminated by sea water is well known for its deleterious effect when used in mortar.

DUST AND GRIT. Wind-borne grit in arid zones where high winds may be expected usually has an abrasive action on exposed surfaces of such materials as stone, glass, and plastics. Dust can be harmful, irritating, or just a nuisance. Precision-instrument manufacturers, clinics, and food-processing establishments demand rigorous dust control, as do hospitals, particularly those dealing with eye and lung complaints. Mining areas, quarries, foundries, etc. are big dust creators. Artificial corrosion is also an enemy of industrial floors, where impact loads, abrasion, oils, urine, blood, and similar liquids ruin the surface. Wet industrial processes may also attack concrete; and floor slopes need to be laid to a suitable fall, in order to drain off liquids while still preventing slipping.

EROSION. The durability and decay of cementitious materials is as much influenced by the internal structure and nature of the material as by the external elements themselves. Decay can occur through the crystallization of salts, causing fragmentation; calcium sulphate can be washed out of limestone; or the material may fail because of the presence of ground water, or through wind erosion, organic growth, or chemical attack. Much depends on the quality of construction and the mortar used.

SOILS. Soils and foundations play an important role in building, and affect both the type of structure to be erected and its cost. Problems likely to be encountered are underground water, made-up ground, quicksand, coastal and tidal hazards, subsidence, earth movement, landslips, and the nature of the soil itself. Foundations will be discussed further in Chapter 4.

VIBRATION. Seismic prospecting for earthquake and volcanic activity is now quite an advanced science, and paths of probability have been charted for most regions. Special care needs to be taken when building in these areas, though unfortunately this is not always done. Expansive clays and heaving or cotton soils, rifts, and faults also need special treatment which will be discussed in later chapters. In growing industrial areas machinery-vibration can be troublesome, especially where laboratories, delicate trade processes, high buildings, or office accommodation are concerned. Forge hammers, heavy traffic, blasting, aircraft, and pile driving have to be considered, in addition to earth tremors.

NOISE AND SOUND INSULATION. Sound insulation qualities are usually measured by the grading of the materials. Grade curves for impact-noise and airborne-sound insulation are obtainable from a number of publications of the B.R.E., England. Levels of decibels may be measured from tests with impact machines.

Lightweight concrete is one of the newer forms of construction used to combat noise and sound penetration. Normally the most influential factor is the weight of the structure but, with new developments in industrialized building, sound insulation factors other than weight have to be explored. Double-leaf walls provide insulation in excess of their weight contribution, but, to be effective, air gaps should be closed to prevent sound-movement. This can sometimes prove difficult with dry construction now popular in concrete and timber-framed buildings.

Roofs also need special attention. Pitched roofs are seldom airtight, but flat roofs can generally cope with normal noise, depending on their construction. It should be possible to provide for the effect of noise before the building is erected by measuring exposure, but allowance has to be made for the increase in sound resulting from pressure at the face of the new building after it is put up. Where new roads or highways are to be built near the structure provision must be made for the gradient

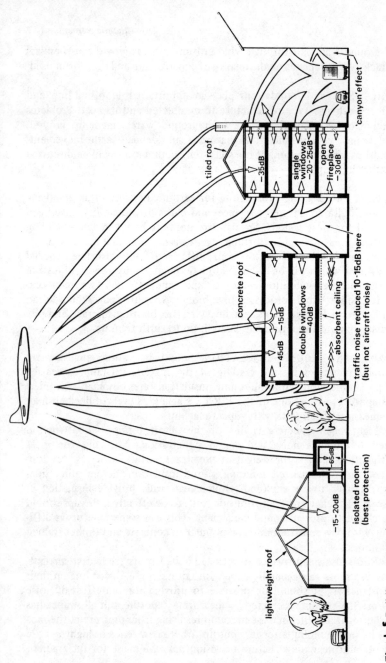

Fig. 1.5 Buildings exposed to outdoor noise

lightweight roof

isolated room
(best protection)

-15-20dB

-65dB

concrete roof

-45dB
-15dB

double windows
-40dB

absorbent ceiling

traffic noise reduced 10-15dB here
(but not aircraft noise)

tiled roof

-35dB

single
windows
-20-25dB

open
fireplace
-30dB

'canyon'effect

of the road, ground absorption, and traffic frequency; and soft grass shoulders or noise screens should be provided if necessary.

1.4 Humidity and temperature

The quantity of water vapour in the air is referred to as *humidity*. Saturated air is that holding the maximum amount of water vapour. When air is saturated any drop in temperature causes it to lose some of its moisture as condensation or dew. The point at which this happens is known as the *dewpoint*, and any rise in temperature permits air to hold more water vapour; until it again becomes saturated. *Relative humidity* is the ratio of the amount of water vapour in the air to the amount it could hold at saturation point at any given temperature. Cooling of the air causes the relative humidity to rise.

CONDENSATION. This is the moisture deposited on surfaces cooler than the ambient air. In temperate zones it invariably occurs in kitchens and bathrooms when the warm moist air strikes the colder outside walls of the room. As walls warm up the condensation evaporates; but cold metal service pipes in warm surroundings retain condensation, and so tend to corrode. In equatorial coastal climates with little seasonal or diurnal variation in temperature there is not a great deal of heavy condensation, though humidity can be high. Hotels and public buildings are often equipped with air conditioning; if left on for long periods it reverses the process normal to temperate zones, causing heavy condensation to form on the outside of windows and thus spoiling visibility and speeding deterioration of the frames.

Drying out is mainly a matter of the time taken for water vapour from cellular pores to reach the surface of the material: the rate of evaporation reduces as the temperature falls. Buildings when drying out are thus cooler than the ambient air, and this is the principle on which the traditional earthenware water cooler works. No rule of thumb on drying out is possible; but premature decoration can be disastrous, especially in damp regions. Dehumidifiers are sometimes used for this purpose if circumstances warrant it. Two types are in general use, dessicant and refrigerant. These can extract up to 6 litres of water per hour depending on the R.H., though effectiveness falls off at lower temperatures. Chemical dessicant dehumidifiers rely on hygroscopic properties of various chemicals: either the chemical dissolves in the water it attracts, to form a solution later discarded, or hygroscopic paper impregnated with chemicals holds the water like blotting paper. Refrigerant dehumidifiers are described in Volume 2.

RELATIVE HUMIDITY (R.H.), expressed as a percentage, is dependent on dry bulb temperature and the concentration of moisture in the gaseous mixture of water vapour and air. There are several ways of determining this ratio, one of which is with the whirling hygrometer. Two thermometers are mounted side by side; one is a dry bulb and the other takes moisture from a small reservoir by means of a wick. As evaporation from the wick takes place the wet bulb is cooled and records a lower temperature. The instrument is whirled for some seconds and the R.H. read off from the scale provided. Humidity-meters of extremely sensitized metal are also available, giving direct readings in percentages from a dial.

Though meteorological charts and records are available on a global scale they give only a rough average figure derived from official weather stations, and cannot be relied upon implicitly. In the same way air temperatures and relative humidity are not in themselves infallible guides to human comfort. The wide range of individual response combined with the variation that can occur in environmental conditions makes it impossible to suit everyone all the time. Attempts have been made to produce a comfort-index, and experimental work has been carried out in this field with some success.

1.5 Resistance and absorption of heat

The rate of conduction, i.e. conductivity, depends mainly on the density of the material; but heat transfer by radiation depends on the temperature and nature of the surface. Thus, while some of the sun's radiation is absorbed by the material, some of it is reflected back. As light coloured surfaces have a high reflectivity, the use of aluminium is widespread. Porous materials on the other hand absorb heat, and so reduce heat transmission through the structure. Convection in warm countries has its main use in air circulation, so when cavities are used to prevent heat transfer they should not be ventilated. Cavities can of course play an important part in thermal control.

The interdependence of relative humidity, dry bulb temperature and concentration of moisture in water vapour and air mixture has an effect on condensation: this can be seen in chart form in Fig. 1.6. From this may be determined the extent of surface condensation on walls, the interstitial condensation at any point in the wall, and also the vapour-pressure drop. Two examples of the calculation of temperature differences are given. The first is taken, with permission, from B.R.E. Digest No. 110, suitable for Europe; and the second shows calculations for warm climates as a comparison.

THERMAL PROPERTIES OF MATERIALS. The expressions and abbreviations used in Examples 1 and 2 may be explained thus:

Thermal transmittance (U) is a measure of heat flow through the construction of a building expressed in watts per square metre of surface area for each degree C rise or drop in temperature, i.e. W/m^2 $^{\circ}C$. Standard U-values are calculated from resistances of component parts necessary for comparing different constructions in varying types of climate.

Thermal conductivity (k) is a measure of thermal transmission per metre thickness of the material, i.e. Wm/m^2 $^{\circ}C$, or more usually W/m $^{\circ}C$.

Thermal resistivity ($1/k$) is the reciprocal of conductivity, i.e. m $^{\circ}C/W$.

Thermal resistance (R) is the product of resistivity and thickness of the material, i.e. m^2 $^{\circ}C/W$.

Abbreviations

Total R . . . sum of thermal resistance of components
    i.e. $R_1 + R_2 + R_3 +$ etc. or thermal resistivity $\times$ width.
T.D.. . . . . . temperature drop between surfaces.
V.R.. . . . . . vapour resistance i.e. vapour resistivity $\times$ width.
V.P. . . . . . . vapour pressure in millibars.
* . . . . . . thermal resistance.
** . . . . . . surface resistance.
V.D.. . . . . . vapour pressure drop between surfaces (mb).

CELLULAR PLASTICS. Foamed and expanded plastics for thermal insulation may be of expanded polystyrene with a thermal conductivity of about 0·036 W/m $^{\circ}C$, or of similar materials: see Properties of Materials, p. 13. Most are used in tropical conditions, though polystyrenes do have a high coefficient of linear expansion. Cellular plastics are light, attractive, easily handled, and resistant to fungal attack and termites. To prevent undue build-up of heat a white finish is desirable; though even this protection may not suffice in arid zones. To avoid loss of thermal insulation vapour barriers may be used where condensation is likely to occur, though this is not normally a problem in tropical countries.

The insulation of roofs is important. Reflective surfaces are more effective when used horizontally rather than vertically, but should always be protected from the sun's rays, particularly when facing the sun directly. Hot bituminous coverings over expanded polystyrenes are also to be avoided. The cooling effect on cloudless nights which usually follows a drop in temperature can be evened out by the stored heat in insulation; such heat, however, affects the rate of degradation, so that the useful life of these plastics tends to be shorter in the tropics than in temperate zones.

Some rigid cellular plastics are used in sandwich panels, with sheets of plywood, plasterboard, etc. bonded to core with adhesives. They are light and have good thermal qualities, but are used mainly as partitions. Glass-fibre reinforced plastic cladding panels are now in general use in developed countries, and are being imported to the larger tropical towns. They consist of glass-fibre reinforcement impregnated with resin and sulphur catalysed to harden. They were first used as corrugated roofing sheets, being light and mouldable in various colours. They can also be used as a sandwich coating, with a core of cellular plastic fixed with steel to the building framework.

AERATED CONCRETE. This is now used as a loadbearing material, and has thermal qualities suitable for warm climates. Sound absorption is higher than for dense concrete and good rain-resistance can be obtained when suitably rendered. This material is more fully described in Chapter 3.

WINDOW HEAT. Window heat and solar gain create problems in warm climates. After the difficulties of daylighting, visual effects, loss of privacy, noise, and condensation have been overcome there still remains the question of overheating, and also of chilling, when the temperature drops. The necessary data on the transmission, absorption, and reflection of solar radiation by sun-control mechanisms are available, and internal temperatures and their diurnal variation can be computed and compared with any collected survey data which may exist. The publications lists of B.R.E., England are a useful guide.

Thermal transmission (U) through glazing varies according to the system and the degree of exposure, i.e. whether sheltered, normal, or severe. Double glazing with a 6 mm air space, for example, gives a U-value of 3.4 $W/m^2 \, ^{\circ}C$ (without frames); but with heat-reflecting glass and a special metallic surface-coating this may be improved. Heat-resistance through windows must be considered in relation to the overall heat-resistance of the building; and when calculating cooling-loads in air-conditioned buildings it is necessary to know the solar gain through windows. Sound-insulation cavities, unlike those for heat-insulation, require a gap of not less than 100 mm.

HEAT AND GROUND FLOORS. Where solid floors are in contact with the earth in areas subject to seasonal and diurnal swings in temperature an insulating layer may be advisable, together with provision of a vapour

(*continued on page 18*)

Fig. 1.6.

EXAMPLE 1.

Consider point A. This represents an air condition of 0 °C and 90% R.H. From the right-hand scales it can be found that such air contains 3·49 grammes of moisture per kg and that the vapour pressure is 5·4 millibars.

Point B indicates air with the same moisture content and therefore the same vapour pressure, but as it is now at 20 °C its relative humidity has changed to approximately 23%. This shows what happens to the air after it is warmed, if no other change occurs.

Point C indicates air still at 20 °C, but with moisture content raised to about 10·3 g/kg. Vapour pressure therefore also rises, to about 16·5 mb. The increase in moisture without change in temperature means that relative humidity has risen, and the curved lines show this now to be about 70%. This is what might occur when the air has picked up moisture from the atmosphere.

Reading horizontally to the left from C, point D indicates when saturation would occur, i.e. when the air is cooled to a temperature of about 14·6 °C.

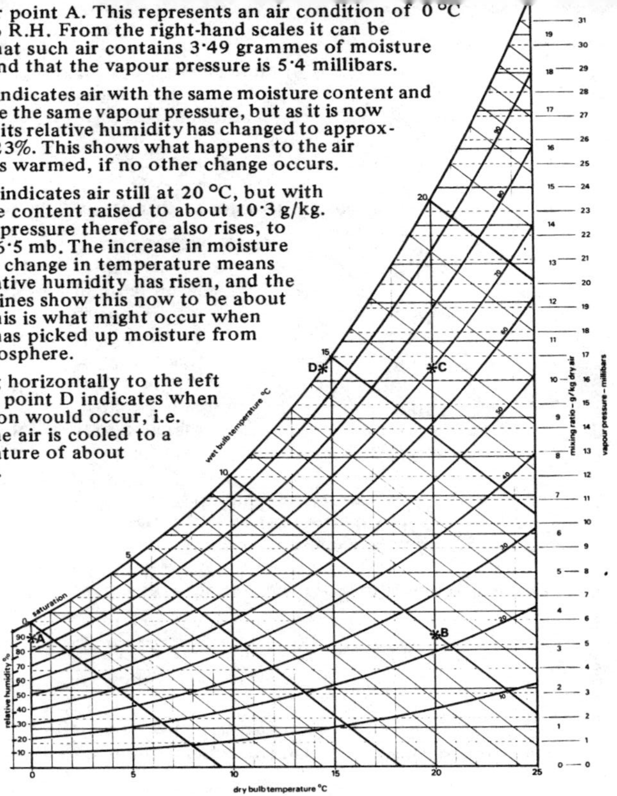

PROPERTIES OF MATERIALS

Typical Values	Thermal Resistance (R)
Surfaces	
Inside wall surface	0·123
Outside wall surface	0·053
Inside roof (or ceiling) surface	0·106
Outside roof surface	0·044
Internal air space	0·176

Materials	Thermal Resistivity $\left(\frac{1}{k}\right)$	Vapour Resistivity
Brickwork	1·04	200 to 20
Concrete	0·69	96 to 29
Rendering	0·83	96
Plaster	2·08	58
Timber	6·93	73 to 46
Plywood	6·93	5800 to 1450
Fibre building board	18·7 to 15·2	58 to 14
Hardboard	6·93	730 to 460
Plasterboard	6·24	58 to 44
Compressed strawboard	11·8 to 9·7	73 to 46
Wood wool slab	8·66	41 to 14
Expanded polystyrene	27·72	580 to 145
Foamed urea formaldehyde	27·72	32 to 19
Foamed polyurethane (closed & open cell)	27·72	960 to 29
Expanded ebonite	27·72	58 000 to 11 600

Table 1.1.

SURFACE CONDENSATION

Object: to determine temperature drop between external and internal faces to compare dewpoint. (Air temp diff. = 6·5 °C.

Temperature drop

$$= \frac{0.123 \times 6.5}{TR}$$

where 0·123 = thermal resistance of inside wall surface

Total $R = R_1 + R_2 + R_3 + R_4 + R_5$

where R_1 = external surface resistance = 0·053
R_5 internal surface resistance 0·123
R_2, R_3 etc. = thermal resistance of components

R_1 = external face resistance	=	0·053
R_2 = R of concrete		0·052
= 0·69 × 0·076		
R_3 = R of insulation		0·693
= 27·2 × 0·025		
R_4 = R of concrete		0·052
as above		
R_5 = internal surface resistance	=	0·123
Total resistance (TR)	=	0·973

Temperature drop of internal surface

$$= \frac{0.123 \times 6.5}{0.973}$$ 0·8 °C

Inner surface temperature 4·7 °c
= 5·5 −0·8

This is too close to dewpoint (4 °C)

EXAMPLE 1 Table 1.2

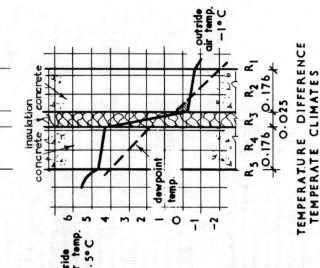

Points

inside air temp. 5·5°C

insulation
concrete   concrete

outside air temp. −1°C

dewpoint temp.

R₅ R₄ R₃ R₂ R₁
0·176 0·176 0·025 0·176

TEMPERATURE DIFFERENCE
TEMPERATE CLIMATES

Fig. 1.7

INTERSTITIAL CONDENSATION

Object: to plot temperature drop from point to point

$$TD = \frac{\text{Air temp. diff.} \times R}{TR}$$

$$\frac{6 \cdot 5 \times R}{0 \cdot 973} = 6 \cdot 68 \, (R)$$

Internal air temperature $= 5 \cdot 5\,°C$
Inside air to Point 1
$= 6 \cdot 68 \times 0 \cdot 123 = 0 \cdot 82$
$\overline{4 \cdot 678}$

Temperature drop to
Point 1. to 2. $= 6 \cdot 68 \times 0 \cdot 052 = 0 \cdot 347$
$\overline{4 \cdot 331}$

Temperature drop to
Point 2. to 3. $= 6 \cdot 68 \times 0 \cdot 693 = 4 \cdot 629$

Temperature drop to
Point 3. to 4. $= 6 \cdot 68 \times 0 \cdot 052 - 0 \cdot 298$
$\overline{0 \cdot 347}$

Temperature drop to
Point 4. to outside air
$6 \cdot 68 \times 0 \cdot 053 - 0 \cdot 646$
$- 0 \cdot 354$

Temperature drop to $- 1 \cdot 00\,°C$

Temperatures plotted on Fig 1.7 (heavy line)

Table 1.3

VAPOUR PRESSURE

Object: to compare temperature drop with dewpoint

$$VD = \frac{\text{Total VD} \times VR}{\text{Total VR}}$$

Total VD $= 8 - 5$ mb.

Total VR
Point 1 to 2 $96 \times 0 \cdot 076$ $= 7 \cdot 29$
Point 2 to 3 $145 \times 0 \cdot 025$ $= 3 \cdot 62$
Point 3 to 4 $96 \times 0 \cdot 076$ $= 7 \cdot 29$
Total VR $= \overline{18 \cdot 20}$

$$\frac{\text{Total VD}}{\text{Total VR}} = \frac{3 \text{ mb}}{18 \cdot 20} = 0 \cdot 164$$

		Correspdg. dewpoint.
Internal VP	$8 \cdot 00$ mb	$\ldots 4\,°C$
Pt. 1 to 2	$0 \cdot 164 \times 7 \cdot 29 = 1 \cdot 19$	
	VD to $\ldots$ $6 \cdot 81$	$\ldots 2\,°C$
Pt. 2 to 3	$0 \cdot 164 \times 3 \cdot 62 = 0 \cdot 60$	
	VD to $\ldots$ $6 \cdot 21$	$\ldots 0 \cdot 5\,°C$
Pt. 3 to 4	$0 \cdot 164 \times 7 \cdot 29 = 1 \cdot 19$	
	VD to $\ldots$ $5 \cdot 02$	$\ldots -2\,°C$

EXAMPLE 1

Table 1.4

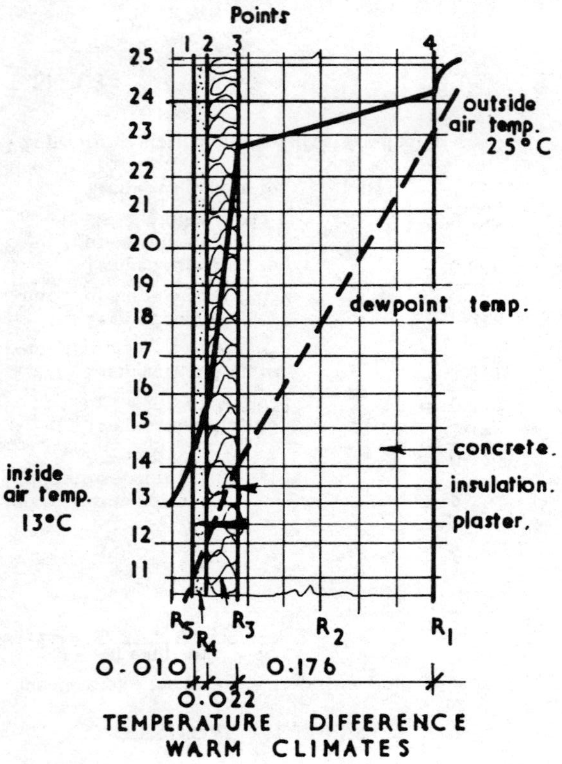

SURFACE CONDENSATION

| Inside temp. . . . 13 °C | Diff . . . . 12 °C |
| Outside temp. . . . 25 °C | R.H. . . . 90% |

Note: As outside temperature is the greater, external R.(0·053) will be taken for temperature drop calc.

$$\text{Temp. drop} = \frac{0 \cdot 053 \times 12}{\text{Total TR}}$$

R_1	External face resistance		0·053
R_2	R of concrete		0·121
	0·69 × 0·176	=	
R_3	R of insulation		0·594
	27·2 × 0·022	=	
R_4	R of plasterboard		0·062
	6·24 × 0·010	=	
R_5	Internal face resistance		0·123
	Total resistance . . .		0·953

Temperature drop of external surface

$$\frac{0 \cdot 053 \times 12}{0 \cdot 953}$$

0·667

Outer surface temperature

$$25 \cdot 00 - 0 \cdot 667$$

24·33

As dewpoint is almost similar (23·35 °C) condensation is likely to occur on outside.

EXAMPLE 2

Table 1.5

Fig. 1.8

INTERSTITIAL CONDENSATION

$$TD = \frac{Total\ TD \times R}{Total\ TR} = \frac{12 \times R}{9.53} = \qquad 12\cdot6R$$

External air temperature
Outside air to Point 4   = 25·00 °C
  = 12·6 × 0·053   = 0·7
Temperature drop to
Point 4 to 3 = 12·6 × 0·121   24·3
  = 1·5
Temperature drop to   22·8
Point 3 to 2 = 12·6 × 0·594   7·5
  = 15·3
Temperature drop to   0·8
Point 2 to 1 = 12·6 × 0·062   14·5
  = 1·5
Temperature drop to
Pt. 1 to inside = 12·6 × 0·123   13·0 °C
Temperature drop to

Temperature difference plotted on Fig. 1.8.
(heavy line)

Table 1.6

VAPOUR PRESSURE

Outside temp. 25 °C   R.H. 90%
External V.P. (from chart)   = 28·80 mb.
             Total VD   13·5
     = 15·3

Total VR.
Point 4 to 3 = 96 × 0·176   = 16·89
Point 3 to 2 = 145 × 0·022 = 3·19
Point 2 to 1 = 44 × 0·010   = 0·44
                20·52

$$\frac{Total\ VD}{Total\ VR} = \frac{15\cdot3}{20\cdot52} = 0\cdot745$$

Correspdg.
dewpoint.
28·80 mb. 23 °C
External VP   ∴   = 12·59
P. 4 to 3 = 0·745 × 16·89 =   16·21 … 14 °C
  VD to   2·38
P.3 to 2 = 0·745 × 3·19 =   13·83 … 12 °C
  VD to   0·33
P.2 to 1 = 0·745 × 0·44 =   13·50 … 11 °C
  VD to

As dewpoint temperature is low, no vapour barrier is needed.

EXAMPLE 2

Table 1.7

barrier. Expanded polystyrene, laid as boards on a concrete base with a damp proof membrane between them, may be used for this purpose. Sound-insulation will also be improved by this method. Expanded polystyrene may also be stuck to inner faces of walls and casings where required.

1.6 Exclusion of water

The effect of soils on water and *vice versa* needs consideration except in the dryest areas. The shrinkage and swelling of soils, the effect of moisture on vegetation, and the water-retaining properties of soils all affect foundation design. Combinations of soils and water are responsible for the settlement of buildings, cracking, shrinkage, decay, damp, disease, chemical attack, subsidence, and landslips.

SURFACE WATER AND DRAINAGE. The two main sources of moisture affecting building are precipitation in the forms of rain or condensation, and capillary attraction in the form of rising damp. The frequency of rainfall is well recorded by weather stations everywhere, and, despite local variations, the designer usually knows what to allow for. In determining open channel and gutter sizes the annual rainfall is not so important as the actual period of the storm and its intensity. Run-offs can also be hampered by insufficient fall, blockage, damage through traffic, or a sewerage system which is simply too small. Soakaways should be large enough to accommodate sudden storms, and should also have a separate drainage system.

Sewerage systems accept only foul sewage in wet regions: elsewhere partially separate systems accept some surface water, and study should be made of roads, footpaths, yards and hard standings in the vicinity. The construction of soakaways may be governed by local authorities, who sometimes have a standard layout. Rainfall statistics in the tropics rarely give an adequate picture of the combined effects of rain, wind and freak storms. Achieving a perfect match of weather and building is not often possible, owing to variations both in exposure and in the performance of the structure itself. These latter may be due to detailing, materials, the size of components, workmanship, and other such factors. And again, rain does not wet walls in the absence of wind. Storms, however, frequently drive rain horizontally, and even upwards under balconies and ventilators.

RISING DAMP. Rising damp can now be overcome by means of the many types of barriers available. Surface membranes are placed under floors, usually as a sandwich between concrete and screed, and a well-constructed building now has adequate water-stops around openings and cavities, in addition to standard d.p.c.s, Figs. 1.2, 1.9.

ROOF DRAINAGE. The most likely source of trouble from water is the roof. Built-up felts, flat roof overflow, rainwater wall streaking, expansion, imperfect flashing, felt blistering, inaccurate flow-load calculations, incorrect sizes of large valley gutters, outlets, and r.w.p.s are common causes of failure. Plastic pipes and gutters also need flexible joints to accommodate thermal movement. Areas subject to sudden storms should have adequate disposal capacity. Ground aprons which take roof discharge direct should have suitable collection channels with sufficient fall, and should be free of blockage and provide easy access to soak-aways and drains.

FLOODING. In coastal areas especially flooding is a common occurrence. The extent of damage will depend on the nature of the foundations, the degree of buffeting by the elements, the degree of expansion of soils, erosion, settlement, and cracking. Services, including electricity, may be impaired, plasterboard ruined, timber and plywood split and warped. After draining or pumping, it may be necessary to punch holes in the lowest areas and remove water trapped in underfloor ducts, pits, or cavities. Ventilators and similar openings may have to be cleared of mud. Sea-water flooding can cause erosion of metals, concrete, and lime. Drains should be inspected before use; and one unpleasant effect could be due to backwash, particularly where cesspits and septic tanks are in use. Having cleaned up as well as possible, doors, windows, and vents should be opened and everything should be done to create maximum air flow through the building.

DROUGHT. In hot dry climates where living conditions have been adapted to the environment low annual rainfall is not considered as drought. In such circumstances a building is made to combat excessive sunshine, heat, wind, and dust storms. Sanitary services are either designed to be able to cope using the water-supply available, or other means of disposal are used.

Drought in areas of reasonable rainfall and humidity can have serious effects. Shrinkage and settlement, with resulting cracking and bleaching of exposed finishes can occur. The building and its environment will deteriorate where maintenance depends on a supply of water, not to mention the unpleasantness and distress caused by the depletion of the supply. What is sometimes described as a drought is often a decrease in supply resulting from increased demand, particularly in growing towns.

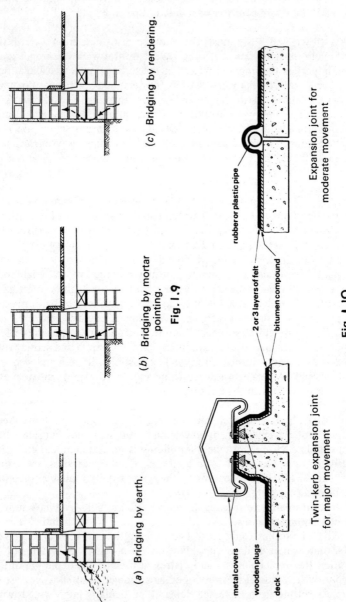

(a) Bridging by earth.

(b) Bridging by mortar pointing.

(c) Bridging by rendering.

Fig. I.9

Twin-kerb expansion joint for major movement

metal covers
wooden plugs
deck

2 or 3 layers of felt
bitumen compound

rubber or plastic pipe

Expansion joint for moderate movement

Fig. I.10

DECAY. Decay in timber is mainly due to the presence of moisture. Wood-destroying fungi attack moist timber and, unless it is naturally resistant, preservation treatment may be necessary. The main sources of damp and moisture content need to be known before the wood is treated. Much of the timber used in humid climates is unseasoned, and decays after only a short time of use. Corrosion of metals, especially of steel in coastal areas, is quite common. Lack of preparation and chemical reactions tend to shorten the useful life of metals and replacement is expensive, particularly of imported commodities. The cost of adequate protection is often recovered many times over in terms of repair bills, maintenance expenses, and inconvenience. Much deterioration is due to lack of skilled supervision.

JOINTS. As joints in building materials and components are often subject to structural and thermal stress as well as moisture, it is necessary that they should be designed for the job they have to do. Glues, sealants, gaskets and open channel joints need to be designed to cope with the local environment. More will be said about this in Chapter 5.

1.7 Wind and air pressure
Wind blowing square against a building is slowed up, with a consequent build-up of pressure against the face; at the same time it is deflected and accelerated around the end-walls and over the roof. As this can cause damage to glazing and cladding, local authorities are insisting increasingly on adequate calculations deemed to satisfy the requirements of the region.

Roofs with a pitch of less than 30 degrees may be subjected to severe suction, but above this positive pressure can be built up, though even here suction can develop at ridge level. Suction can be in excess of the dead weight of the roof, which would then require firm anchorage to an adequate foundation. Distribution of wind over a roof is not uniform; wind may blow along a ridge, or vortices may be produced along the edge of eaves and verge when the wind blows against a corner. High suction can thus occur at all edges and corners of square buildings.

WIND LOADS. Details of calculating wind loads are given in a number of documents, the best know of which is the British Code of Practice 3: Chapter V: Part 2. This states the various factors which should be taken into account when designing structures and components. The following is a simplified version of how loads may be determined. The first task is to calculate the designed wind speed, which is found from the following formula:

$$V_s = V \times S_1 \times S_2 \times S_3$$

where $V_s =$ designed wind speed

V ...   basic wind speed based on 3-second gusts estimated to be exceeded not more than once in every 50 years.

S_1 ...   topography factor taken as 1·0 for open country
                                  1·1 exposed hills and valleys
                                  0·9 steep sided enclosed valleys.

S_2 ...   roughness factor depending on height and size of building combined with nature of ground and wind gusts. A normal office building in a small town or suburb would have a factor of about 0·74.

S_3 ...   statistical factor based on probability of violent storm over the years and permanency of the structure. Normally taken as 1·0.

Given velocity (V_s) the dynamic pressure of the wind (q) may be found from the conversion chart, Fig. 1.11. As an example, a wind force (V_s) of 40 mph gives a wind pressure (q) of 200 N/m^2 or 4·4 lbf/ft^2. This enables structural stresses on the area and members to be calculated directly.

The above example explains the general principles only. The value of q may have to be modified to take account of the size and shape of the building and the angle of the wind. This is done by using pressure coefficients to allow for the fact that the total force on a wall or roof depends on the difference of pressure between the inner and outer faces. Thus force F is derived from:

$$F = (C_{pe} - C_{pi})qA$$

where A Area of surface

C_{pe} External pressure coefficient

C_{pi} Internal pressure coefficient

Different values of C_{pe} and C_{pi} are used for walls and roofs of rectangular buildings, with allowances made in the form of local coefficients to deal with edges and corners. Open doors, windows and ventilators on the windward side will increase the air pressure inside the building; this will increase the loading on those points of the roof and wall which are subjected to external suction. The C_{pi} coefficient was introduced to allow for these variations.

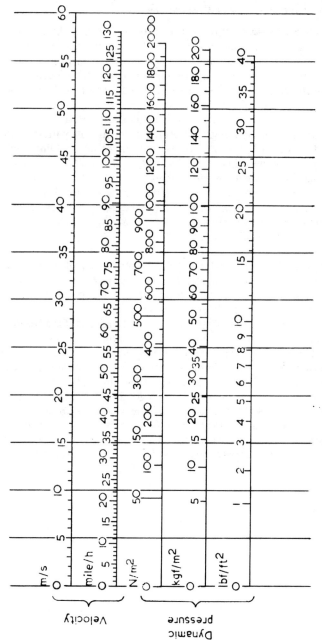

Fig. 1.11. Conversion chart for wind speed and dynamic pressure

WIND ASSESSMENT AROUND TALL BUILDINGS. Tall buildings deflect wind downwards, creating unpleasant high-speed currents in pedestrian areas. Movement of the atmosphere high above the earth is controlled by large weather patterns, but frictional drag on the surface slows it down. Such wind speeds, measured on the Beaufort Scale, are sometimes available from local authorities.

When wind flows over rows of low buildings the area is sheltered; when it meets taller buildings, however, a different flow pattern occurs. Below, it descends to form a vortex in front of tall buildings and to sweep around windward corners. If a building is on stilts or has an open ground plan the descending wind passes beneath. If the structure is not too high it can be refreshing provided it is free from dust. Unpleasantly high speeds in the environment of the building must be avoided, but where buildings have already been erected it may be possible sometimes to alleviate the worst effects.

WIND AND SANITARY VENTILATION. In the design of main soil and waste pipe ventilation systems, problems can be encountered as a result of the wind's creating suction across the tops of vent stacks, which then tend to suck water from traps and break the seal. As wind suction is usually heaviest near the eaves and parapets it is advisable to avoid placing soil and vent pipes in these positions. Problems can also arise with mechanical ventilation systems. As the outside pressure on the windward side of a building is greater than the inside pressure this can have an effect on air extraction from internal bathrooms and w.c.s, particularly as there is only a single duct from the w.c. to the outside air. With sufficient fan power and resistance to air flow the extractor system can be made immune to external air changes. It is wise, however, to provide duplication of equipment or an alternative power source, together with regular servicing to prevent breakdowns.

2 | Preliminaries and setting out

2.1 Introduction

The scope of this chapter will be confined to domestic housing and medium-scale building. The latter will consist of a multipurpose hall for meetings, functions and entertainment. As it is necessary for the reader to become acquainted with sites larger than individual plots, the setting out will cover a portion of a bigger housing development with road layout and a community centre. It could also be considered as a single phase of an extensive estate to be developed in stages. The construction of this part of the overall plan will be undertaken in accordance with the sketch plan, Fig. 2.1. This shows the layout of semi-detached and terraced houses and a central hall, shops, etc. to serve the area. The types and methods of construction will be discussed in later chapters. A part of a sewage and drainage layout has also been added, Fig. 2.9.

2.2 Authorities and undertakings

Before any large building project is begun, detailed information is usually required in order to establish whether the development is suitable in every way, legally, physically, and economically. On obtaining the contract, the first item of information to be circulated is usually the Contract Details sheet as shown in Volume 2. Following this a list will be needed of people, bodies, authorities, organizations, suppliers and others who will be concerned with the project. This will contain names, addresses, references, and other necessary details of clients, vendors, engineers, surveyors, architects, water boards, transport, electricity, and telephone departments, insurance brokers, lawyers and other legal representatives and government departments concerned with construction. Each will deal with matters arising from the development in his own capacity: the client with user requirements, planning consent etc.; planning authorities with factors liable to affect permissions, restrictions and approvals; the engineer with foundations, structure, calculations, soils, drainage, roads, environmental influences; service departments with electricity, water, telephones, and so on. Site environmental problems have been discussed in Volume 1.

Problems may also arise over boundaries, adjacent buildings, works

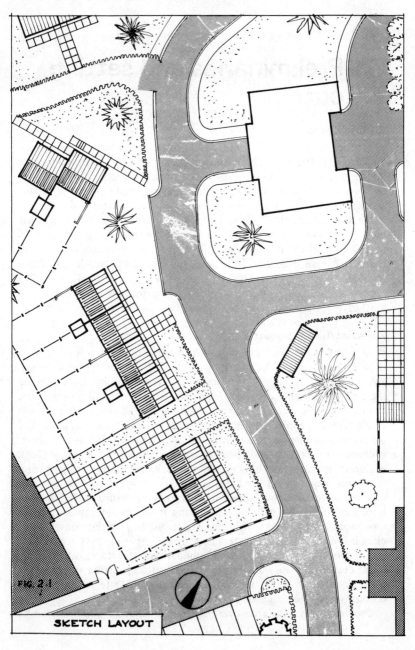

FIG. 2.1

SKETCH LAYOUT

on site, basements, roads, encroachments, rights of way, existing buildings, holes, and structures which may have to be dismantled and re-used. Legal matters arising from the scheme, other than those dealt with under the Standard Form of Contract, are usually handled by specialists and principals and are outside the scope of this work. Following a site study a feasibility report and programme will be drawn up for the consideration of all concerned.

REGULATIONS, PLANNING, AND LOCAL BY-LAWS. Where building schemes are to be developed by the local authority, who may be both client and architect, the interpretation of regulations and procedures becomes easier. Apart from improved communication a measure of priority is often given where otherwise delays might occur. Local authorities, however, are often much stricter with contractors than independent principals would be. The main considerations affecting preliminary works are:

Type of development which is concerned with the nature or purpose of the structure(s) and effect upon the community and environment.

Materials which are examined for fire resistance, weatherproofing properties, strength, insulation, durability, or special qualities such as ultra-violet ray resistance.

Site itself which must be acceptable as to location, population-density, convenience, health, amenity, and accessibility.

Foundations which must be designed to carry the structure, and be large enough to avoid overloading the permissible bearing capacity of the soil. They must also be capable of resisting any malfunction of the soil itself.

Structure which must conform to the requirements of the appropriate standards and codes of practice currently in force and required by building regulations, and be able to carry all loads and stresses demanded of it.

Drainage which must be of approved material, laid to adequate falls, and constructed according to regulations.

Building regulations in some countries are adapted from those of temperate zones but incorporate appropriate clauses and local by-laws to meet special needs. Strictness of interpretation often depends on local requirements or climatic needs such as water conservation. See Chapter 14.

2.3 Site considerations

Water supply and disposal is a problem in many countries where low annual rainfall is often combined with heavy periodic storms. Low rainfall is of course experienced in humid as well as dry climates. Some humid coastal towns, such as Accra in Ghana, have a yearly rainfall of only 680 mm — no more than that of London. Sometimes, on large developments such as universities, hospitals or industrial complexes, when supply is likely to be intermittent, a large permanent storage tank is installed at the beginning of the contract. This is filled during off-peak periods — usually at night when demand is reduced. This can later be used in times of emergency or connected up to small domestic tanks to augment the normal supply. Building operations may have to be postponed during the dry season. Alternative sources of water on site could be investigated before the work commences.

DRAINAGE. The method of drainage and sanitation for the area will need consideration and the local authority may have to be consulted quite early in the development. Use of septic tanks, cesspits, sewage disposal plants or main drainage, either wholly or in part, will have to be decided upon. Frequently individual institutions such as military camps, hospitals, or industrial installations have their own sewage disposal systems, with rivers or ponds to drain off the main surface-water. In dry zones with short rainy seasons the question of waterless night-soil disposal may have to be considered. Chemical closets in such circumstances could be preferable to w.c.s without water.

BOUNDARIES. Difficulties of demarcation are a familiar problem and questions of crossings and rights of way may arise from lack of knowledge of traditional customs and usage. In such matters legal authority may not automatically take precedence and the needs and privileges of the local inhabitants could well become the subject of long and patient negotiation. This applies equally to owners of adjoining properties. Boundaries need to be set out with care and Fig. 2.2 shows a typical triangulation method using a chain or steel band. Offsets are then taken as required from the main sight lines. If convenient the same triangulation or part of it could be used as base lines for setting out later.

OBSTRUCTIONS. These can often affect site work as has been mentioned in Volume 1. Consideration may also have to be given to old workings, filled-in excavations, made-up ground, old clay-pits, ground-water, subterranean faults, wells, flooding, and erosion of physical features.

SITE PRELIMINARIES. Having obtained the necessary contract information and possession of the site, the engineer or supervisor will check

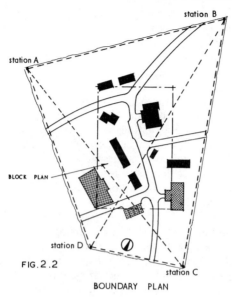

FIG. 2.2

BOUNDARY PLAN

such details as those given in Volume 1. Larger developments will need a wider approach making allowances for adequate storage areas, access, information on existing and proposed infrastructure, services, permanent roads and areas, preservation of trees and natural features, and security and protection from the elements.

Before preparing a plan of site activities and operations he will first clear the site of scrub, rubbish, derelict buildings, unwanted physical characteristics such as hillocks, ant-hills, and top-soil. The site plan is then prepared, showing the position of existing buildings, with features and services clearly marked. After clearing, and filling and levelling old holes and workings, setting out and gridding may commence.

2.4 Trial holes

On all but the smallest sites it would be advisable to obtain some idea of what may be encountered below ground. The science of soil mechanics is now well developed; and site exploration would clearly be the first priority when dealing with heavy industrial structures, tall buildings, deep basements, power stations, and the like. Even with shallow foundations up to 3 metres deep, investigation is necessary to identify types of soil, and their physical properties, strength, behaviour under load, and stabilization.

Where hand labour is available manual digging with basket and hoe is best, though rather slow, particularly if it is desirable to dig, inspect, and fill-in during a single day. A hole of about a metre square is usually big enough to give the engineer an opportunity to inspect the sides and bottom. Machine digging is quicker and, provided no boulders, lateritic lumps or water are encountered, can be done in a matter of minutes. A suitable digger for this purpose is shown in Volume 2. The position of the trial hole should be shown clearly in the plan; this could be done if necessary, by triangulation before the site plan is prepared. Trial holes should be kept clear of the foundations and of any existing services, if their positions are known. The local authority should be told when the holes are to be dug as an inspector present at this stage could save trouble later on. Having left the hole open for an hour or so to check for seepage, the soil may then be returned, filled, and rammed and a report prepared. A simple but satisfactory *pro forma* for shallow foundations is shown in Fig. 2.3.

In addition to machine or hand digging, drilling may also be used; a suitable type of machine is shown in Volume 2. Alternatively, in reasonably soft soils, holes may be drilled by hand-auger, using a four-man capstan. Field testing and soil stabilizing suitable for medium-sized contracts is described in Volume 2. Physical tests on soils, however, are rarely necessary for medium-sized contracts, though samples may have to be checked for the presence of sulphates harmful to concrete. This is normally done in a laboratory.

Physical tests of soils to determine chemical properties and moisture content are generally unreliable and local experience could be the safer guide. The strength of soils on site could be measured by using the vane test which consists of rotating a four-bladed vane of about a 100 mm diameter and 100 mm deep at the bottom of the hole and measuring the torque required to rotate it. The advantage of this method is that it may be used in a hole drilled either by hand or machine. Volume 2 gives further information on simple foundation equipment. Other methods are available for complex or deeper foundations.

2.5 Survey and setting out

The equipment normally needed for starting site operations is given in Volume 1. In addition to rods, lines, battens, profiles, ranging rods, square, dumpy level, steel tapes, boning-rods, compass and tools, he may well need a folding surveyor's rod, chain, steel band, optical square, standard metric staff or levelling rod and possibly a theodolite.

Levelling rod. While the old Sopwith staff is still much in use it is being superseded by the metric levelling rod. The metric rod shown in

TRIAL HOLE SITE REPORT

| Site. Address | | Date. Today's date. |
| Location. (Quote Drg. No.) | | No. of holes. . 4. |

Hole	Depths	Remarks
Hole No. 1. Loamy top soil. Some roots and scrub.	O — O 225 225	Hand digging of holes throughout.
Red sandy clay. Some stones. Little gravel.	525 750	
Red clayey sand. Football size laterite lumps. Soil firm but damp. Clay and small stones present.	2475. 3000	Some trace of water in sides and bottom after one hour. Not serious
Hole No. 3. All as above.		For holes no's 2 and 4 see separate sheet.

REPORT

Average soil conditions for this area. No foundation problem expected. Building inspector reports no sulphates present in soil. Unmarked drains may exist at N.W. corner. Care to be taken.

PRESENT

. . . . Engineer. Structural Dept. H.O.

. . . . Building Inspector.

. . . . Site Agent.

Signed. . . . .

FIG.2.3

Engineer.

Volume 1 is not as widely used now as the British Standard version shown in Fig. 2.4. The readings are in metres, 100 mm and 10 mm. The staff illustrated is available for both trisectional and telescopic use and the readings have a yellow background for high definition.

Theodolite. This is the most widely-used instrument for horizontal and vertical angles. It can be used for normal levelling and may be supplied with a compass needle if required. A reasonably-priced instrument suitable for building work and small surveys is shown in Fig. 2.5. The vertical angle is useful for such tasks as setting out foundation base pads in basement areas or on undulating ground where the horizontal angle alone would be inadequate. It can also be used for measuring heights of towers and tall buildings. For horizontal work it can also be used as a check on physical triangulation measurement using trigonometrical ratios.

GRID SURVEY. Having obtained the sketch layout of the site from the architect or local authority, the next step is to impose on this a grid survey or contour set in relation to a known datum or setting-out point. Fig. 2.1 shows a portion of the site to be developed. The whole plan would of course show an inset block plan giving exact location of the site and the surrounding areas. This would normally be traced from an Ordnance Survey map to a scale of, say, 1 : 2500 (25 in to a mile) or 1 : 1250 (50 in. to a mile). Should none exist then a block plan would have to be drawn in relation to permanent roads, buildings or landmarks in the area, Fig.2.2.

First it is necessary to decide on a suitable baseline. This may be a boundary line as shown in Fig. 2.6, or a building line of the local authority, or it may be necessary to set up one's own. Should a peg or mark exist relating to a previous development or phase it could be used as a starting point. Assume that a point O.D. 48.35 has been placed permanently near the existing hall, Fig. 2.6. From this a baseline is produced and a setting-out peg, preferably a 10 mm reinforcement rod, is placed conveniently near to the proposed community hall. This done, the survey framework may be laid out by using a grid of, say, 5 m to 10 m squares. The co-ordinates must be related to the existing setting-out points and triangulated as shown to ensure accuracy of position. In addition to the two stations given, a third is desirable; and for this the corner of an existing school has been chosen (51.00). This enables a triangulation check to be carried out. A thin r.c. rod bedded in concrete with the top levelled smooth and trowel cuts crossed at dead centre makes a good peg. *continued on page 36*

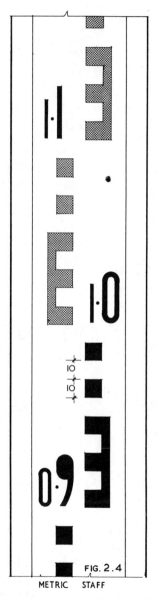

FIG. 2.4

METRIC STAFF

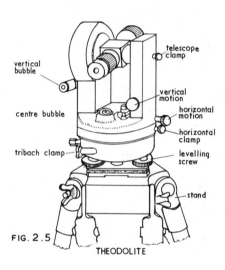

FIG. 2.5

THEODOLITE

INSTRUMENTS — TRIANGULATION

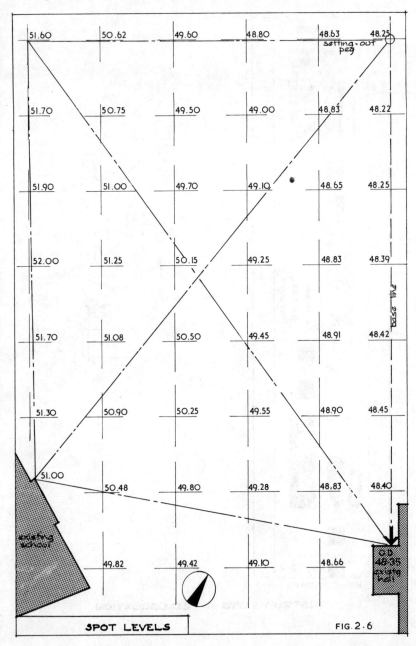

SPOT LEVELS   FIG. 2·6

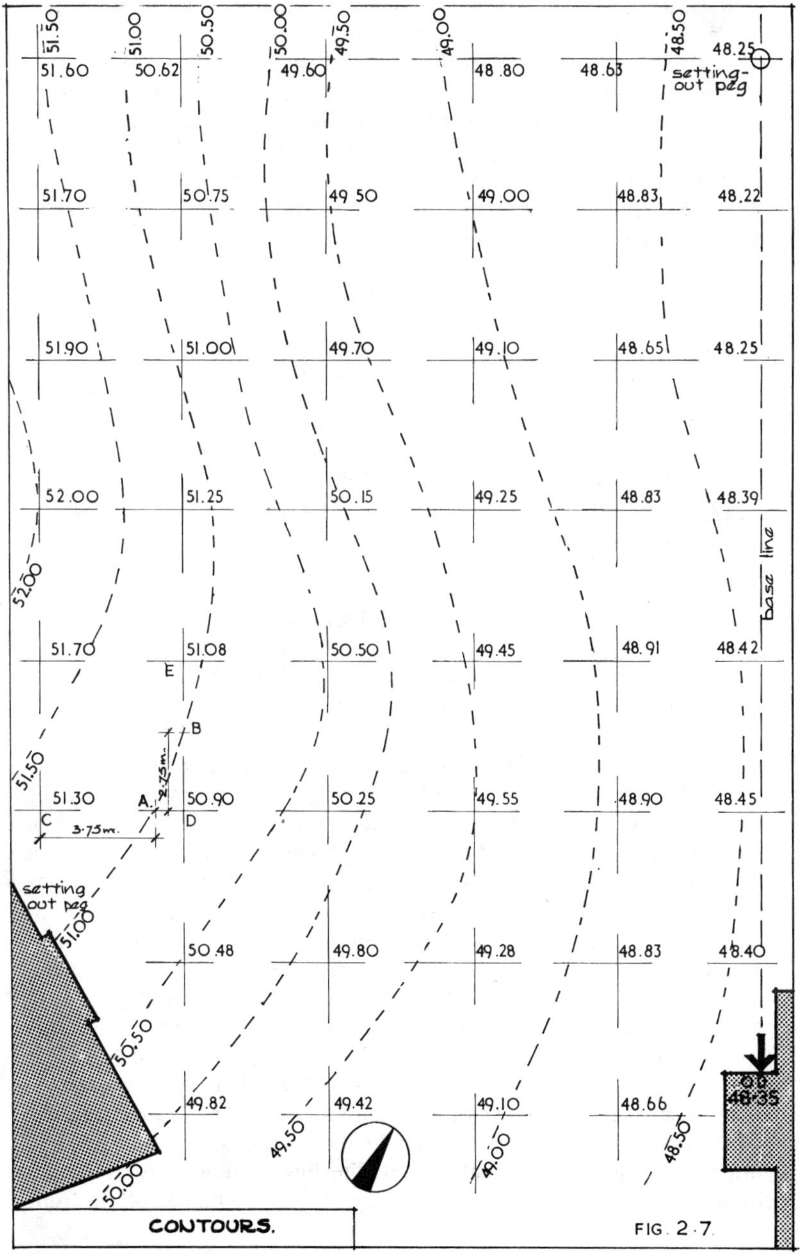

51.50	51.00	50.50	50.00	49.50	49.00	48.50	48.25
51.60	50.62		49.60		48.80	48.63	setting-out peg

51.70   50.75   49 50   49.00   48.83   48.22

51.90   51.00   49.70   49.10   48.65   48.25

52.00   51.25   50.15   49.25   48.83   48.39

52.00

51.70   51.08   50.50   49.45   48.91   48.42

base line

E

51.50

B

2.25 m.

A

51.30   50.90   50.25   49.55   48.90   48.45

C   3.75 m.   D

setting out peg

51.00

50.48   49.80   49.28   48.83   48.40

50.50

49.82   49.42   49.10   48.66   O.O 48.35

49.50   49.00   48.50

50.00

CONTOURS.

FIG. 2·7.

CONTOURS. As it is difficult to visualize the slope of a site from spot levels alone, contours can be added. To do this, first hatch in contours in pencil at approximate positions on the grid, Fig. 2.7. Contours may be taken at vertical intervals to suit the site gradient and the degree of accuracy required. In this case 500 mm intervals have been taken.

More accurate plotting can be done by simple calculation using similar triangles. A pocket calculator is useful for this. Consider contour 51.00, Fig. 2.7, point A and B. With grid co-ordinates at 5 m centres the data could be set out as follows:

A. Easting

Cont.	Co-ord. (C)	Diff. (C & A)	Diff. (C & D)	Distance (C to A)
51.00	51.30	0.30	0.40	$\dfrac{0.30 \times 5}{0.40} = 3.75\,\text{m}$

B. Northing

51.00	(D)	(D & B)	(E & D)	(D to B)
	50.90	0.10	0.18	$\dfrac{0.10 \times 5}{0.18} = 2.75$

Fig. 2.8 Contour schedule

To avoid confusion always interpolate from left to right to give 'Easting' and from bottom upwards to give 'Northing'.

SETTING-OUT SHEET. The working drawing is prepared by overlaying the grid contour on the sketch layout to give Fig. 2.9. Careful scaling will then give all the dimensions needed for setting out the buildings. This is sufficiently accurate for most purposes if the work has been properly done. The next stage is to decide on finished floor levels. The community centre can be omitted for the moment.

Consider the pair of semi-detached houses Nos 14 and 15 on contours 51.50 and 51.00. These levels would suggest a step of 500 mm to maintain the same relationship of finished floor level (F.F.L.) to ground in each house. Now let the F.F.L. be 150 mm above ground level. This would give the F.F.L. for house No. 14 as 51.65 and that of No. 15 as 51.15. Only a small amount of cut and fill would then be necessary.

Now assume that the architect would prefer a stepped and staggered condition in this case. This may be for aesthetic reasons, privacy, orientation, lighting, ventilation, building-line requirements or simply to accommodate the slope. Then house No. 15 would be both in front

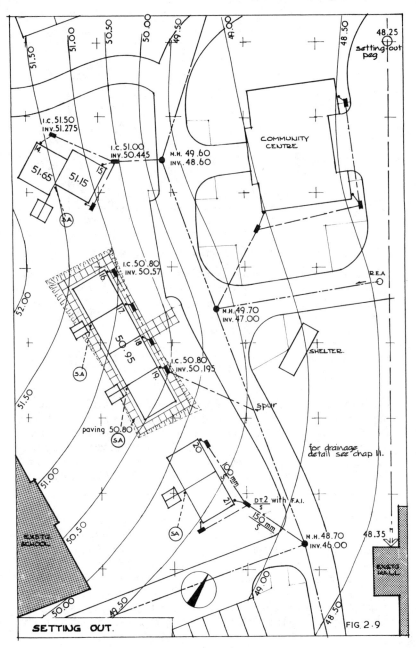

SETTING OUT.

FIG. 2·9

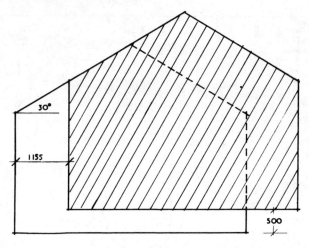

STEPPED & STAGGERED ROOF. FIG. 2·10

of and lower than its neighbour; the amount of stagger could be varied within reason. Supposing it would be desirable to avoid stepping the roof, then the lower house could be brought forward horizontally until the roofs line up, Fig. 2.10. With a roof pitch of 30 degrees, for example, the stagger between houses would be:

$$\frac{\tan 30°}{500\,\text{mm}} = \frac{0.577}{500} = 1.155\,\text{m for a step of } 500\,\text{mm}.$$

CUT AND FILL. Assume that the four houses Nos. 16, 17, 18, and 19 are all to be on one level and that it is desirable to balance the cut with the fill as far as possible. From the contours the east corner paving would be at 50.00 level and the west corner at 51.60. With an embankment cut at 45 degrees the plan would be as shown. It should be noted that the entrance paving would be stepped up from the sidewalk, say at 50.25 level, in which case each step would be at 130 mm.

SETTING OUT DRAINAGE. For this example we will assume that the normal practice of surface-water disposal will be adopted viz. open concrete channels of a size determined by local rainfall. Rainwater from roofs of small buildings will fall on to the apron or footpath, with open channels provided as necessary. Channels, for instance, would probably be needed at the foot of the embankments to house No. 16. A channel could also be provided across the open space, to discharge into the main channel at the roadside. Waste from bath and sinks would normally discharge into the foul drain via inspection chambers, though this would depend on local practice. On some estates, where there is no main drainage, foul drains discharge into septic tanks, and waste into soakaways.

The example shows inspection chambers to each house with covers flush with house paving. Falls of drains should be about 1 in 40 minimum if possible, though care must be taken to ensure that pipes are always at least 225 mm below the ground at the highest point. The lowest inspection chamber connects to the sewer or spur inserted by the local authority at the time of building. A suggested layout with inspection chamber and invert levels is given in Fig. 2.9. Information regarding sewerage levels and other services is supplied by the local authority — usually on their own drawings, though this can only be counted on in large towns. On completion of the drainage plan a schedule of inspection chambers is prepared with each one numbered and levels of cover and invert given. A few levels are given as a guide.

COMMUNITY CENTRE. A study of Fig. 2.9 and the methods outlined in this chapter should enable the setting out of the community centre together with F.F.L. and drainage to proceed without difficulty. The height of F.F.L. above ground will have already been decided by the architect and the location of the inspection chambers will depend on the sanitary appliances within the building. Foundations and floors are discussed in later chapters.

3 | Materials

3.1 Introduction

In recent years technology has made considerable strides in the development of building materials for practically all purposes. The use of natural substances, soil, stone, timber, and metal in elemental form is becoming less common than hitherto, though this development is not yet very noticeable in some growing countries. But it can now be seen almost everywhere else with the use of concrete, glass, synthetic and composite laminates, plastics, asbestos and metal alloys. Space does not permit development of this subject but further reference will be made to materials as they apply to construction in appropriate chapters.

CONCRETE

3.2 Properties

The four functional properties of concrete are strength, durability, fire-resistance and thermal insulation. To obtain the particular degree of each of these properties required for any specific purpose it is necessary both to select the correct materials and to mix them in the appropriate proportions.

CEMENT. This is a generic term covering limes, cements, plasters and sometimes adhesives. It will be used here, however, to denote Portland cement only. Of the wide variety of cements available such as ordinary, rapid-hardening, sulphate-resisting, water-repellent, hydrophobic and others, only those in common use in warm climates will be mentioned here.

The bulk of cement used in developing countries is of ordinary quality. It is adequate for all normal purposes, though its resistance to acids is low. As cement's rate of hardening is assisted by the high temperature in tropical zones, the need for rapid-hardening cement is restricted to special purposes. Sulphate-resisting cement is used where ground-water is suspect; waterproof and water-repellent cement make concrete and renderings less permeable to water. Low permeability also depends on the density of the concrete used.

Hydrophobic cement, briefly referred to in Volume 1, prevents partial hydration during storage and is useful in humid climates and conditions. Manufacture is undertaken only to special order and it is therefore more expensive. But normal cement is hygroscopic and so attracts moisture, and in areas of high humidity, especially coastal areas, it is liable to premature set. Hydrophobic cement for high-humidity zones may therefore be worth investigation. One problem, however, with using several different types of cement could be that of storage, especially when trained storekeepers are not available. Distribution difficulties could arise, but for large individual contracts special cements could be a distinct advantage. In order to reduce premature set with ordinary cement, special 6-ply sacks with vapour-proof linings can be specified with the order.

High alumina cement is not to be recommended for use in tropical countries except for very special cases. Although it has good resistance to oils, chemicals and acids and can develop high early strength, it can deteriorate under high temperature and humid conditions. It also requires special care in mixing and placing, since any surplus water in the mix leads to eventual failure.

AGGREGATE. The choice of aggregate is often limited to the immediate locality, for long hauls over inferior roads are impracticable except where large civil engineering contracts make it necessary. The strength of the concrete depends on the grading of the aggregate, its surface texture, and the extent of the impurities in the stone. Aggregates are loosely classified into coarse and fine, the latter referring to particles passing through a 5 mm sieve. Details of grades are given in B.S. 882. As aggregates become finer the surface area of the particles increases and more water is needed to produce concrete with the required workability; and this results in lower strength concrete. The effect of using sand grading on 1:6 concrete at constant workability may be seen in B.R.E. Digest 150.

ADMIXTURES. Properties of concrete in both plastic and hardened states can be modified by the use of admixtures. Some air-entraining admixtures, for instance, introduce air bubbles into concrete and so increase its workability, though with some reduction in compressive strength. Better workability means that the water/cement ratio can be reduced so that a more durable concrete may be obtained. Other types of admixture such as accelerators, are available, but since their main purpose is to work the concrete at low temperatures they are of little concern here. The need for careful grading and control of concrete mixing and placing is not diminished by using admixtures.

STRENGTH. Transverse strength is necessary where bending is more important than compressive strength, and denser concretes are needed to produce this property. In plain concrete walls the permissible stress is related to the cube strength of the concrete of which the wall is made. More will be said of this later but detailed structural calculations, though worked out for some examples given, will not be discussed here.

LIGHTWEIGHT CONCRETE. Structural lightweight concrete is used to provide improved thermal insulation and fire-resistance. It reduces weight by about a third and can produce concrete which combines cube strength of up to 55 MN/m^3 with a density of 30 to 40 per cent less than ordinary cement concrete. Large precast concrete storey panels have been used in countries where lightweight aggregates are available and used in conjunction with normal *in situ* frames.

Lightweight aggregates may be difficult to obtain in some countries as they depend on the production of well-burnt surface clinker, foam blast-furnace slag made by treating molten slag with water, exfoliated vermiculite, expanded clay or shale, pumice, sintered powdered fuel-ash (P.F.A.). Further information may be obtained from appropriate British Standards. Lightweight concretes have higher shrinkage movements than normal concretes though their fire-resistance is high owing to low thermal expansion, but they often provide less protection to steel reinforcement. Lightweight precast concrete blocks are widely used where suitable materials are to be found. Densities for lightweight concrete are usually taken to be less than 960 kg/m^3 for coarse aggregate and 1200 kg/m^3 for fine aggregate.

No-fines concrete is made by omitting the fine aggregate and small particles so as to leave voids. It has a weight of only about two thirds that of dense concrete and is usually cast *in situ* for both loadbearing and non-load-bearing walls. It is cheaper and lighter than normal concrete, saves material and has better insulation, but care must be taken to check its structural and functional properties before use.

Aerated concrete is usually produced by introducing air into a cement so that when the mass has set a uniform cellular concrete is formed. It is also sometimes 'cured' by high-pressure steam — this process is known as autoclaving. Air-cured concrete has only about half the strength of autoclaved and it shrinks very much more. When used in the form of precast building blocks and reinforced units, contraction joints are necessary about every 6 metres. The mortar used in jointing should not be too strong. These blocks could be used as shown in Volume 2 as an alternative to glass-fibre blocks.

DURABILITY. The durability of concrete depends on many factors such as the nature of any impurities present, the level of the water table, seasonal variations of climates and the form of construction and type of concrete. Sulphates and other impurities reach concrete by the movement of their solutions in ground-water, and concrete wholly above water is unlikely to be affected. Low permeability is essential to resist such attack and sulphate-resisting cement is necessary in some zones. Resistance to attack from organic impurities in ground-water is best achieved by a low water/cement ratio, thorough compaction and curing. Mass concrete deteriorates less than thin-walled sections but the concrete needs to be fully compacted. Water for mixing must of course be free from impurities.

Salts in aggregates and sea spray can also cause erosion and spalling especially in reinforced concrete; atmospheric gases have a similar effect in industrial areas. Grit and sands in latitudes of high winds and sparse vegetation can cause scour by impact and abrasion. Solar radiation can also cause evaporation leading to poor curing.

APPEARANCE. Durability and function in themselves are not enough to satisfy the client and appearance plays a part in his demands. Concrete discolours quickly in polluted or smoky areas; but weathering may be due to poor detailing and execution. Streaking can occur where horizontal projections have not been undercut with drips; lichens and weeds grow in damp crevices. Decay, often due to poor specification of mix and materials, becomes evident in cracking and crazing leading to spalling and shrinkage. The latter combined with thermal movement, creep, or settlement introduces the risk of water penetrating to the reinforcement and setting up corrosion and eventual disintegration of the concrete surrounding the steel.

Finishes due to the cement, aggregate, or sand do not always weather attractively, particularly in dirty atmospheres. Carelessness can also spoil the finish − for instance when red laterite earth becomes mixed with the concrete when placing. Insufficient care at the end of the day's work leaves untidy laitance lines, which may also let in water. Staining of concrete finish can also occur through rusting reinforcement or metal piping, or through rust-contaminated water splashing the surface. Where corrosion of reinforcement is likely to occur, zinc-coated steel could be specified − a method of protection obtained by hot-dip

Table 1 Timber species and properties

Standard name	Sapwood distinct	Heartwood durability†	Resistance to impregnation with preservatives	Woodworking properties			Amount of movement on re-wetting	Paint performance in service	Remarks
				Resistance in cutting	Blunting effect				
Afrormosia	Yes	Very durable	Extremely resistant	Medium	Moderate		Small	Good, provided paint is not affected by oily exudations	Non-ferrous fitments necessary
Afzelia				High	Moderate		Small	*	May discolour other building materials
Iroko				Medium	Fairly severe		Small	*	
Kapur				Medium (variable)	Moderate (occasionally severe)		Medium	*	Non-ferrous fitments necessary
Makoré				Medium	Severe		Small	*	Non-ferrous fitments necessary
Teak					Fairly severe (variable)		Small	Good, provided paint is not affected by oily exudations	Has been known to cause discoloration of granite
Agba	Moderate	Durable		Medium	Moderate		Small	Moderate, but experience is limited	Log core may contain brittleheart. Resin exudation may occur with some parcels, especially if the timber has been air-dried
Idigbo	Moderate				Mild		Small	*	Log core may contain brittleheart. May cause staining of other building materials
Sweet chestnut	Yes			Medium	Mild		Small	*	Non-ferrous fitments necessary
Utile					Moderate		Medium	Good	

Douglas fir	Yes	Moderately durable	Resistant	Medium	Medium	Small	Moderate	Non-ferrous fitments advisable
African mahogany			Very resistant	Medium (variable)	Moderate (variable)	Small or medium	*	Log core may contain brittleheart
Gurjun Yang			Resistant	Medium (variable)	Moderate (occasionally severe)	Large	Moderate	Resin exudation occurs with some parcels of these timbers (especially keruing); select material free from resin for painted work
Keruing						Medium	Moderate	
Red meranti Red seraya	Moderate					Small	*	Log core may contain brittleheart
Sapele	Yes		Extremely resistant	Medium	Moderate	Medium	Good	
Redwood	Yes	Non-durable or perishable	Sapwood permeable Heartwood moderately resistant	Low	Mild	Medium	Good	
Western hemlock	No		Sapwood and heartwood resistant	Low	Mild	Small	Good	
Whitewood								
American red oak	Yes		Sapwood permeable Heartwood resistant	Medium	Moderate	Medium	Poor	Non-ferrous fitments advisable
Beech	No		Permeable	Medium (variable)	Moderate (variable)	Large	Good	
Ramin				Medium	Moderate	Large	Moderate, but experience is limited	

† The sapwood of all species is either perishable or non-durable.

* Seldom painted; there is therefore little painting experience.

galvanizing. It may be worth the extra cost for industrial and other high-corrosion areas.

MOISTURE PENETRATION. The question of damp-proofing must always be considered when designing structures of reinforced concrete. Whilst penetration of moisture in itself may not be harmful, its effect on finishings and coverings such as wood-block or insulation linings could be damaging. Damp can cause lifting and warping of blocks and impair the insulation value of linings.

TIMBER

3.3 Properties

In most warm countries the availability of timber depends on the humidity of the climate; its use is therefore sparse or prodigal according to the environment. The type of timber used locally also depends to some extent on the export market and the demands of the importer. This often leads to neglect of good secondary species which it is not thought profitable to fell. With the well-known species of hardwoods strength depends usually on density and the properties of woods can in most cases be related to cost.

Conversion into specific sizes of felled logs is not adhered to rigidly but Table 3.2 gives a few examples of sizes usually quoted for export purposes.

Source	Exported condition	Species	Width mm	Length m
West Africa	Logs	Various	Min. dia. 559	3.7 to 9.1 m
	sawn		Min. dim. 152	1.8 min.
			Min. av. 178	Min. av. 3.7 m
Malaya	sawn	Keruing	Min. dim. 152	1.8 min.
			Min. av. 178	Min. av. 2.7 m
	sawn	Ramin meranti	Min. dim. 152	1.8 min.
			Min. av. 229	Min. av. 2.7 m

Table 3.2

THERMAL INSULATION. Timber is a good insulator; its conductivity (k) is rated as 0.144 W/m °C and transmittance (U) for 102 mm thickness is approximately 1.19 W/m² °C for timber weighing 481 kg/m³ with 20 per cent moisture content. It has a low thermal movement and expansion joints in structures are not normally required.

In almost all cases strength of timber depends on its density; the greater the weight the stronger the wood. Where movement is restrained after loading in conditions of humidity or alteration of moisture-content the final dimensions may vary from those originally obtained. This is sometimes an advantage, as in the case of, say, tongued and grooved flooring's becoming clamped together after nailing. Usually, however, distortion takes place where timber cannot shrink equally in all directions and leads to cupping of plain sawn boards or diamonding of square timbers. Distortion is also caused by a combination of tangential and radial movement (T + R) which for good class hardwoods is less than 3 per cent.

DURABILITY. The efficient maintenance of timber depends largely on how often it is carried out, though this is not the sole criterion. Deterioration may be caused by fracture through loading; abrasion by sand and/or wind; leaching caused by flowing water; alternate wetting and drying causing expansion and contraction; sunlight causing fading and embrittlement of the surface; bacteria and fungi causing superficial discoloration; wet and dry rot; and beetles and termites causing disintegration. Durability classifications of hardwoods are available from B.R.E., Princes Risborough.

Moisture is the greatest single factor affecting the properties of timber. Moisture-content expressed as a percentage of dry weight is most simply measured by an electric moisture meter; otherwise it is necesssary to dry a sample of timber in an oven to test its loss of weight. The level of moisture-content depends on the nature of the building, ambient humidity, method of cooling, diurnal temperature changes, percentage of moisture present at time of assembly, and relative humidity.

To ensure protection of timber it is necessary to convert it immediately on felling and then season it as described in Volume 1. Movements of timber caused by moisture at various R.H.s have been recorded, and details may be obtained from B.R.E., Princes Risborough. Where timber is likely to be damp in service, whether durable or not, a preservative should be used.

PRESERVATION. Preservation of non-durable hardwoods is especially necessary where secondary species of locally-grown woods are used externally. Processes of preservation vary from superficial treatment of limited protective value to pressure-impregnation in plants of high efficiency. The latter are still comparatively rare in the tropics. Air seasoning of such timber is desirable where time permits.

For external woodwork which will not need painting the liquid should be flooded over the surface to enable it to absorb as much as possible, particularly at joints and end grain. Some hardwoods are resistant to any preserving process. Preservatives can also be used to resist attack by fungi and insects and can be specified as non-toxic to plants and animals or as odourless, or as not affecting painting processes, or as incapable of being washed out by rain. Creosote is the best-known and cheapest but cannot be painted over and has a strong odour at the time of application. Waterborne preservations are widely used and the W.B. 2 grade can meet most requirements.

Painting. The need for good priming treatment of hardwoods is often overlooked; many woods receive only a top coat of paint and deteriorate after only a short time. Hardwoods, even dense varieties, frequently have large pores which permit paint to penetrate but this alone is not enough and a heavy primer or filler may be necessary. As a smooth surface is essential for good performance, sanding is often resorted to but this can reduce paint penetration, especially with small-pore species. Some woods such as teak contain oil which interferes with adhesion, while unseasoned timber allows dimensional changes to occur which cause the paint to crack. External condensation, where it occurs, will also penetrate window- and door-frames through joints and putty. Sunlight too, can affect colour by causing fading and clear finishes fail even more rapidly. Painting is not a guaranteed defence against fungal spores or harmful organisms, as they may be present before painting starts.

Aluminium primers are usually considered best for hardwoods, though lead-based fillers may be needed for sills and thresholds. Stopping and filling should always follow priming. Priming should always take place before assembly or, better still, complete painting of components should be carried out before fixing, especially window- and door-frames. Priming is best carried out at times of lowest humidity.

DEFECTS AND DECAY. Natural defects may be found in timber which have occurred while it was growing. Brittleheart (softheart) is common with hardwood, leading to brittle fracture and thus reducing strength. Checks, i.e. small splits in timber left in the sun after felling, are fairly common. Splits and shakes are due to stresses in trees while growing; and timber is often stressed by too rapid seasoning through exposure to the sun. Rind gall is also common where trees have been defaced. Wood-rotting fungi are also apt to attack timber if it is left wet for long periods, though exterior timber not in contact with the ground, if of moderately-durable heartwood, is unlikely to decay. (Sapwood is vulnerable, and preservative treatment may be necessary.)

Condensation is not a serious problem in warm climates except as discussed in Chapter 1. Vapour barriers may have to be provided where condensation can be expected e.g. cold stores and areas subject to high diurnal change. As untreated timber is susceptible to fungal attack condensation may accelerate the process.

Fungi and insects. Durability embraces resistance to fungal attack. Destructive fungi are those which cause dry rot and wet rot, whilst non-destructive fungi are those which cause moulds and staining. Common pests are the marine borers, wood-destroying insects, termites, beetles etc., which have been discussed in Volume 1. Boring in moderation is not always regarded as a defect in structural work, provided it has been treated, and if exit holes are not too numerous or conspicuous.

Exposure to weather. Any wood exposed to sunlight and rain will in time lose its original colour. This may be due to loss of water-soluble extractives, but is mainly due to the ultra-violet rays in sunlight. The surface may become pitted or striated or roughened by wind-borne abrasives. It can also be affected by algae attracted by alternate wetting and drying.

Prevention of decay. Decay is caused by timber's becoming sufficiently moist to allow wood-destroying fungi to grow. Prevention depends on the selection of timber naturally resistant to decay and/or protected by preservative. Moisture control is obtained by good storage, careful fabrication, good detailing of joints, thorough painting, regular maintenance and prompt attention to repairs. Woodwork exposed to prevailing rain or wind needs special care. Decayed timber, where discovered, must be cut away, and paint stripped from the surrounding area so that it can dry out. Wood exposed by cutting must be treated with preservative or primer. Enclosed woodwork in floor joists or structural timber should have through ventilation and the ends should be treated with preservative.

Drying out buildings. Excessively fast drying can create problems, though artifical means are not usual in warm countries. If dehumidifiers are used in zones of high humidity then the building must be enclosed for them to be effective. Hot, dry zones are not normally affected. If necessary, moisture in timber can be measured by a moisture meter: this is done by inserting probes into the wood and reading the electrical resistance between them.

Chemical resistance. Compared with metal, wood has a good resistance to alkalis and weak acids. The former are rarely a problem while the latter may be encountered in growing industrial areas. But this is not yet a cause for concern in most warm countries.

Fire. The hazard of fire is growing as building in towns and industrial areas continues to expand. Its behaviour is different from forest or bush fires to which most rural inhabitants are accustomed. In zones where relative humidity approaches zero it can be potentially dangerous, particularly where water is in short supply. Where hardwood is more plentiful the large solid sections normally survive better than steel sections of equivalent strength. Precautions, however, are rarely taken, especially where regulations tend to interfere with bodily comfort: see Chapter 14. An example of such precautions would be the provision of fire checks at staircases and landings, which will reduce cross-ventilation and through draught, unless air conditioning has been installed.

Where fire-doors are required these will create a check of half an hour, or up to one hour in the case of special types of door. Existing doors may be converted by covering with asbestos insulation board to improve their resistance. Rigidity of construction is essential otherwise the top and bottom corners of the doors not stiffened by the hinges would distort and so cause draughts through the gaps. Effective stops are to be provided especially where rising butts are used. Intumescent strips should be inserted in the edge of the doors and deeper stops used than is normal.

GLUES. The choice of glues for wood depends on a number of factors such as surface preparation, required adhesive application and extent of durability needed. There are four main types:

Type W.B.P. (weather, boil proof). These are highly resistant to climatic changes, micro-organisms and heat, and are suitable to warm, humid zones.

Type B.R. (boil resistant). These are suitable for normal exposure but not as strong as W.B.P. Good for external situations.

Type M.R. (moisture resistant). These are not as strong as either of the above and will not withstand full exposure to tropical elements. They are quite good for protected interior situations.

Type INT. (interior). These are not suitable for humid tropical conditions. One of the dangers of providing such a choice of adhesives is that they may be interchanged, especially in times of shortage – thus glues intended for internal use could be used externally, particularly on, say, locally-made doors, which then disintegrate quite quickly.

TIMBER-BASED BOARDS. With the introduction of improved glues and adhesives the range of building boards and slabs has increased considerably. The most commonly used are:

> Chipboard
> Plywood
> Hardboard
> Blockboard and laminboard
> Fibreboard
> Woodwool slabs.

Plywood is the most commonly used as it can be manufactured in a number of local sawmills, and has thus largely eliminated hardboard in some countries. Unfortunately not all mills state the quality on the sheet, with the result that some plywoods disintegrate rapidly. Imported plywood is supplied in grades and durability according to requirements.

Blockboard and laminboard are similar products and are used mainly in fittings and furniture. Laminboard is usually of superior quality. Particle and chipboards are also used for flooring decks, roofing, linings and furniture; they have now become popular and tend to supplant block and laminboard.

Standard hardboard can be supplied in a number of forms; perforated, enamelled, P.V.C. faced, wood veneered, moulded etc.

Fibre building boards include insulating fibreboard, hardboard, and wallboard, which is slightly denser than insulating board and is used for room linings.

Woodwool slabs are popular in tropical countries, particularly for insulating purposes in both walls and roofs. They have a low flame spread and are resistant to termites and fungi. For a full range of sizes and other details of wood-based boards application can be made to the Fibre Building Board Development Organization (FIDOR) or Chipboard Promotion Association (C.P.A.).

METALS

3.4 Iron and steel

Apart from the normal effects of corrosion discussed in Volume 1, the chemical and physical effects of water on ferric metals need consideration. Salt, whether present in water or as sea spray, will erode, as will salts dissolved in the normal water supply. Corrosion of pipes, tanks, and boilers is usually initiated by the presence of oxygen, and can be accelerated by bimetallic contact.

PROTECTION. Apart from having a high U-value these metals become uncomfortable to touch when exposed to the sun, and also need special provision to counteract thermal movement. Stainless steel, though expensive, has properties of strength and corrosion-resistance that make it attractive not only for services but as a finish in high-quality building. It is useful as a cover for protection to columns, street furniture, window sections, water tubes, exposed flashings, and cladding. It is widely used for anchors and dowels in parapets and for fixing precast concrete slabs to the structural frame.

Treatment of steel is a wide subject and before proceeding in industrial areas it is advisable to consult such sources as the B.R.E. publications and those of the British Steel Corporation, where new methods and applications are always being investigated. When applying protective methods special attention should be paid to instructions and also to accessibility, with inspection and control at all stages of the process.

3.5 Other metals

Aluminium. In order to reduce glare, chromate phosphate coating may be used to produce 'architectural green' as it is known in the industry. It also provides a good base for painting. Chromate oxide coating, produced by immersion in a hot alkaline metal chromate solution, gives a grey coating suitable for decoration and anti-glare purposes.

Copper. Copper surfaces are painted merely to match the surroundings. The surfaces can be roughened with fine abrasive paper, preferably used wet or with white spirit, before painting. Aluminium-pigmented primers are also satisfactory. Special lacquers can also be used to preserve the original appearance and prevent discoloration.

The composition of natural waters from lakes, rivers, and reservoirs and deep wells can in some instances have a deleterious effect on metals. This applies both to internal action in service pipes and external corrosion. Local conditions are worth studying before large-scale projects are embarked upon. It is necessary to select materials which will resist the action of the external environment as well as internal attack.

PLASTICS

3.6 Introduction

Though space does not permit a much wider expansion of this subject than was given in Volume 1, further information will be given as appropriate in other chapters. Forms of plastics include tubes, mouldings, sheets, films, cellular forms, paints and coatings, plumbing and service fittings; though not all have their uses in tropical countries.

When considering the use of plastics it is necessary first to study their properties and function. These could come under the headings of strength, reactions to temperature, fire-resistance, thermal conductivity, insulation and resistance to solar effect. The main concern is that of durability, usually in relation to sunlight, warmth, and moisture. The photochemical action of solar radiation can cause polymerization and oxidation leading to embrittlement, crazing, cracking, and discoloration. Thermal action causes a decrease in tensile strength and elasticity, leading to embrittlement and warping. In some instances permeability to water vapour increases with temperature.

Some plastics absorb moisture more readily than others, causing swelling and shrinkage as the moisture content changes. The combined effects of moisture, longer hours of sunshine, high indoor temperature, and exposure to climate and the elements tend to produce degradation as much as three or four times as fast as in temperate climates. Plastics exposed in this way, particularly foamed plastics, are subject to abrasion by wind-borne particles and grits; insects and micro-organisms also can attack organic fillers and plasticizers, causing embrittlement. It is possible, however, to add anti-oxidants and absorbing pigments to improve durability. As a general rule plastics are better used internally or in shaded situations, unless guaranteed to serve their intended purpose.

Glass-fibre reinforced polyester resin products (G.R.P.) have now found a wide use in the building industry, apart from marine structures and boats. They have high strength and resistance to impact, with a strength/weight ratio greater than that of mild steel. Uses include cold-water cisterns and tanks, translucent sheets and panels, concrete form-work, window mullions, tubes used as fence- and guard-rails, channels and sections for light structural work, and many other uses.

GLASS

3.7 Introduction

Up to the end of the Second World War the use of glass for building purposes was practically unknown in many developing countries, except in capital cities. Here its use was confined mainly to large department stores, government and prestige building, and high-class housing. Smaller towns and villages were – some still are – without glass in doors and windows. In the better types of building, glazing was confined to steel-casement windows and doors, the other windows having either wooden shutters or louvred casements as a means of lighting, ventilation and protection. The introduction of glass louvres was an instant success and transformed building comfort, especially in humid areas. Those with traditional casement windows applied adaptor strips to convert casements to louvres, see Volume 1. The new louvre was comparatively inexpensive, easily imported as a knock-down unit, rot-proof, and could be installed by comparatively unskilled labour. It required only narrow strips of glass which could be fixed without putty or beads. It had certain disadvantages in the early days, but these have now been largely overcome.

TYPES OF GLASS. Many different glasses are manufactured throughout the world, a large number of which are not suitable for the tropics. The main types imported are:

clear sheet	...	3 to 6 mm
clear plate	...	6 mm
wired plate	...	6 mm
wired cast	...	6 mm
rough cast	...	5 to 6 mm

Special glasses such as armour-clad, toughened plate, domed lights, heat-resisting, anti-sun, heat-absorbing, fire-resisting etc. may be ordered if desired. Coloured glass panels, glass mosaics, glass sliding doors, curtain walling, and windows have already been discussed. Natural lighting, sound-proofing, thermal insulation, and double glazing were introduced in Chapter 1.

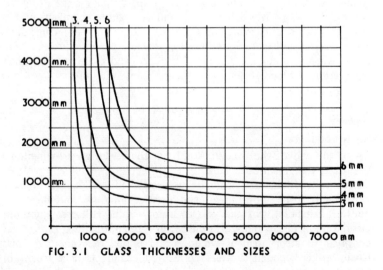

FIG. 3.1   GLASS   THICKNESSES   AND   SIZES

Wind pressure. The growing practice of using larger windows makes the risk of wind damage all the greater, especially in humid zones. Wind impact loads may be calculated from information compiled as described in Chapter 1. Glass, however, is weak in tension and unpredictable in strength. As a result, safe stresses have been computed on a large number of samples to give recommended sizes and thicknesses for clear sheet-glass. These are given in Fig. 3.1 and relate to clear sheet-glass glazed on all four edges where wind loading is 1000 N/m². It could also be shown that clear plate-glass is weaker than sheet, and that wire in glass reduces its strength. (CP 3. 1972, Chap. V). Information on sizes of glass and frames for adjustable glass louvres is given in Volume 1.

BITUMINOUS PRODUCTS

3.8 Introduction

A bituminous product is a durable material if protected from hot sun. It is resistant to acids, alkalis, industrial liquids, and oils. Pitch is even more durable. Bituminous paints like creosote bleed through ordinary paints, so other protective means may have to be used. Bituminous products are also softened by heat and thermal action which cause them to flow, though tropical road research has produced surfaces which can resist high temperatures. One widely-used cut-back bitumen (MC2) can resist temperatures up to 66 °C. This has also been used successfully as a rendering when mixed with sandy top-soil, particularly on flexible or unstable backgrounds.

Asphalts and bitumens are not successful when used on exposed surfaces, especially flat roofs. Thermal insulation below the roof-covering prevents heat from being absorbed by the structure, but, as a result, the covering becomes hotter and so deteriorates. It also suffers from solar radiation as ultra-violet rays are absorbed. This causes brittleness and cracking; and the chemical change permits water-soluble compounds to form. Light-reflecting surface chippings or limewash may prolong the life of the covering, but hot sun will still penetrate to some extent.

Insects and micro-organisms can also attack bituminous surfaces, as the green mould seen on neglected roofs will testify. Termites will also attack such products, particularly when they are deteriorating.

The effect of high winds on flat surfaces has been shown in Chapter 1. Suction, especially at the corners and edges of buildings, causes felt to lift and tear. Felt is still widely used, however, and it is possibly cheaper to replace it at intervals than to employ costlier methods initially. It lasts fairly well in protected places.

INTERNAL PLASTERING AND PLASTERBOARD

3.9 Introduction

In warm countries it is customary to use local materials and to conform to traditional methods, especially in rural areas. Where lime and gypsum plasters are not available, imported cement and sand are sometimes the only materials used for internal plastering and external rendering and mortars. As countries develop, and where lime deposits are being exploited, efforts are being made to introduce quicklime as an additive or internal substitute for cement mortar for housing and community building.

PLASTERBOARD. Where they can be afforded, plasterboard and fibreboard are imported mainly for internal use. Plasterboard is normally a solid or cored plaster surfaced with heavy grey paper. It may also have a cellular core. The board may be used as a lining for walls or ceilings, which can be either plastered or finished direct. The latter is usually preferred where fine coat plastering is not normally a traditional skill. Some manufacturers provide 'egg crate' cellular panels and also gypsum planks made from two boards stuck together. Plasterboards are classed as combustible in some building regulations, owing to the paper liner; but the plaster itself provides a reasonable degree of fire-resistance. Among the numerous types of board manufactured are:

Gypsum wallboard which has an ivory finish ready to receive paint on one side, the other having a grey paper lining suitable for plastering.

Gypsum lath is a narrow plasterboard designed to receive plaster only. For this reason it is not always popular, as extra plastering is involved.

Gypsum plank is not often imported unless additional strength and/or sound-insulation is needed.

Insulating gypsum board is useful where extra thermal insulation is needed. It has aluminium foil attached to one side to reflect thermal conductivity (k) of 0.07 W/m °C for the normal 12.7 mm with adjacent cavity.

Other types of board, such as P.V.C.-faced with smooth or textured surfaces, or perforated boards to increase sound-absorption are obtainable, but usually only to order, the general product being gypsum wall board of 12.7 mm thickness. The usual size is 2400 mm × 1200 mm, though other sizes are available. For details of fixing and general uses see Chapter 5.

Wall linings. These may also be of asbestos insulation or fibreboard. Various methods of fixing and normal usage are described in Vol. 2.

Glass-fibre reinforced gypsum (G.R.G.) has been developed by B.R.E., England for such purposes as for use as partition units; as slabs used instead of flooring boards; and in special shapes, such as pyramidal panels used to form domes, ducting, and other items by bolting sections together.

Glass-fibre reinforced cement (G.R.C.). This is a composite of cement and *Cem-fil*, an alkali-resistant glass-fibre developed by B.R.E. and Pilkington Bros. Ltd. G.R.C. components are lighter and less bulky than r.c. products and do not need the 25 mm protective cover necessary for steel.

A wide range of commodities has already been made using C.R.C., including precast cladding panels, system-built housing based on timber framing, sewer linings, formwork, sheet piling, ducts, street furniture, roofing products, etc. At the time of writing it is not intended that it should be used for any primary structural purpose until more experience has been gained on its performance.

When used as formwork it can be moulded to complex shapes, receive a variety of surface treatments and colour applications. It can satisfactorily withstand fire tests, is totally resistant to fungal growth, termites, etc., and is particularly suited to tropical climates. However, it is expensive; though the cost is offset by its reduced weight, enabling hand labour to be used where otherwise heavy lifting-gear might be needed.

4 | Foundations

4.1 Site exploration

The size and shape of a building has an effect on the type of foundation used; though this is also dependent on soil and site conditions. Site investigation is a complex task for large-scale building and civil engineering, but in small and medium-range work the information gathered from trial holes is usually sufficient to make assessments. Wet and dry seasons can have an effect on soils, particularly at the time of change-over — which in some areas can be quite sudden. When water evaporates from a soil the grain-particles become more tightly packed. In porous soils and laterites such movement is limited, but in clays, especially in hot climates at the end of a wet season, the ground shrinks visibly. Fortunately most buildings in such areas have wide verandas or aprons surrounding the external walls and corners, which protect them from shrinkage and scour.

Sometimes local knowledge or existing records are available to help in deciding on the type of foundation. Mining activities, made-up ground, quarries, ditches, slow-moving slopes, ground faults, silty soils and peats could all produce unstable conditions. Evidence of flooding must be looked for; arid beds in dry seasons could become lakes or rivers when rains commence. Ground-water in underlying strata may be detected when digging trial holes. Local farmers and contractors can also be helpful.

Clays which shrink and swell according to the time of year are common in some zones, and the change effected by vegetation, roots, plants, and grasses can be quite striking. A building on shallow foundations close to vegetation in, say, savanna country is liable to seasonal movement as a result of root growth. However, the building itself tends to protect the ground to some extent from seasonal wetting and drying.

4.2 Types of foundations

Foundation design should aim at keeping differences in settlement to a minimum in order to prevent harmful distortion; the object should be to keep movement as slight as possible. If some change is expected then foundations and structure together should be sufficiently rigid to redistribute the load and keep settlement within limits. Whatever the type of building, the forces carried by the structure at ground level should be calculated. These should include superimposed loads, wind, suction and overturning stresses as well as the dead weight of the building. When the foundations have been designed, their own dead weight should be added to obtain the total bearing pressure on the soil beneath.

All soils will compress to some extent under load, causing cracking and distortion, however slight. Shallow foundations consisting of strip, pad footings, and rafts increase the pressure on uniform soil to a breadth and depth equal to approximately one and a half times the breadth of the foundation. The resulting settlement depends on the increase in pressure and on the soil properties in an imaginary envelope of these dimensions, Fig. 4.1. In shrinkable clays short piles bored by hand or mechanical augers are effective. Pad footings are also equally effective.

STRIP FOOTINGS. Standard types of footings for houses and single-storey buildings are shown in Volume 1. These are used where a sufficient depth of reasonably strong subsoil exists near the surface or, with a basement building, at the level of the lowest floor. The foundation is obtained by calculating the total load per metre run of the building including superimposed load and weight of structure by the safe bearing pressure of the soil. For single-storey buildings a depth of 150 mm is usually adequate, but beyond this it should not be less than the projection beyond the wall face. In districts subject to minor tremors or similar movements a reinforced-concrete strip may be used. In this case all steel at corners and junctions needs to be firmly joined, preferably by welding, including bases at cross walls. A simple strip footing for outhouses and small sheds is shown in Fig. 4.2., for cases where the hardcore does not exceed 600 mm in depth. Above this it is usual to use hollow concrete blocks filled with concrete up to ground level. Where the floor is to be used as a garage or workshop it may be reinforced with light mesh and the wall strengthened with 10 mm rods placed in alternate cavities, Fig. 4.3.

PAD FOUNDATIONS: These are frequently used for isolated columns of concrete, steel, or timber. The area of the base is usually calculated according to the load but the depth and the reinforcement is supplied by the structural engineer. Houses and similar dwellings have long been raised on piers, which permit the shaded area beneath the house to be used for storage, recreation, or livestock. Though this practice is now decreasing, pads are still the normal method of supporting columns in

Soil type and site condition	Foundation	Details	Remarks
Rock, solid chalk, sands and gravels or sands and gravels with only small proportions of clay, dense silty sands	Shallow strip or pad footings as appropriate to the load-bearing members of the building	Breadth of strip footings to be related to soil density and loading (see Fig. 4.2). Pad footings should be designed for bearing pressures tabled in CP 101 : 1963. For higher pressures the depth should be increased and Civil Engineering Code of Practice No. 4. 'Foundations' consulted	Keep above water wherever possible. Slopes on sand liable to erosion. Foundations 0.5 m deep should be adequate on ground susceptible to frost heave although in cold areas or in unheated buildings the depth may have to be increased. Beware of swallow holes in chalk
Uniform, firm and stiff clays: (1) Where vegetation is insignificant	Bored piles and ground beams, or strip foundations at least 1 m deep	Deep strip footings of the narrow widths shown in Fig. 4.3 can conveniently be formed of concrete up to the ground surface	
(2) Where trees and shrubs are growing or to be planted close to the site	Bored piles and ground beams	Bored piles dimensions as in page 65.	Downhill creep may occur on slopes greater than 1 in 10. Unreinforced piles have been broken by slowly moving slopes
(3) Where trees are felled to clear the site and construction is due to start soon afterward	Reinforced bored piles of sufficient length with the top 3 m sleeved from the sur-rounding ground and with suspended floors, or thin reinforced rafts supporting flexible buildings, or base-ment rafts		

Soft clays, soft silty clays	Strip footings up to 1 m wide if bearing capacity is sufficient, or rafts	See page 63 and CP 101 : 1963	Settlement of strips or rafts must be expected. Services entering building must be sufficiently flexible. In soft soils of variable thickness it is better to pile to firmer strata below (See Peat and Fill below)
Peat, fill	Bored piles with temporary steel lining or precast or *in situ* piles driven to firm strata below	Design with large safety factor on end resistance of piles only as peat or fill consolidating may cause a downward load on pile (*see Digest 63*) Field tests for bearing capacity of deep strata or pile loading tests will be required	If fill is sound, carefully placed and compacted in thin layers, strip footings are adequate. Fills containing combustible or chemical wastes should be avoided
Mining and other subsidence areas	Thin reinforced rafts for individual houses with load-bearing walls and for flexible buildings	Rafts must be designed to resist tensile forces as the ground surface stretches in front of a subsidence. A layer of granular material should be placed between the ground surface and the raft to permit relative horizontal movement	Building dimensions at right angles to the front of long-wall mining should be as small as possible

Table 1. Choice of foundation

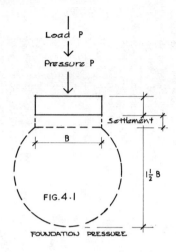

Load P

Pressure P

settlement

B

1½ B

FIG. 4·1

FOUNDATION PRESSURE

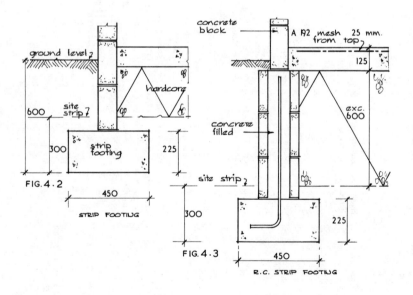

ground level

concrete block

A 142 mesh 25 mm. from top

125

hardcore

600

site strip

exc. 600

concrete filled

300

strip footing

225

site strip

FIG. 4·2

450

STRIP FOOTING

300

FIG. 4·3

450

225

R.C. STRIP FOOTING

SHALLOW FOUNDATIONS.

system or industrial building or where large numbers are required quickly as with housing schemes. In large modern buildings where open space is needed at ground level, isolated piers on pads are widely used.

SLABS. As a combined floor- and ground-beam, slabs are the most common type of foundation. They are convenient for forming verandas around buildings with the edge thickened to contain hardcore and prevent scour. This also serves as a good base for roof posts. The underside of the slab can also be thickened to form a strip foundation for walls and partitions.

RAFTS. Where the filling is unsound or where uneven bearing cannot be avoided, rafts are sometimes used, Fig. 4.4. They are usually reinforced and extend beyond the outer walls of the building to give extra bearing. Where upward pressure due to water may be heavy the raft is designed as an inverted slab-and-beam floor, as shown in Volume 1. Foundations are taken down to firm ground with piers on pads, or piles support the ground beams. Figs. 4.4 and 4.5 show the plan of a typical r.c. slab for part of a community hall or similar building, with *in situ* columns supported by shallow foundations. These could be on piles in place of pads or, where the ground is poor, a raft could be placed over the whole site. In this case pads have been used. Only sufficient detail has been given to show the principle of ground beam-and-slab construction.

Hardcore is usually needed on site to raise the slab to the required finished floor level (F.F.L.). After removing the top layer of soil and vegetation the hardcore is laid, watered, and rolled or tamped in layers of about 200 mm. The materials used should be chemically inert, easily transported, and available in sufficient quantities. If they are to be in contact with Portland cement, soluble sulphate content should be limited to 0.5 per cent. Materials that swell on wetting should not be used. Clinker sintered into hard lumps is as good as rubble; as is gravel, quarry waste, or stone. In some areas, where hardcore is not available, earth-fill is used, rolled and tamped to maximum compaction.

4.3 Piles

These are not yet used extensively in tropical countries, chiefly owing to the difficulties of manufacture and driving, except in capital cities where equipment is available. As their use is growing, however, the more usual types of equipment and methods of driving are given here. Many systems are obtainable and a knowledge of soils and foundations is needed in order to select the most suitable one for the job. Generally

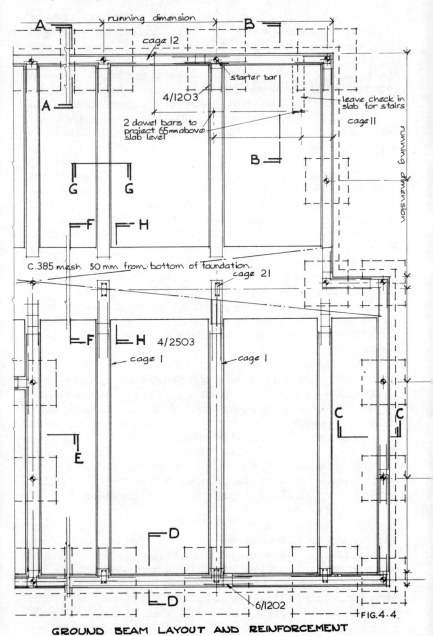

A

running dimension

B

cage 12

starter bar

4/1203

2 dowel bars to
project 65mm above
slab level

leave check in
slab for stairs

cage 11

A

B

G   G

F   H

C.385 mesh 50 mm from bottom of foundation.

cage 21

F   H   4/2503

cage 1   cage 1

C   C

E

D

D   6/1202   FIG.4.4.

GROUND BEAM LAYOUT AND REINFORCEMENT

piles may be supported by friction, or by firm soil at a lower level, or by square-ended piles driven into sand or gravel to act as a column. The choice will depend on the nature of the soil − on whether it is aggressive and liable to attack the pile, particularly in coastal areas where the use of piles is most common.

TYPE OF PILES. Piles may be of timber, r.c., or steel. They can be of the displacement or replacement type. Displacement piles are inserted by driving and forcing the soil away to make room for the pile. They may be made of precast concrete (Volume 1), prestressed, Fig. 4.7, or of hollow tubes with the lower end enclosed, thus displacing the soil as it is driven, Fig. 4.8. They may also be of hollow steel tubes driven with a detachable steel shoe at the bottom; as the tube is withdrawn the shoe or plug is left behind, Fig. 4.10. Concrete may also be poured into the tube and consolidated by vibration or compressed air, while at the same time the tube is slowly withdrawn and the concrete thus packed tightly against the soil. Sometimes tubes are driven and the concrete compacted simultaneously by vibration which keeps the piles moving continuously under the weight of the vibrating hammer and the pile itself, Fig. 4.8.

Replacement piles are usually formed by driving open-ended tubes of steel or concrete into the soil as far as needed. The hole is then cleaned out, a helical cage of reinforcement inserted if required, and concrete poured in as described. Compression of concrete during placing is also done by a drop hammer working inside the helical reinforcement. The usual method of installing a pile is by means of a drop hammer raised by a winch and allowed to drop on the head, Fig. 4.9. This system is gradually being replaced by hammers driven by steam, diesel, or compressed air.

Bored piles. These are often used where ground conditions permit. Mechanically-driven augers drill holes in fairly firm ground at the same time extracting the soil. The holes are then filled as described. Short bored piles are popular for shallow foundations and cohesive soils free from ground-water, and can penetrate beds of clay up to about 4 m in depth. They can be constructed from 250 to 900 mm diameter. A rough guide for calculation purposes is that in stiff clay a 300 mm pile under 4 m long would carry a load of 75 KN. The spacing depends on the structure, loading, and length of the pile. Once they are in position they are spanned, in the same way as precast driven piles, by shallow r.c. beams on which the structure is to be built. Hand augers may be used for simple work, and also for soil testing, lamp standards, flag

continued on page 71

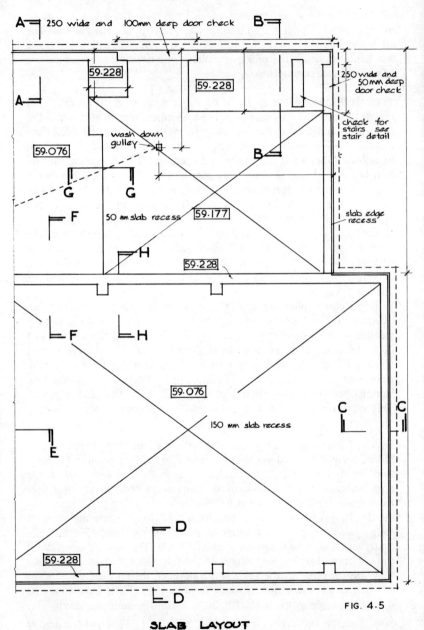

A 250 wide and 100mm deep door check B

59.228

59.228

250 wide and 50 mm deep door check

check for stairs see stair detail

wash down gulley

59.076

G G

slab edge recess

F

50 mm slab recess

59.177

H

59.228

F H

59.076

150 mm slab recess

C C

E

D

59.228

D

FIG. 4.5

SLAB LAYOUT

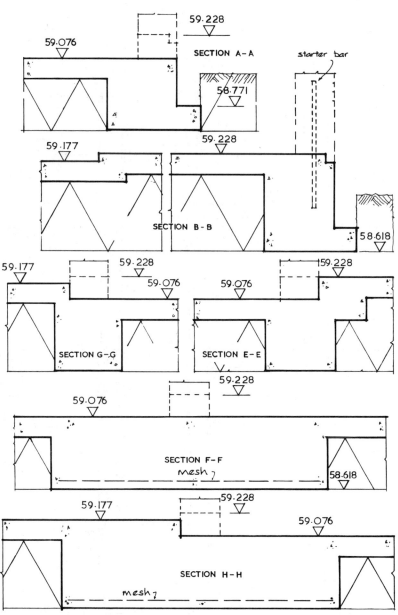

SLAB AND RAFT SECTIONS.

FIG. 4.6

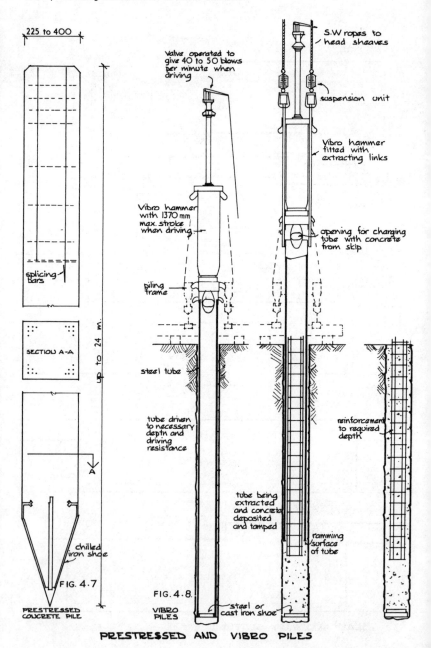

225 to 400

splicing bars

SECTION A-A

up to 24 m.

chilled iron shoe

FIG. 4.7

PRESTRESSED CONCRETE PILE

Valve operated to give 40 to 50 blows per minute when driving

Vibro hammer with 1370 mm max. stroke when driving

piling frame

steel tube

tube driven to necessary depth and driving resistance

tube being extracted and concrete deposited and tamped

FIG. 4.8

VIBRO PILES

steel or cast iron shoe

S.W ropes to head sheaves

suspension unit

Vibro hammer fitted with extracting links

opening for charging tube with concrete from skip

ramming surface of tube

reinforcement to required depth

PRESTRESSED AND VIBRO PILES

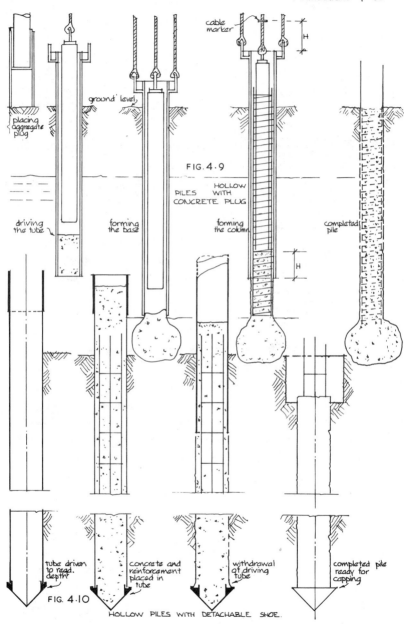

FIG. 4·9

HOLLOW PILES WITH CONCRETE PLUG

placing aggregate plug

ground level

driving the tube

forming the base

forming the column

cable marker

H

completed pile

H

FIG. 4.10

tube driven to reqd. depth

concrete and reinforcement placed in tube

withdrawal of driving tube

completed pile ready for capping

HOLLOW PILES WITH DETACHABLE SHOE.

HOLLOW CONCRETE PILES

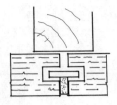 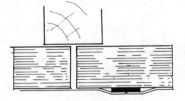

A...Foamed polyurethane strip well compressed in butt joint formwork.

B...Foamed polyurethane strip well compressed in tongued and grooved formwork.

C.: Joint in lining with masking tape to sheet metal or hardboard lining.

FIG. 4.11   LEAK PROOF TIMBER FORMWORK.

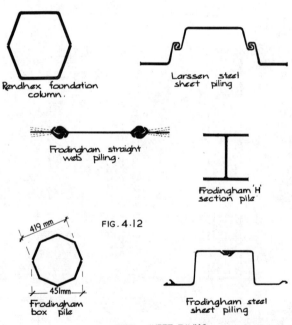

Rendhex foundation column.

Larssen steel sheet piling

Frodingham straight web piling.

Frodingham 'H' section pile

419 mm

FIG. 4.12

451mm

Frodingham box pile

Frodingham steel sheet piling

STEEL SHEET PILING

TIMBER FORMWORK   **STEEL SHEET PILING**

poles, shallow wells, temporary latrines, and gate-posts.

Timber piles are not used extensively, as wood of such dimensions could be more profitably used for other purposes. Where they are employed they are made of sufficient length to enable the tops to be cut off after driving. The same applies to precast and prestressed piles as splicing can be difficult and expensive.

Steel piles of 'H' section are manufactured, but are used mainly by civil engineering contractors on large projects, Fig. 4.12.

Sheet piling is widely used in both building and civil engineering work and may be of wood, concrete, or steel. It can be employed for temporary work on slopes, embankments, cofferdams, streams and culverts, retaining walls, etc. Wood is used where available by driving planks, usually vertically, between guide frames of strong horizontal members and sealing the vertical joints to make them watertight. Permanent sheet piling may be of r.c., precast and driven in the same way by drop or steam hammer. Piles are obtainable in a range of cross-sections and lengths from 300 to 100 mm wide and up to 6 m long. After the tops are cut off level, the ground beam or capping is cast *in situ* along the top to hold them firmly in place.

For ease of placing and general convenience steel sheet piling is most widely used, Fig. 4.12, but it can be expensive if imported. It is employed for both permanent and temporary work as it can be readily extracted. Steel sheet piling for permanent installation should be coated before driving and this should be renewed on exposed parts periodically as required.

Before commencing piling operations it is necessary to consider any information gained from site investigation, records, estimated load to be carried by the pile, degree of permitted settlement, and the need to protect adjacent property from inconvenience and damage.

4.4 Removal of water

Two main problems concerned with water usually arise in foundation work, viz. rising damp and percolation. Where materials used in the building permit, horizontal sawcuts can be made through the wall and a felt or copper strip inserted at the usual d.p.c. level. A d.p.c. can also be made by injecting water-repellent substances into holes at the base of the wall. The best materials appear to be those employing a siliconate solution in water or a siliconate/latex mixture. After injection the material changes to form a water-repellent layer. It may also be used in retaining and basement walls, Fig. 4.13.

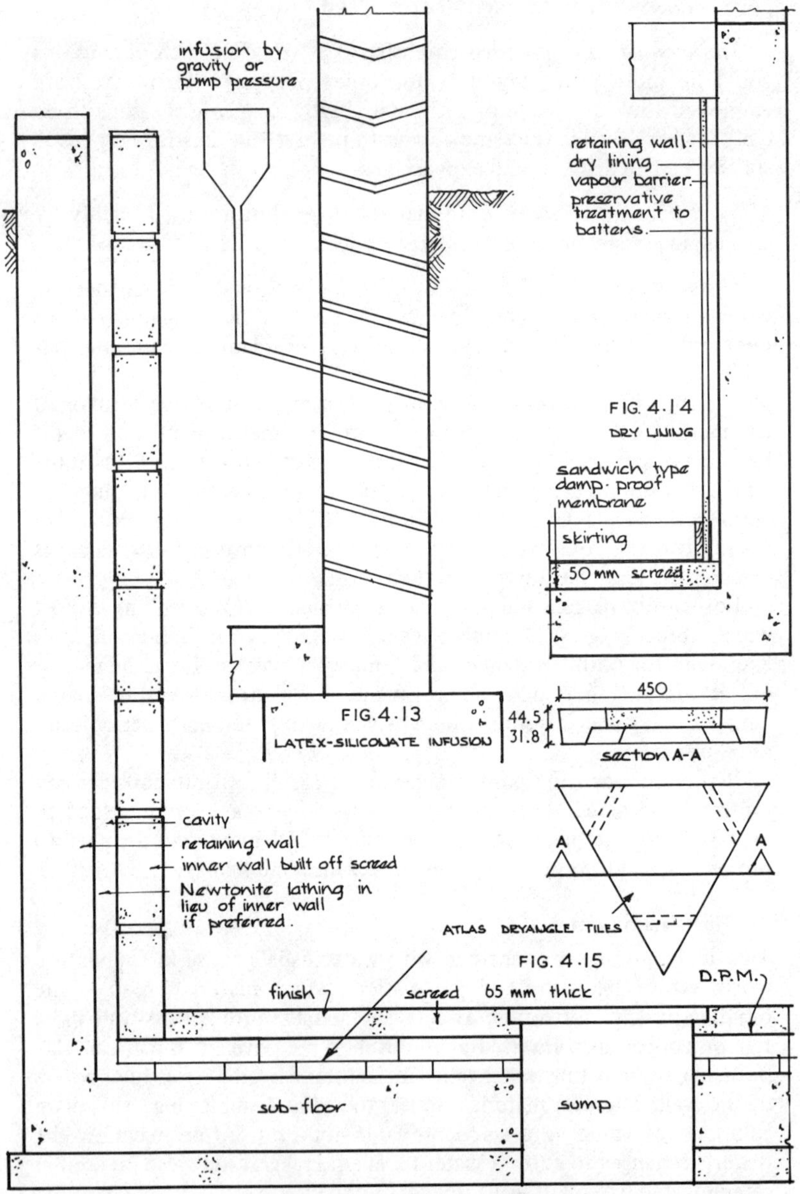

infusion by
gravity or
pump pressure

retaining wall
dry lining
vapour barrier
preservative
treatment to
battens.

FIG. 4·14
DRY LINING

sandwich type
damp·proof
membrane

skirting

50 mm screed

FIG.4·13
LATEX-SILICONATE INFUSION

450

44.5
31.8

section A-A

cavity
retaining wall
inner wall built off screed
Newtonite lathing in
lieu of inner wall
if preferred.

ATLAS DRYANGLE TILES

FIG. 4·15

A   A

D.P.M.

finish   screed   65 mm thick

sub-floor   sump

WATERPROOFING BASEMENTS

Electro-osmosis is another process used to prevent the rise of water in a wall, and this is done through diffusion by electrical means. It is not to be recommended for humid climates because of possible corrosion of the electrodes used to control the rise of moisture.

Ground-water. The ground-water condition of sites needs to be examined not only because of rising damp but because of constructional difficulties which may arise as a consequence. Trial holes should reveal most of the dangers arising from the different strata. Disposition of layers should be noted, such as fine sand below water level which might otherwise have to be excavated when the seasons change. Clay at foundation level could also confine water underneath, which could cause uplift during excavation. Dangers of flooding should also be kept in mind.

A simple way of preventing rising damp, if circumstances permit, is to provide adequate drainage and so lower the ground-water. Aprons, footpaths and other impervious surrounds can help in many cases by protecting the footings. To protect inside walls from damp at ground level, wall linings are sometimes used but cannot be considered a permanent cure. The wall is lined with plasterboard on timber battens, coated with preservative to prevent rotting, and the back of the board treated with fungicide, Fig. 4.14.

4.5. Basement and retaining walls

In buildings of up to about three storeys the use of basements is now unusual. They may be necessary, however, as sumps, lift-wells, cellars, and workshops for heavy machinery. They may also be only partly below ground, for instance on sloping sites. In the event of ground-water being present several methods of construction may be adopted, depending on the severity of the problem. To resist upward pressure, where this is not too great, a normal unreinforced floor may be laid containing a sandwich damp-proof membrane of plastic sheeting, as for ground-floors. Should this be insufficient a reinforced loading floor would be necessary, Fig. 4.14.

Walls may be waterproofed by building a concrete-block wall of sufficient thickness to prevent overturning. The inside is then treated using a waterproof rendering, vertical d.p.m. or other means, and an inner skin of blocks built to hold the d.p.m. in position. It is unwise simply to render the inside wall of the basement as the water pressure could be strong enough to force the rendering off. The same applies to floor screeds.

It is sometimes difficult to make an underground structure waterproof.

Even though it may seem waterproof at first, leaks could occur later. For this reason it is wise to leave all basement walls accessible so that they may be treated further if necessary. If leaks are bad, pressure grouting could be used by injecting grout through small holes drilled through the wall, or latex-siliconate could be infused by pump, Fig. 4.13. Concrete basements without a waterproof membrane, although waterproof, may not be vapourproof. To overcome this linings may be added as described above. The floor can be treated with a polythene d.p.m. placed over the whole floor and covered with 50 mm of fine concrete, Fig. 4.14.

CONCRETE BASEMENTS. Where basements are to be constructed of reinforced concrete properly proportioned and workable mix is essential, Fig. 4.16. The concrete must be of good quality and have a compressive strength of at least 27 MN/m^3. Before starting work the ground-soil must be firm and clean; oversite concrete is desirable as a casting slab for the whole floor. It may consist of a lean mix but should be well-compacted and grout-proof. The placing of concrete should be planned beforehand: it must be poured continuously between predetermined positions, Fig. 4.17, and protected from the sun's rays. It should be vibrated if possible and all joints constructed as described in Volume 1. When working in waterlogged soils it is essential that the water-pump should continue working.

Formwork. When casting basement walls *in situ* it is necessary that the formwork should be properly constructed with leakproof joints to prevent loss of grout. Three methods of sealing formwork are shown in Fig. 4.11. The use of well-compressed foam-plastics is desirable at all joints liable to open because of water-pressure or shrinkage.

Construction joints are usually made with the aid of strips or wood-stops nailed to the formwork. Extra care needs to be taken to remove laitance, preferably by spraying the concrete with a mist of clean water just before the concrete hardens. Construction joints in the floor slab are made as shown in Fig. 4.17. Vertical joints are made in the same way as horizontal ones, and the joint between base slab and wall is made as in Fig. 4.16. The formwork must be strong enough to resist any movement in the slab-concrete after laying. A poker vibrator is essential for sound concrete in this position.

One method of casting a pad foundation in a basement is shown in Fig. 4.19. To prevent grout-loss the pad-base formwork is placed on oversite concrete and sealed with mortar. The base is then poured, together with the column kicker. After the next lift of the column is

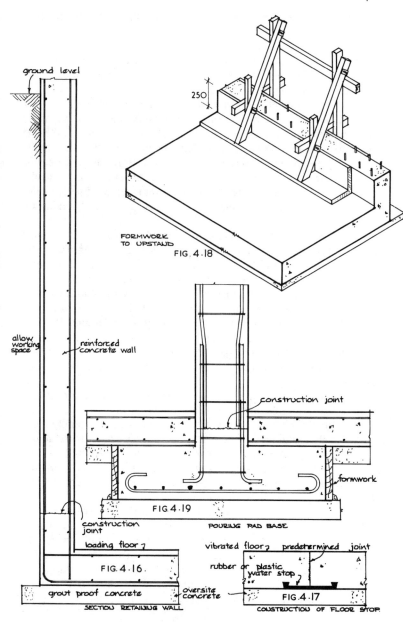

ground level

250

FORMWORK
TO UPSTAND
FIG. 4.18

allow
working
space

reinforced
concrete wall

construction joint

construction
joint

formwork

FIG. 4.19

POURING PAD BASE

loading floor

FIG. 4.16.

grout proof concrete

oversite
concrete

SECTION RETAINING WALL

vibrated floor   predetermined   joint

rubber or plastic
water stop

FIG. 4.17

CONSTRUCTION OF FLOOR STOP.

REINFORCED CONCRETE BASEMENTS

cast, the base formwork is removed and oversite concrete or floor completed.

CAVITY BASEMENT WALLS. The normal practice when building retaining walls in basements is to make the wall completely impermeable and to seal off all seepage. This can sometimes be difficult. An alternative method is to erect a normal concrete wall, either cast *in situ* or made of hollow block, concrete-filled for stability, Fig. 4.15. The floor is then covered with special grooved triangular tiles. On top of these is placed a d.p.m., and the whole screeded. The inner cavity is then built. Any water penetrating the outer wall now runs down inside the cavity and across the floor between the tiles into a collecting sump. Special care is necessary to ensure that the sump water is removed periodically as it could become a breeding place for cockroaches and malarial larvae.

RETAINING WALLS. The most common materials for retaining walls are brick, stone, and concrete. The nature, shape, and density of the material used dictate the type of wall to be built; in this respect local custom is a good guide. The traditional design is the wedge shape, Fig. 4.20, the size and shape of which is determined by the stability of the soil it has to support. In order to prevent a build-up of hydrostatic pressure behind the wall, loose rubble is usually packed against it as the soil is being returned, filled, and rammed, thus providing an outlet for water percolation. This discharges either into a channel at the bottom of the wall or through weep holes placed at regular intervals in the face of the wall.

There are a number of types of retaining wall. Two cantilever designs, both of which are built in concrete, are shown in Figs. 4.21 and 4.22. Fig. 4.21 shows the more economical design, but 4.22 is generally more suitable for the building of basements, particularly where these adjoin main roads. They are often built with a check to prevent sliding, Fig. 4.23. Typical reinforcement is shown; the size of bars depends on the stresses involved. Retaining walls are sometimes of r.c. counterfort pattern, with buttresses built in as required, Fig. 4.24.

Where the length and purpose of the wall justifies it, precast concrete 'L' shaped units can be made into cantilever form and lifted into position. They may be used for earth-retention or bulk storage as required. Provision could be made in the base of the unit to prevent lateral movement and to allow the wall to be anchored in place. The weight of the earth or storage material helps to resist overturning. Where the demand is sufficiently great these sections can be supplied by proprietary firms.

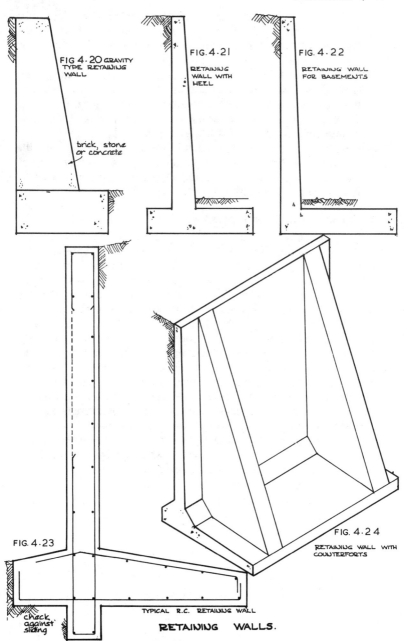

FIG 4.20 GRAVITY TYPE RETAINING WALL

brick, stone or concrete

FIG. 4.21 RETAINING WALL WITH HEEL

FIG. 4.22 RETAINING WALL FOR BASEMENTS

FIG. 4.23

check against sliding

TYPICAL R.C. RETAINING WALL

FIG. 4.24 RETAINING WALL WITH COUNTERFORTS

RETAINING WALLS.

Another type of wall consists of precast concrete hollow-block units built to a battered profile from one to two metres thick where the backfill is allowed to enter the interstices, thus reducing the hydrostatic pressure and adding to the stability.

Perhaps the most common method of retaining soil is by using sheet piling, which acts as a vertical cantilever when driven into the subsoil. The tops of the piles are then encased in concrete capping, which also acts as a base for a crash-barrier or handrail, or as a kerb to the enclosed area. Where the piling is insufficiently stable to retain soil cables may be anchored to concrete stays embedded in the soil itself, so holding the piles in place. Sheet piling used in conduits or against banks of narrow streams can be braced across by flying buttresses; or the stream can be wholly encased in a box culvert if preferred.

4.6 Stepped and staggered foundations

Referring to the setting out drawing Fig. 2.9, it will be seen that the semi-detached houses Nos. 14 and 15 have stepped foundations which are also staggered, Fig. 4.25. The difference between finished floor levels (F.F.L.) is 500 mm. It has been assumed that the houses are part of a precinct to be composed of prefabricated components, either of precast concrete or timber frame. These would be placed on a base comprising concrete edge-beam and floor, Fig. 4.26. With no strip footings as such, the weight of the components is carried by pads, the size and position of which depend on the component design. The purpose of the edge-beam is only to contain the hardcore and to form a plinth below the superstructure, which is of fairly light construction.

In order to determine the depth of the edge-beams and the supporting pads, consider the table below which has been prepared from the setting out plan, Fig. 2.9. From the contours shown, the ground-levels at each corner of the house plan have been interpolated, and it is these levels, together with the F.F.L.s, which govern the edge-beam depths. There are two conditions to be observed: first, that the minimum depth of the edge-beam is to be 300 mm from F.F.L.; and next, that the bottom of the edge-beam is to be not less than 150 mm below site strip level. The depths have been worked out as follows:

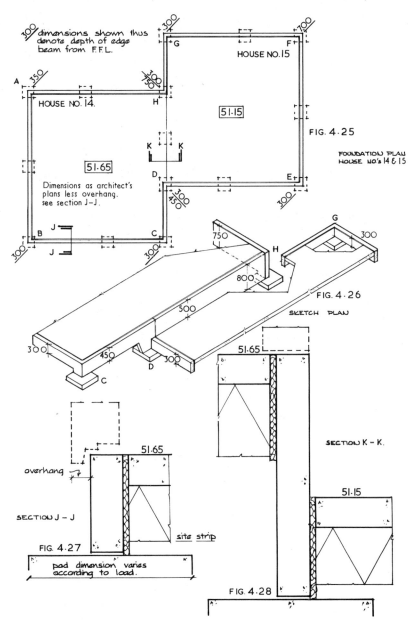

300 dimensions shown thus denote depth of edge beam from F.F.L.

HOUSE NO.15

HOUSE NO.14.

51.15

FIG. 4·25

FOUNDATION PLAN
HOUSE NO's 14 & 15

Dimensions as architect's plans less overhang. see section J–J.

51.65

750

300

800

FIG. 4·26

SKETCH PLAN

500

300

450

300

51.65

SECTION K – K.

overhang

51.65

51.15

SECTION J – J

site strip

FIG. 4·27

pad dimension varies according to load.

FIG. 4·28

STEPPED AND STAGGERED FOUNDATIONS

Hse.No. & F.F.L.	Location	Actual level	Strip site level	Bottom of beam level	Depth from F.F.L. (min)
14.	A.	51·60	51·45	51·30	350
51·65.	B.	51·83	51·68	51·53	300
	C.	51·65	51·50	51·35	300
	D.	51·50	51·35	51·20	450
	H.	51·20	51·05	50·90	750
15.	D.	–	–	50·85	300
51·15.	E.	51·15	51·00	50·85	300
	F.	50·75	50·60	50·45	700
	G.	51·15	51·00	50·85	300
	H.	51·20	51·05	50·90	300

Only the corner locations have been given. The intermediate pads may be used when a change of depth in the edge-beam is necessary. Note also that the depth of the pads is not always the same as the edge-beam depth at the corners. (See location 'C'.) When depths of beams exceed 600 mm it is usual to provide light reinforcement to prevent the beams from bulging outwards. This may consist of two 10 mm bars placed on the outside edge of the beam, one above the other, to take up tensile stresses. It should be noted that on sites subject to scour, care must be taken to protect shallow foundations.

5 | Walls, beams, and columns

5.1 Introduction

Traditional building usually makes a clear distinction between walls, beams, and columns. Such structures are usually unframed, and the functions of each are clear and understood. In modern construction the distinction is less clear. A wall may still be defined as a vertical bearing member whose width is more than four times its thickness — which distinguishes it from a column. But the introduction of reinforcement into a wall can convert it, in part, into a beam or a column, and some knowledge of the theory of structures is now essential. For convenience, however, these headings will here be maintained.

WALLS

5.2 Concrete

Walling has already been discussed in Volume 1, and further examples of block bonding are given in Figs. 5.1, 5.2, and 5.3, which apply to both brick and blockwork. Cladding and concrete panels will be dealt with later in the chapter.

Concrete is possibly the most widely-used material in building today, and the appearance of the finished product is nowadays given close attention, whether precast or cast *in situ*. Much depends on the type of formwork used, which may be of hardwood, plywood, sheet metal, or plastic-faced plywood. (Plastics are quite widely used in Europe.) Inset patterns can be formed by securing plastic or rubber discs or other shapes to the formwork before casting. When a first-class finish is required it is advisable to seal the formwork joints as shown in Fig. 4.11. Formwork may also be secured by the method shown in Volume 1, or by using the many proprietary methods available through builders' merchants. Wire ties are often used as a cheap and effective method of securing formwork, though this should be avoided where there is a risk of rust stains.

Precast concrete slabs and similar units are sometimes cast on expanded polystyrene to form a permanent face anchored to the

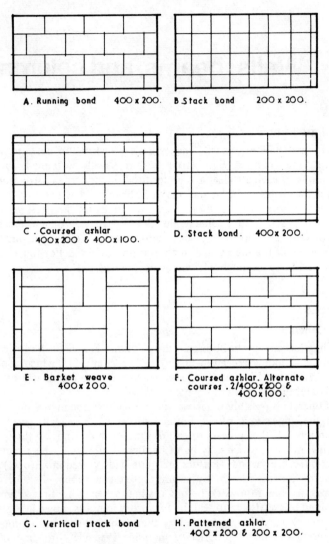

A. Running bond 400 x 200.

B. Stack bond 200 x 200.

C. Coursed ashlar 400 x 200 & 400 x 100.

D. Stack bond. 400 x 200.

E. Basket weave 400 x 200.

F. Coursed ashlar. Alternate courses. 2/400 x 200 & 400 x 100.

G. Vertical stack bond

H. Patterned ashlar 400 x 200 & 200 x 200.

FIG. 5.1. BLOCK BONDING

concrete; or sculptured forms can be cast into the fabric of the structure. As well as reinforced plastics, polypropylene panels may also be used as moulds, though these are mainly for coffered floors. Polythene film over formwork is excellent for precast work and for lining wooden moulds. As this is leakproof, close watertight joints are not so important, though wrinkles have to be avoided if a smooth face is required. Expanded polystyrene may also be reshaped for prestressed hollow members but, is not strong enough for repetitive casting.

Apart from formwork, variations in finish can also be effected by use of cements of different colours, or by treating the surfaces to reveal the colour of the aggregate. White, black, red, or other colours can be imported but they are expensive and need care in storage. The texture of the surface is also a factor affecting appearance; texture can be conditioned by the grading of aggregates, cement content, or water/cement ratio. A poor texture is also often due to poor workmanship, which may result in rough formwork, gaps in shuttering, irregular construction joints or faults due to slipshod pouring. Poor grading of concrete also causes deterioration, even with normal weathering. Lack of quality-control could result in shrinkage-cracking, pyrite-staining, sulphur attack, and corrosion from salt and sea water. Failure of concrete may be due to inadequate cover, poor aggregate, badly-made formwork, poor compaction, and poor construction joints.

Though it is difficult in the tropics to obtain a concrete finish derived from the grain of the wood, it is possible to vary the width of the boards, using different kinds of timber, or to line the formwork with hessian or similar material. Timber or rubber strips may also be nailed to the formwork before pouring. Shot-blasting or bush-hammering is also popular, but care must be taken to maintain sufficient cover to protect the reinforcement. Precast or natural stone can also be secured to the concrete face as shown in Volume 1. Sprayed cement or other rendering could also be applied.

The design and appearance of both concrete and blockwork can be marred by streaking, crazing, pattern staining, penetration by water, metal corrosion, and around areas of timber decay. Dirty surfaces are normally cleaned by water spray, or by grit blasting in severe cases. Walls can also be damaged by ground movement, shrinkable clays, expansion caused by corrosion, unsound materials and manufacture, and salt. Behaviour could also be affected by solar radiation, temperature, rainfall, humidity, termites, earthquakes, and hurricanes.

STRENGTH AND STABILITY. It is desirable that the principles of structural mechanics should be applied where there may be a risk of

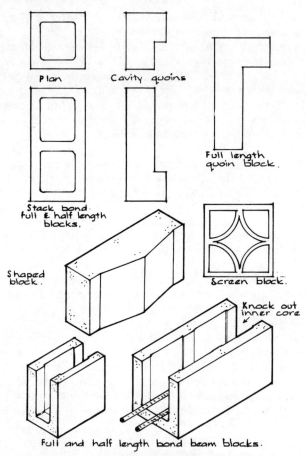

Plan

Cavity quoins

Full length
quoin block.

Stack bond
full & half length
blocks.

Shaped
block.

Screen block.

Knock out
inner core

Full and half length bond beam blocks.

FIG. 5.2   SHAPED BLOCKS

failure. Retaining walls are a typical example. It is also necessary to take into account those factors affecting stability, such as foundations, subsoil, estimated wind load, estimated live and dead load, local stresses, mortar jointing, and strength of materials. There may be other factors to consider such as scour, insulation, movement from external sources, eccentric loading, effective height, and whether the wall is free-standing or supported. When heavy beams or girders bear upon a wall it is usual to provide support with pads of stone or concrete to spread the load. Wall plates serve the same purpose for roof-rafters and floors.

Investigation of wind pressures may be undertaken in accordance

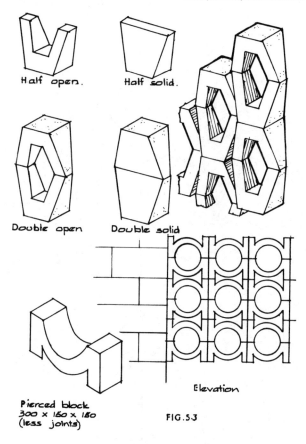

Half open.

Half solid.

Double open

Double solid

Pierced block
300 x 150 x 150
(less joints)

Elevation

FIG.53

SCREEN WALLING

with C.P.3, Chap. V, part 2 as described in Chapter 1. Effective height,
lateral support, and effective thickness may be determined by using
C.P. 111 where it applies. Structural walls need to be tied, and free-
standing ones designed to withstand wind pressure. Walls may fail when
they cause the actual ground beneath the base to exceed the permissible
ground pressure; or when the panel between the buttresses is too thin;
or when the weight of the wall itself is insufficient to prevent overturning; or when the wall fractures at ground or d.p.c. level. The strength
of the brick or block walls in relation to their size, thickness, or height,
together with bonding of plain or reinforced brickwork, has been

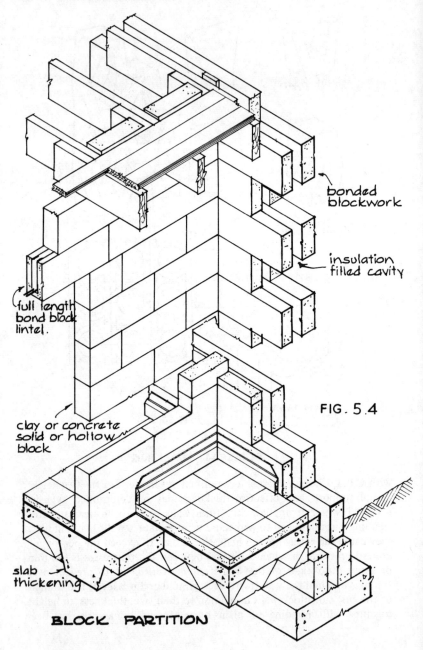

bonded
blockwork

insulation
filled cavity

full length
bond block
lintel.

clay or concrete
solid or hollow
block

FIG. 5.4

slab
thickening

BLOCK PARTITION

analysed by building and engineering laboratories. C.P. 111 and similar publications give details of structural strength and dimensional stability of walls. Movement, for instance, could be caused by elastic and plastic deformation, chemical action, temperature change, or moisture content. Panels should be separated by straight joints so that tensile stresses can be relieved in predetermined positions. Reinforcement should be placed in areas of high concentrated stress such as corners of door and window openings.

High-strength bricks need 1:3 mortar and the slenderness ratio of the wall should not exceed 18 except in dwellings of two storeys, when it may be taken as 24. Wind loads greatly affect the stability of the walls, as much by suction as by wind pressure. Traditional designs may be used in exposed areas if reinforcement is placed in the bed joints as shown in Figs. 5.2 and 5.6. Thin wall sections of both brick and concrete can create problems of dimensional stability, particularly thermal expansion. Expansion joints are needed in long walls at, say, ten metre intervals to avoid cracking. In earthquake areas and where ground conditions are unstable, foundations may have to be strengthened and wall reinforcement added. One well-known method is that of the Quetta bond, Fig. 5.6, which consists of a one-and-a-half-brick Flemish bond, built to permit vertical reinforcement to be placed, and filled from above. Another method is to use a rat-trap bond, i.e. a Flemish bond of bricks laid on edge, with gaps filled as before. Examples of reinforced walls are given in Figs. 5.6b and c.

WALL PANELS. Although traditional brick- and block-laying is still the standard method of wall building in the tropics, there is a growing tendency in Europe to prefabricate brick panels by hand in jigs, and reinforce as required. Prefabricated brickwork includes blocks. Three modes of assembly are used: first, jig laying *in situ*, i.e. laying by masons but using jigs to preserve accuracy; second, horizontal casting in jigs; and third, vertical casting. Jig laying has been widely used in areas where skilled craftsmen are scarce; and jigs may be made locally. The simplest form consists of vertical posts or steel scaffolding tubes placed at each corner, with nails or sawcuts inserted at brick-course heights. These keep the perpendicular and horizontal joints true and even. Another simple device is a light metal frame which slides horizontally along the course below to keep the wall plumb and the courses regular. Doors and windows are built in the normal way.

Where a first-class finish is required, panels may be prefabricated and lifted into position. This method is quick and clean, but not the most economical, as a crane is usually needed to lift the panel into position.

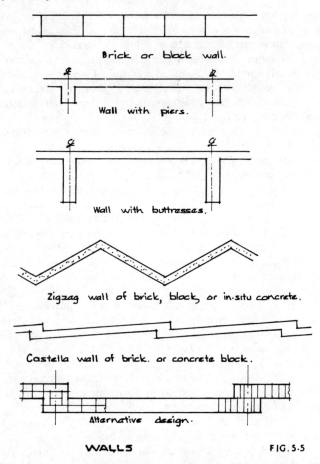

Brick or block wall.

Wall with piers.

Wall with buttresses.

Zigzag wall of brick, block, or in-situ concrete.

Castella wall of brick. or concrete block.

Alternative design.

WALLS   FIG. 5.5

Horizontal casting is usually done by means of a space-grid of simple design, on which bricks are laid, mortar inserted between the joints and the slab added. The process can be refined by using a foam-rubber base on which the bricks rest face downward, which prevents the mortar staining the face. The panel can also be vibrated, and horizontal and vertical reinforcement added if required.

Vertical casting is also used as a proprietary system whereby bricks are placed in position between upright formwork lined with foam-rubber to protect the face, and the whole clamped together to hold the bricks in place. Liquid mortar is then poured from the top of the panel,

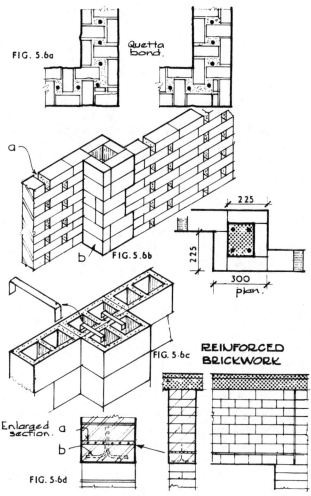

FIG. 5.6a   Quetta bond.

FIG. 5.6b

225
225
300
plan.

FIG. 5.6c   REINFORCED BRICKWORK

Enlarged section.

FIG. 5.6d

which has already been reinforced if necessary.

Concrete panels. There are a number of industrial systems which cater for the production of housing units of twenty or more houses, and also for medium-sized community halls and public buildings. Attempts have been made to set up factories in tropical countris to ease housing problems in large towns. To remain viable, however, such plants need continuity of orders, together with assured supplies of raw materials, equipment, and good organization. Transport is becoming less of a problem as communications improve and it is possible that industrialized production may be more widely adopted. The finished product is of standard quality, erection is quick and the problem of unskilled labour is largely overcome. Systems vary, but a fairly standard type of low-rise housing and community hall consists of concrete wall-panels of single storey height delivered to site with glazed windows complete if desired. The external walls are coffered on the inner face, and a lining of expanded polystyrene bonded to plasterboard is fixed after erection. The amount of insulation would depend on climatic conditions, see Figs. 5.10 to 12.

A number of building firms in Europe now produce wall-panels in a wide variety of shapes, sizes, and finishes which enjoy particular favour as external covering to buildings where repeated use of the same design is possible. Fig. 5.8 shows one system. In this case the main wall panels (panel C) are 8 metres long and weigh 6 tonnes. They are hoisted into position by means of cast-in lifting hooks, later cut off if necessary, and held to the main concrete frame by means of non-ferrous cramps. In areas of high pollution or corrosion it is better they should be of stainless steel which, though more expensive than bronze alloy, will last indefinitely. The window openings are formed by means of spacer units, panels A and B. Another type of cladding has a ribbed finish, Fig. 5.9. This section shows a typical coping unit, the length depending on the unit module used. With irregular shapes such as these, lifting points have to be worked out in order that the panel should remain vertical during hoisting and fixing. Such panels as these are made to a close tolerance and to a designed mix of 35 MN/m^3.

Wall blocks. The most widely used form of precast concrete is the block. When manufactured under an entirely automatic process with proper control these can be very good, but locally-made products can be inferior. Aerated blocks are now made in some areas, though they are not yet in general use in warm climates. These are special high-pressure steam-cured blocks consisting of an inorganic cementing agent, usually with the addition of fine aggregate; and the aerated structure is formed by the incorporation of air and other gases during manufacture. The blocks are graded into three types: type A, weighing not less than 1500 kg/m^3, may be used for foundation walls and superstructure; type B for foundations to internal walls; and type C for internal non-loadbearing walls. Sometimes pigments are added to impart special colours, and other substances are added to improve workability and accelerate hardening. Waterproof compounds can also be added where blocks are likely to be exposed.

With the introduction of metrication and dimensional co-ordination minor changes have been made to block sizes. This can cause disruption when blocks are produced by different machines, and care must be taken to ensure that blocks match up for size. A good concrete block should be capable of providing a sound building structure, with adequate levels of thermal and sound insulation, adequate resistance to rain and damp penetration, good fire resistance, and clean internal and external finish. It should be of excellent durability. Mortar should not be stronger than necessary for loadbearing and durability, and should be able to absorb as much movement as possible without failing. A reasonable mix would be 1:6 for protected areas and 1:2½ for exposed faces. If plasticizers are added the sand content could be increased by a quarter. A useful leaflet on the design of concrete blockwork may be obtained from the British Concrete Federation, Data sheet No. 121. Various types of blocks and bonds are shown in Fig. 5.1 to 5.5. Most large towns have facilities for producing these, together with slabs, planks, posts, beams, and columns.

5.3 Walls, timber

Timber, concrete, and metal are all used as cladding in some form, and timber in the tropics, where growth is plentiful, could be more widely utilized. It is a flexible material which permits plenty of scope in design. Panels, being lighter and cheaper than other cladding, can be transported from sawmills quite quickly, and erected without the use of heavy lifting gear. However, continuity of demand is essential. During the Second World War standardized 'Lagos Hutting' was prefabricated in Nigeria and shipped to many parts of West Africa. It proved simple and effective. Fibre-glass mats can now be prepared in widths to fit between timber studding, and faced with Kraft paper or polythene, which is fixed to the studs. This also acts as a vapour barrier should the need arise. A fairly typical example of a proprietary timber frame is shown in Fig. 5.7.

INDUSTRIAL SYSTEMS. Prefabricated timber houses, designed to resist earthquakes and typhoons or unstable ground conditions, are also made in Japan. They look extremely individual, and can be made in a number of different designs, with floor areas from 10 to 800 m². Bungalows, twin storey and split-level units are made from standard panels by securing plywood to stud frames. These units are used for walls, floors, and roofs. Foundation walls are built above ground level after the sub-floor has been laid and holding-down bolts have been cast into the

Frameform House

a. Insulation to roof space
b. Gable ladder
c. Gable infill
d. Roof trusses at 600 mm centres
e. Top plate carried over panel junction
f. Junction of two panels
g. Window panel incorporating lintel
h. Moisture barrier breather paper
j. Plywood sheathing applied during fabrication
k. 25 mm insulation battens
l. Foil-backed plasterboard vapour barrier
m. Plywood sheathing applied on site
n. Flexible brick ties
o. Brick or block veneer
p. Damp proof course
q. Plywood floor decking in 2400 × 1200 mm sheets
r. Top plate, site applied
s. Joists at 400 mm centres
t. Perimeter header
u. Concrete slab in two thicknesses
v. Bottom plate secured to slab
w. Footings to suit site conditions

Fig. 5.7 Construction details

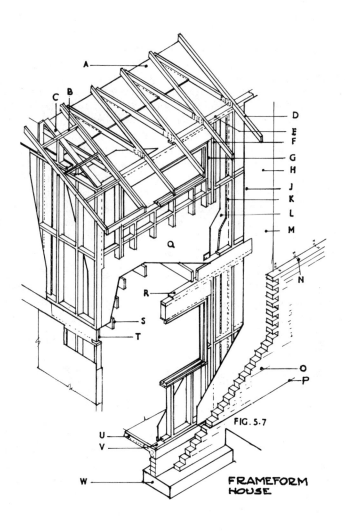

FIG. 5.7

FRAMEFORM
HOUSE

foundation to hold the flooring and wall-panels. One such system is known as the Misawa, and details can be had from the Building Centre, London.

With all forms of timber construction adequate protection with preservatives is essential. The main ones are coal-tar products, water-borne preservatives, and organic solvents. In all cases goggles should be worn when applying. Coal-tar oils include creosote; but water-borne preservatives are generally cleaner. They may be applied by dipping or steeping. Organic solvents are used where a decorative finish is required. Natural oils are used on some exterior hardwoods, but exterior varnish is also popular, though at least four coats are needed, and even then they can only be used in shaded positions. Staining agents that fade or wash off are not recommended. The exposed end-grain of the wood should be sealed with end-sealing compound.

Timber frames are usually dry lined internally with either plywood, hardwood, or plasterboard. The latter has a special ivory-faced finish which is used for decoration. The plasterboard is usually fixed with dabs of plaster, on to which the board is pressed. For Artex finish the butt-edged board is used and the joint filled out, and then covered with paper tape. This makes a neat finish.

Party walls. The current trend towards prefabrication using dry construction techniques adds to the acoustic problem as it is difficult to seal all the air paths against the passage of sound, particularly in terraced housing such as is shown in Fig. 2.9. A typical solution for a timber-framed party wall, which is known to give satisfactory performance in normal service, is given in Fig. 5.7a.

5.4 Joints and connections

Most walls of brick or concrete are prone to cracking, for reasons already discussed. Initial drying, expansion caused by moisture content, settlement, change of temperature, overloading, soil movement, and vibration are some of the causes. Movements caused by temperature changes and moisture are possibly the chief causes of cracking in warm climates, though diagnosis cannot always reveal the cause with accuracy. New walls are sometimes damaged by moisture expansion when, say, the d.p.c. acts as a slip joint, or when the infilling masonry panels between columns are without expansion joints. Thermally-caused movements in flat roofs exposed alternately to tropical sun and night sky tend to push walls over.

Mortars for blocks and masonry should combine good working properties with early working strength. Unnecessarily strong mortar

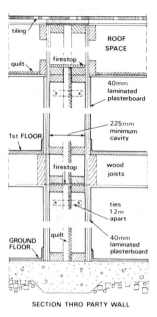

SECTION THRO PARTY WALL

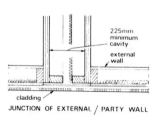

JUNCTION OF EXTERNAL / PARTY WALL

Fig. 5.7a Typical construction of party wall in isolated house system.

creates fewer but wider cracks, but weaker mortars allow greater distribution and the cracks are less noticeable. Lightweight materials require weaker mortars. Plasticizers can provide a useful alternative to lime where this is not obtainable. For external block walling a good mix would be 1:1:5-6 (one part cement, one lime, five to six parts sand) or 1:5-6 (one cement, five to six parts sand with plasticizer).

Accuracy in precast concrete construction is essential for satisfactory performance. Traditional building usually allows for defects to be corrected as the work proceeds, but cutting and fitting precast elements

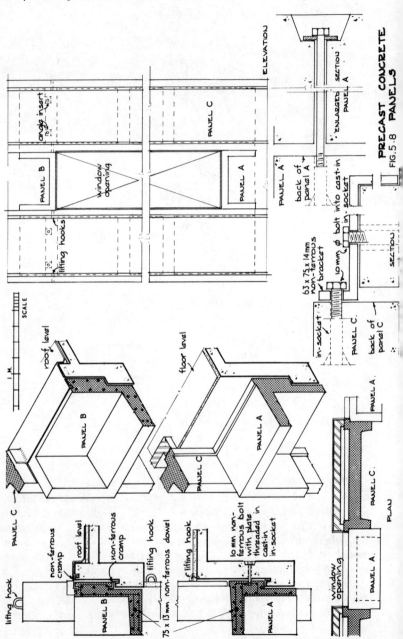

PRECAST CONCRETE PANELS

FIG. 5.8

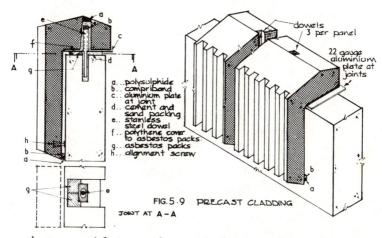

a .. polysulphide
b .. compriband
c .. aluminium plate at joint
d .. cement and sand packing
e .. stainless steel dowel
f .. polythene cover to asbestos packs
g .. asbestos packs
h .. alignment screw

FIG. 5·9   PRECAST CLADDING

JOINT AT A – A

can be an unsatisfactory and expensive business. Tolerances must be dealt with explicitly according to designer's instructions. Information on maximum deviation is usually included on the drawings, but deviations range from ± 0·3 mm for 100 mm components to ± 25 mm for large panels. Simple edge-moulds in casting panels should be capable of being reset after each casting. The drying shrinkage of concrete must also be allowed for. In good concrete this is usually small, but aggregate in some countries is not of good quality. Weathered rock, granite, shale, etc. can cause shrinkage leading to spalling and disintegration. 'Fit' includes not only the requirement of matching for size, but also the capacity of the complete assembly to function satisfactorily. The limits of tolerance, accordingly, will depend not only on deviation caused by man-made inaccuracies but inherent deviation as a result of inadequate resistance to movement caused by temperature changes or moisture.

JOINTING WITH MASTICS. Flexible seals and mastics of rubber/bitumen are commonly used to fill and seal joints against rain penetration, and are usually applied by trowel or by pouring, but sometimes also by tape or strip. Guns, pressure operated by lever or compressed air, can also be used to extrude soft compounds. Compounds supplied in drums, from which they are dug out, are normally applied with trowels. Poured mastics which are heated to make them fluid — usually based on bitumen — are used for horizontal surfaces. Cold poured mastics are also used. Tape or strip is usually stiff enough to handle, and is protected by plastic wrapping until ready for use. Mastics should be easy to apply, should retain their form without slumping, must not bleed or stain and must be resistant to alkalis in concrete.

Glazing-joints in openings and curtain walling should be flexible enough to permit thermal and moisture-caused movement so that the working load does not cause distortion. They should exclude wind and rain, and should be durable, easily repaired, and capable of preserving thermal and sound insulation. They should also allow the components to be easily assembled. The estimated maximum and minimum, temperatures for the locality should be ascertained, and their effects on the particular type of building concerned should be considered. As an example take a light-coloured masonry wall or slab, and an external temperature of 50 deg. C.:

Then for similar construction in dark colours   add 30 per cent

     for black glass on metal insulation behind   add 60 per cent

     for white glass on metal insulation behind   add 30 per cent

     for black metal tray exposed behind clear

         glass and insulated behind   add 160 per cent

     for clear glass in front of dark insulation

         background   add 60 per cent

     for aluminium mullion in curtain wall   add nil per cent

From these figures the effects of jointing in curtain walling may be judged.

Component joints. The larger the component the more difficult it is to position it accurately, and to provide for structural, thermal, and moisture movements; impermeability creates its own problems, as rainwater is discharged over joints. Joints are broadly divided into two types; filled and open, and these may be either lapped or butted. Filled joints depend on mortars, sealants and gaskets, but mortars shrink and cannot absorb movement. Sealants must be selected according to the relationship between joint width and movement, as fracture allows water to enter. Gaskets can be shaped to suit the internal profile of the joint, but may be designed with emphasis on ease of application rather than on waterproofing. Doubtful accuracy in manufacture and assembly can thus create limitations in efficiency.

With horizontal joints, upstands should be provided with sloping sills to reduce water penetration, Fig. 5.10a. An efficient air barrier at the back of the joint is the best defence, Figs. 5.10b and c. Vertical joints need an effective air seal behind. Vertical grooves in the joint do not act as a drainage channel, but do provide a sharp edge which reduces penetration, whereas inclined grooves help to throw water off which is preferable: see Fig. 5.10c. Baffles and ventilated cavities are now in common use but should be in rolling shear to be effective,

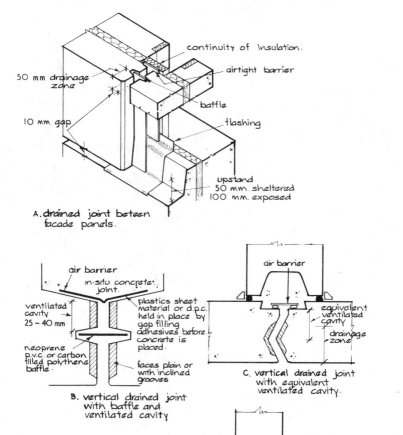

A. **drained joint between** facade panels.

continuity of insulation.
airtight barrier
50 mm drainage zone
baffle
10 mm gap
flashing
upstand
50 mm. sheltered
100 mm. exposed

B. vertical drained joint with baffle and ventilated cavity

air barrier
in-situ concrete joint.
ventilated cavity 25 – 40 mm
plastics sheet material or d.p.c. held in place by gap filling adhesives before concrete is placed.
neoprene p.v.c. or carbon filled polythene baffle.
faces plain or with inclined grooves

C. vertical drained joint with equivalent ventilated cavity.

air barrier
equivalent ventilated cavity
drainage zone

D. vertical drained joint (Reema)

weather strip held in position after insertion by spring clips.
aluminium weather strip
mastics bedding

FIG. 5·10. OPEN DRAINED JOINTS

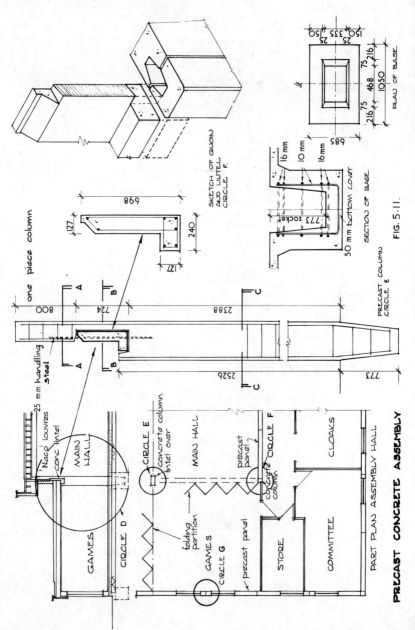

SKETCH OF QUOIN AND LINTEL CIRCLE F.

PLAN OF BASE

SECTION OF BASE

773 socket

50 mm bottom cover

16 mm   10 mm   16 mm

PRECAST COLUMN CIRCLE E

FIG. 5.11.

one piece column

SKETCH OF QUOIN AND LINTEL

25 mm handling steel

Nacp louvres

conc lintel

CIRCLE D

GAMES

MAIN HALL

CIRCLE E

concrete column
lintel over

MAIN HALL

folding partition

precast panel

CIRCLE F

concrete column

precast panel

GAMES
CIRCLE G

STORE

COMMITTEE

CLOAKS

PART PLAN ASSEMBLY HALL

PRECAST CONCRETE ASSEMBLY

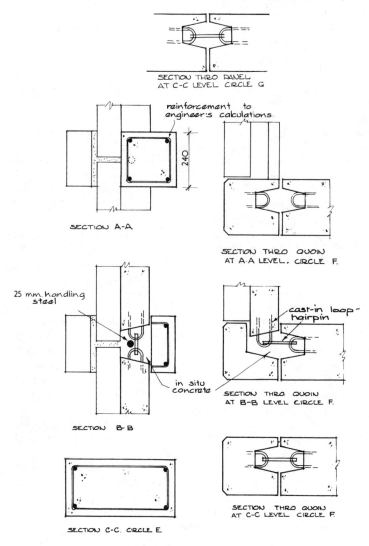

SECTION THRO PANEL
AT C-C LEVEL CIRCLE G

reinforcement to
engineer's calculations.

240

SECTION A-A

SECTION THRO QUOIN
AT A·A LEVEL. CIRCLE F.

25 mm. handling
steel

cast-in loop-
hairpin

in situ
concrete

SECTION B-B

SECTION THRO QUOIN
AT B-B LEVEL CIRCLE F.

SECTION C-C. CIRCLE E.

SECTION THRO QUOIN
AT C-C LEVEL CIRCLE F.

PRECAST CONCRETE DETAILS   FIG. 5.12.

Fig. 5.10b. In low buildings where joints are backed by *in situ* concrete a simple barrier or d.p.c., held temporarily in place with a gap-filling bitumen adhesive, works quite well, Fig. 5.10b.

5.5 Partitions

A wide variety of internal partitions is made, each one being designed for a specific purpose. They may be either loadbearing or non-loadbearing; the latter are lighter in weight and thus possess a higher resistance to heat-flow. Loadbearing partitions can be of timber, concrete, clay block, or proprietary materials. For housing units where timber is available stud partitions are widely used, Figs. 5.13 and 5.14. Plasterboard facing is usually added to both sides, though the finish may be of other material if desired. Clay blocks, where obtainable locally, are popular and so are hollow concrete blocks which are usually rendered on both sides, Fig. 5.4.

Non-loadbearing partitions could be in any of the above materials, but are limited in function. Most are selected according to what is required of them under one or more of the following headings:

demountability   acoustic properties

provision of services   finish   cost

Degrees of demountability vary from semi-permanent wet construction to sliding doors, folding doors and screens. Though greatly improving facilities by its flexibility, demountability must be considered in terms of accessibility and simplicity of demounting. Re-erection must also be considered, and of course the partition should not contain service lines – though many of them do, especially where change takes place only infrequently; in which case removable panels are sometimes provided. When erected below false or suspended ceilings, additional support is needed, for strength as well as for protection against fire and sound. This problem is overcome by the insertion of a barrier above the suspended ceiling and in line with the partition. Demountable partitions must have suitable trim at all edges of the panels, to prevent damage on removal. Where the finish is of permanent laminate it should be of a neutral colour which blend with a variety of surroundings.

A popular type of light partition is the Paramount, Fig. 5.15. It consists of two sheets of plasterboard held apart by a metal or stiff paper diaphragm. It is available in various sizes, the most usual being 2400 × 1200 mm. The partition is fixed to the floor, ceiling, and wall by battens as shown. Openings or cavities are quickly inserted as required. Full details of fixing are available from the makers.

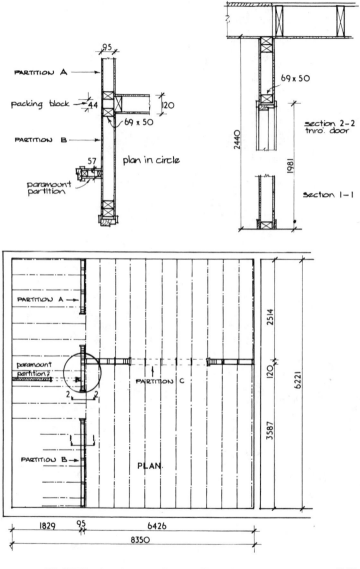

PARTITION A

packing block

PARTITION B

paramount
partition

95

44

69 x 50

57

120

plan in circle

69 x 50

2440

1981

section 2-2
thro'. door

section 1-1

PARTITION A →

paramount
partition

PARTITION C

PARTITION B →

PLAN.

2514

120

3587

6221

1829   95   6426

8350

TIMBER LOAD·BEARING PARTITION.   FIG.5·13

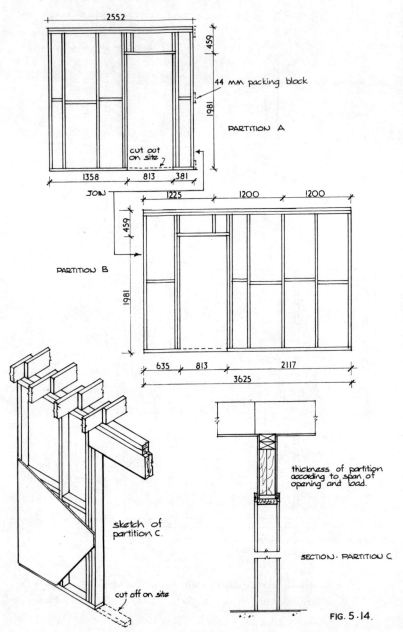

2552

459

44 mm packing block

PARTITION A

1981

cut out
on site

1358   813   381

JOIN   1225   1200   1200

459

PARTITION B

1981

635   813   2117

3625

sketch of
partition C.

thickness of partition
according to span of
opening and load.

cut off on site

SECTION · PARTITION C

FIG. 5·14.

LOADBEARING PARTITIONS. Where a number of houses are to be built at one time, and where timber is available, the stud partition is the most economical, Figs. 5.13 and 5.14. These can be made up in a jig at the sawmills or the joiners' shop and transported direct to site. The studs should be accurately spaced to suit the width of the facing board which, if of plasterboard, cannot usually be cut except at internal angles. This is important on repetitive contracts where taper-edged boards are being used. The board height should be a little less than the floor to ceiling height, to allow movement in fixing. Rust-resisting nails should be used in damp climates and/or the boards fixed by adhesives. For ease of transport and fixing the partition studs should be made up in lengths not exceeding 4 metres if possible, Fig. 5.13, circle.

For stud partitions fibreboard or plasterboard is usually used as a facing. These facings are available in several sizes and thicknesses, the most common being 2400 × 1200 mm and 12·7 mm thick, and 9·5 mm thick for ceilings. When setting out the partition care must be taken to ensure that the positions inside the partition of service inlets and power points (including T.V and telephone etc.) taken from floor conduits do not clash with stud locations; and also that holes are drilled correctly in the bottom plate to allow the entry of cables.

Fibreboard can be decorated like any other internal wall surface, though, where fire hazard warrants it, it may be painted with flame-retarding paint. Other dry partitions are discussed in Chapter 2. It is advisable when fixing hardboard to brush about half a litre of water, depending on humidity, onto the back of the board before fixing so that the subsequent drying out causes the board to tighten after fixing.

Standard insulation board made from uncompressed wood and sugar-cane fibres, giving a low density of not more than 400 kg/m^3, is a popular imported product. It has a low thermal conductivity (k) of not more than 0·06 W/m °C, which, though not important in partitions, can assist in controlling temperature when used as a wall lining to an external wall, particularly when timber cladding is used with an insulating sandwich. Insulating materials and properties have also been discussed in the previous volumes.

5.6 Moisture penetration

Though desirable for the sake of health and comfort the elimination of dampness from buildings is sometimes a difficult thing to achieve. It is also necessary to exclude rain and ground moisture, and to protect the structure from decay and unsightly effects. Solid brick and block walling sometimes offers little resistance to rain penetration, but hollow blocks

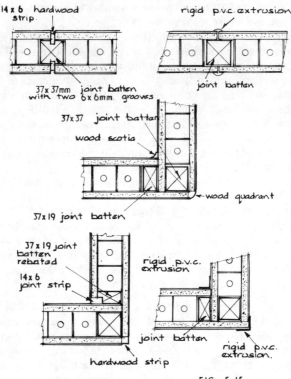

FIG. 5.15

PARAMOUNT PARTITION.

are better in this respect. Normal concrete hollow blocks, when rendered with cement and sand, will resist percolation in most areas.

Before deciding on the treatment of external walls it could be helpful to have information on the effects of driving rain in relation to the temperature and thermal expansion of the material. As this is difficult to obtain in most tropical areas, it is often necessary to resort to such arbitrary classifications as 'for use in sheltered (or 'moderate', or 'severe') conditions', which are in frequent use. High rates of run-off from walls and windows of high buildings can be of concern to designers, though such tests as have been carried out show little effect apart from streaking. Rainwater is normally removed by natural evaporation, the significant climatic factors being solar radiation, wind speed, and relative humidity. Cracking causes up to two-thirds of the problems in walls, for reasons already given; faulty materials and execution make up the remainder.

Water-repellents are intended to improve resistance to rain. There are many waterproofing admixtures to concrete which include plasticizers an and hardeners, some of which are claimed to have all three properties of repellence, hardening, and workability. Some liquids are claimed to be suitable as retarding plasticizers, some to be suitable for hot weather concreting or large pours, and others to be unaffected by extremes of temperature. For a comprehensive list the reader is referred to Specification.

Repellents now in use are often based on silicone resins in the form of a solution of silicones in organic solvents, a water solution of siliconates, or silicone emulsions. The treatment has to be renewed from time to time particularly in coastal zones, but is effective. Before water-repellents are applied the surface must be clean, and cracks or other defects made good.

Paints. Where timber is plentiful, preservatives of different kinds are widely used. Before assembling joinery or securing surface fixings it is necessary to ensure that the backs and edges of components, including sheets, boards, and laminates are treated with appropriate preservatives and primers. Steelwork, especially in coastal zones, must be protected. Where long-lasting results are desired factory processes such as zinc spraying and galvanizing should be applied, even though they may be expensive originally. Good primers may also be used (such as calcium plumbate, iron oxide, lead chromate, red lead, etc.) provided that an absolutely clean surface is first prepared. A thick coating of bitumen provides excellent protection for exposed metals.

Painting of brick, stone, or concrete is sometimes resorted to as a means of waterproofing, but is not always successful. Water can gain access through structural defects, and defects caused by weathering, wrong choice of paint, or unsuitable surface conditions. Precautions against alkali attack should also be taken and the wall dried out before application. For early decoration, cement or emulsion paints should be used; for concrete or rendering, co-polymer and acrylic emulsion paints are best.

Below ground level or where no decorative effect is needed bituminous paints are useful; but they lack durability when exposed to sunlight. Some bituminous paints can be used as waterproofers prior to the application of decorative paints. New paints are continually being manufactured to meet specific needs; these include paints which are anti-condensational, fire-retarding, fungicidal, heat-resistant, multi-coloured, odourless, quick-drying, chemically-resistant, etc.

Damp-proof courses. These are usually of bituminous felt in warm countries. Where movement is not likely to be severe, a 50 mm screed of rich cement mortar is sometimes laid in the wall, which is reinforced with expanded metal mesh; and any slight capillary percolation through cracks can usually be ignored. Where thought desirable a P.V.C. water-stop can be inserted in predetermined positions, though this is not usual. Apart from upward movement of moisture at ground level, protection at roof level is also necessary, at places such as parapets, chimneys, and vents, and also at lintels and over cavity walls. Horizontal movement of water must also be prevented at window and door openings. When using cement/sand renderings to prevent moisture penetration low-sulphate cement is usually available if needed, though in some areas it is not obtainable.

Impregnation. The use of moisture-impervious chemicals in masonry and the inhibition of moisture movement by physical control has been mentioned in Chapter 4, Fig. 4.13.

Efflorescence. Deposits of soluble salts may appear on brickwork or blockwork as an unsightly bloom which is difficult to eradicate. This is often due to the use of salt water and sand during construction, or their presence in the materials themselves. The water dissolves the soluble salts and, as evaporation takes place, the coating is left behind. In rural or coastal areas, where salt water is normally used or where fresh water is difficult to obtain, it has to be accepted; but in more sophisticated building it is guarded against from the outset. Efflorescence is often confused with lime deposit, which leaves white stains on walls after rain; it can also be caused by faulty d.p.c.s, which allow water to be drawn up from the ground.

BEAMS AND COLUMNS

5.6 Fabrication

A wide variety of structural frames are made in metal, timber, concrete, or a combination of all three — outlines have been shown in Volumes 1 and 2, and further examples are given in more detail in Chapter 8. Structural frames are popular in countries where the demand for proprietary products makes fabrication worth-while. Although the current stage of development in some tropical lands would not justify the cost of setting up plant, the reader should become acquainted with assembly and system-building. As is well known the site, with its extremes of heat and exposure to wind and rain, is not the best place to build. Architects,

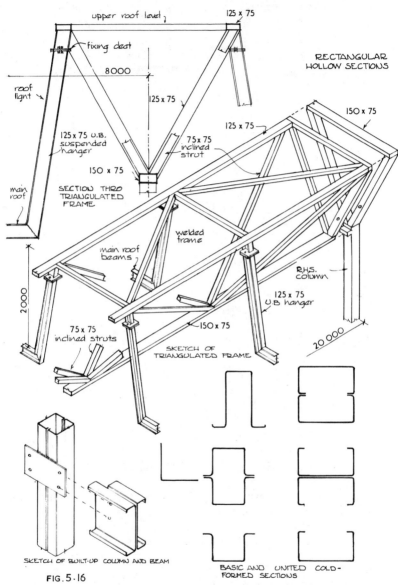

upper roof level
125 x 75
fixing cleat
8000
125 x 75
roof light
125 x 75 U.B. suspended hanger
75 x 75 inclined strut
150 x 75
SECTION THRO TRIANGULATED FRAME
main roof
RECTANGULAR HOLLOW SECTIONS
150 x 75
125 x 75
welded frame
main roof beams
R.H.S. column
125 x 75 U.B hanger
2000
75 x 75 inclined struts
150 x 75
SKETCH OF TRIANGULATED FRAME
20 000

SKETCH OF BUILT-UP COLUMN AND BEAM
BASIC AND UNITED COLD-FORMED SECTIONS

FIG. 5·16

BUILT-UP TUBE AND COLD-ROLLED STEEL SECTIONS

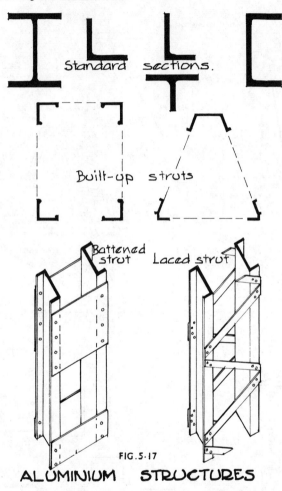

Standard sections.

Built-up struts

Battened strut   Laced strut

FIG.5.17

ALUMINIUM STRUCTURES

in conjunction with structural engineers, could in some instances produce structural frames without full factory facilities. Tropical experiments in this field have already been carried out by the writer and his colleagues with full-scale precast concrete frames.

Where demand justifies the outlay some interesting designs can be carried out. One such example, shown in Fig. 5.8, is capable of both horizontal and vertical repetition, as is the case with most prefabricated systems. The finish could be varied to suit the client's requirements.

Fig. 5.9 shows a ribbed finish now popular in Europe. These precast panels are suitable for office blocks, flats, and places of public assembly.

A design for a prefabricated hall in precast concrete, where columns, beams, and walls are preformed and secured by means of *in situ* vertical joints, is shown in Fig. 5.11. The system is flexible and can be adapted to most requirements. The design could also be used for *in situ* columns and walls if desired.

Steel-framed structures of the heavy type are not widely used in developing countries, except for civil engineering work. However, a number of light cold-formed sections, square and round tubes, and aluminium structural sections have been produced, and this makes the construction of medium-sized social and industrial building in warm countries perfectly viable. Fig. 5.16 shows an example of a triangulated frame made up of steel rectangular hollow sections produced in a wide range of sizes. The main columns are erected first and the triangular frame then bolted in position. The main roof of standard hot-rolled sections is welded together, complete with hangers, and bolted on site to cleats provided. Quite large spans can be covered in this way. A further development has been made in the use of pressed or cold-formed steel and aluminium sections, as shown, which could increase the range and reduce transport costs. A sketch of a built-up column and beam is shown.

Prestressed concrete sections are also used in Europe; but apart from floor and roof beams (Chapters 7 and 8) their use is not likely to become widespread in tropical countries in the near future.

6 | Doors and windows

6.1 Introduction

Standard types of doors, windows and frames have been dealt with in Volumes 1 and 2 under 'External coverings' and 'Internal finishes' respectively and may be found under the appropriate index. The examples given in the following pages have been adapted as far as possible to suit warm climates, though many are manufactured by specialist firms. These designs are usually the outcome of long experience and it is advisable to follow the makers' instructions as far as possible.

DOORS

6.2 Timber doors

There is a tendency today, particularly in good domestic building and the smaller types of administrative and community building, to depart from the ubiquitous flush door and revert to the panelled type popular in earlier years. An example of such an ornamental door is shown in Fig. 6.1. The elevation can be adapted to any of the sectional plans shown: (a) gives a solid 100 X 44 mm hardwood stile with a raised and fielded panel inset; (b) uses a modern technique with W.B.P.-grade plywood, built-up stiles, and raised flat panels with bolection moulding; and (c) is of similar construction but with a raised decorative panel fixed to one face. Such doors are usually imported from specialist manufacturers.

Functional doors. For doors of clean design with a hard-wearing surface for use in hospitals and colleges, the type shown in Fig. 6.2 is available. It is used mainly in dry wall partitioning with a glazed screen supplied to match. The door has a lightweight cellular core with a sapele finish for general purposes or else laminated-plastic facing, if desired. Medium- and heavyweight semi-solid cores, and also laminated cedar cores, which provide the checks required by fire regulations and also improve sound insulation, can be supplied. Asbestos-reinforced laminated doors also meet fire regulations, and can be obtained either fully-finished

continued on Page 116

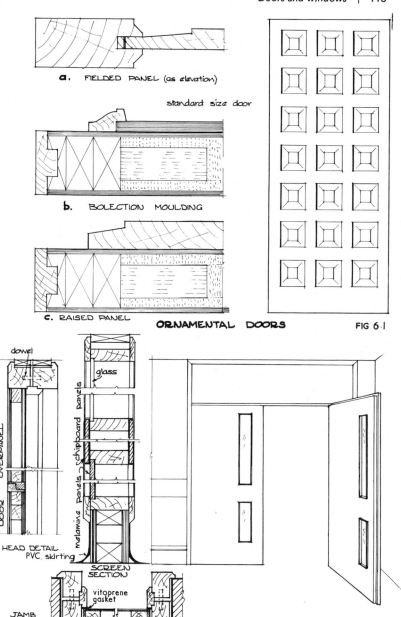

a. FIELDED PANEL (as elevation)

standard size door

b. BOLECTION MOULDING

c. RAISED PANEL

ORNAMENTAL DOORS

FIG 6·1

dowel

glass

DOOR OVERPANEL

melamine panels

chipboard panels

HEAD DETAIL

PVC skirting

SCREEN SECTION

vitoprene gasket

JAMB DETAIL

HOSPITAL DOORS FIG.6·2

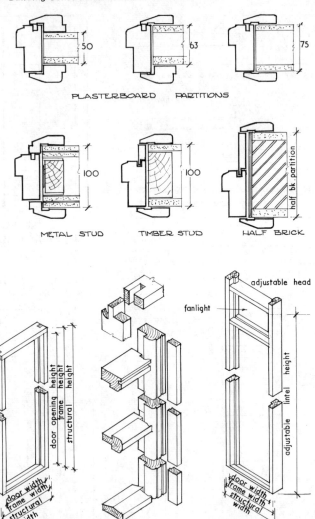

PLASTERBOARD PARTITIONS

METAL STUD   TIMBER STUD   HALF BRICK

DOOR HEIGHT FRAME

STOREY HEIGHT FRAME

DOOR FRAME COMPONENTS   FIG. 6·3

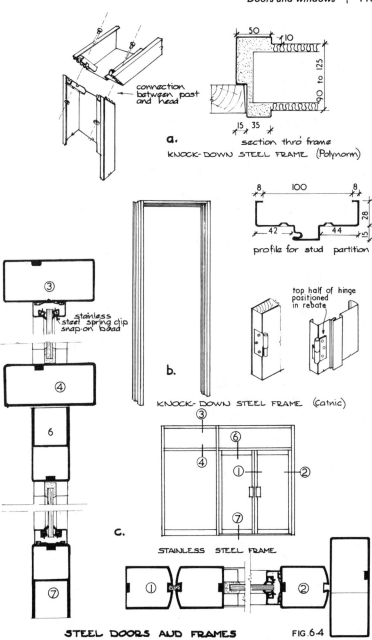

a.
section thro' frame
KNOCK-DOWN STEEL FRAME (Polynorm)

profile for stud partition

top half of hinge positioned in rebate

b.
KNOCK-DOWN STEEL FRAME (Catnic)

stainless steel spring clip snap-on bead

c.
STAINLESS STEEL FRAME

STEEL DOORS AND FRAMES

FIG.6.4

Folding shutter door

1. Welded m.s. suspension bracket
2. Top track
3. Mild steel cover plate
4. Mild steel end panels
5. Mild steel 225 mm shutter plates
6. Hinging strip
7. 138 mm rigid stile
8. 19 mm steel pickets supporting door
9. Bottom track
10. Mild steel sump box with lid
11. Shutter leaves rolled around 3 mm wire reinforcement

Fig. 6.5 Construction details

or primed with paint.

When obtaining doors from Europe it is advisable to have the recesses cut and the hinges, mortice locks, and furniture fitted accurately at the factory. To obtain units complete with frames and fittings usually proves cheaper and more satisfactory than transporting each item separately.

TIMBER DOOR FRAME COMPONENTS. With the advent of mass-production of joinery some firms adopt a system whereby one type of door frame may be used for any internal wall or partition, whatever its size. This is done by using extension linings as shown in Fig. 6.3, with rebated architraves to suit the width of the partition. This could be used, for instance, with the two partitions in Fig. 5.13.

With a door size of 2040 X 826 mm, for example, the following dimensions would obtain:

Door opening height	. . . . . .	2047 mm
Frame height	. . . . . .	2089 mm
Structural height	. . . . . .	2095 mm
Door opening width	. . . . . .	+6 mm
Frame width	. . . . . .	+64 mm
Structural width	. . . . . .	+74 mm

Other door sizes vary by the same amounts. Whatever the system — and there are several — frames must be made accurately, with set allowances to suit door sizes.

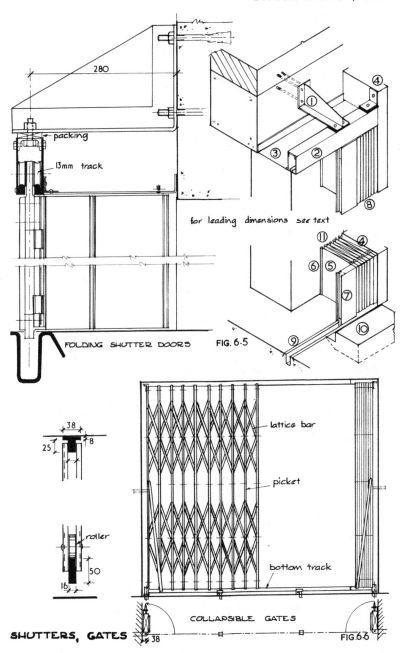

280

packing

13mm track

for leading dimensions see text

FOLDING SHUTTER DOORS   FIG. 6.5

38
25   8

roller
50
16

lattice bar

picket

bottom track

COLLAPSIBLE GATES

SHUTTERS, GATES   38   FIG.6.6

6.3 Metal doors and frames

Steel frames have been used in certain areas of the tropics for many years. They are rot- and termite-proof, require little skilled labour to install, and can be supplied k.d. for easy transport. Fig. 6.4a shows a plan and sketch of a mitred joint which simply needs bolting together. Another type is shown in Fig. 6.4b. The hinge is supplied with the frame together with simple assembly instructions. The groove in the stop provides for a rubber sealing-strip fixed after the frame is in position. Various clamp anchors are used to secure the frame to the structure, one of which is a countersunk screw used with timber stud partitions. Others include wire anchors and clamps fixed after the wall is built.

A typical stainless-steel frame is shown in Fig. 6.4c, with sections marked to make it self explanatory. The doors are factory-glazed and arrive with wool-pile weathering at meeting-stiles and jambs for dust protection, and sealing for air conditioning. Installation is by wood screws concealed by glazing beads and fixed to hardwood plugs.

Folding shutter doors. For transport garages, airports, and industry generally, steel folding shutter doors are used. They may be hand or electrically operated and can be made to practically any length and heights of up to 12 metres or more. Fig. 6.5 shows a typical sketch and section.

Collapsible gates. Such gates are in use all over the world. In warm countries they have the additional advantage of being rot-proof, permitting ventilation and requiring little attention apart from greasing. There are a number of types, some with spear heads and others with floor grooves. Fig. 6.6 shows a typical pair, hinge aside, with a fixed top track and folding bottom track. When the gates are bunched they can be swivelled back against the wall out of the way. The sections vary according to the size of the opening.

Aluminium entrance doors to glass-fronted and other buildings of modern design are popular, particularly for public and commercial building in city centres. A great deal of thought has gone into their design and Fig. 6.7 shows a typical section. The frame takes glass from 6 to 9 mm. Silicone-treated polypropylene-pile weatherstrip is inserted at door stiles, interlocked with the extrusion, and secured with adhesive. The frame is of anodized finish, and all hardware is supplied as required, e.g. completely-sealed overhead door-closer, security locks, quick-action flush bolts and pivot-closing as part of the door frame. Where two or more entrance doors are required they may be joined by timber distance

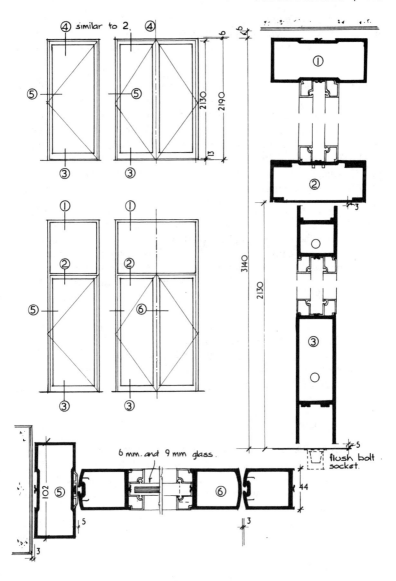

ALUMINIUM DOORS AND SECTIONS FIG.6·7

pieces covered with aluminium cladding. Waist rails (not shown) 233 mm high are obtainable which take glass framed to match the doors. Rubber finger-guards at stiles are also available for schools and nurseries if required.

Revolving doors for modern offices are of similar design — though hardwood is still used for public building, banks, etc., to match internal joinery. A complete typical revolving door, which can be supplied as a standard product by some manufacturers, is shown in Fig. 6.14. These are useful in air-conditioned premises and for keeping out fine dust and wind. They may be folded in the centre to permit two clear passages and through ventilation, or may be rolled away to one side for un-obstructed passage of goods and packages.

Fire-doors. Most building regulations and by-laws have some pro-visions for the prevention of fire, though these are not always rigidly adhered to. Where large numbers of people are likely to congregate, as in stadiums, department stores, and halls, fire doors are often of steel 6 mm thick and secured to a frame, as in Fig. 6.9. The frame is normally of 51 × 51 × 6 mm angle. Up to 1060 mm in width single doors are usual; but over this size they must be of two leaves. Sliding doors of one leaf are in frequent use. They are suspended from rollers mounted on an inclined overhead track and held open by a wire with a fusible link, which when melted causes the door to close.

An example of a typical standard timber fire-door, which gives a one-hour fire check, is shown in Fig. 6.10. The frame containing the door must also have a one-hour resistance to the passage of flame while re-taining its stability.

6.4 Composite doors

Shutter and garage doors. Whereas roller shutters of steel and timber have been in use for many years it is only fairly recently that laths of plastic have been introduced. Where bright modern colours are desirable, as in market stalls or department stores, coloured shutters of plastic are popular: see Fig. 6.11. They are supplied in a range of shades, and multicoloured stripes can be provided if specified. Such shutters can be used externally or internally, and are suitable for large or small openings. The laths are made in several sizes, two of which are shown. The bottom rail is usually of aluminium alloy. The size of coil D depends on the clear height; a complete table may be had from the manufacturers. As an example, using a lath of 60 mm with a clear height 2400 mm, the table shows that a coil diameter 'D' of 290 mm would be needed.

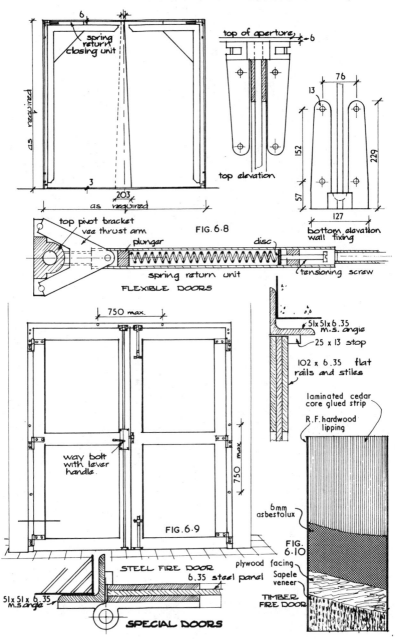

6

spring
return
closing unit

3

203

as required

as required

top of aperture

6

top elevation

76

13

152

229

57

127

bottom elevation
wall fixing

top pivot bracket
vee thrust arm

FIG. 6.8

plunger

disc

spring return unit

tensioning screw

FLEXIBLE DOORS

750 max.

51 x 51 x 6.35
m.s. angle

25 x 13 stop

102 x 6.35 flat
rails and stiles

laminated cedar
core glued strip

R.F. hardwood
lipping

750 max.

way bolt
with lever
handle.

FIG. 6.9

6mm
asbestolux

FIG.
6.10

plywood facing

STEEL FIRE DOOR

6.35 steel panel

Sapele
veneer

51 x 51 x 6.35
m.s. angle

SPECIAL DOORS

TIMBER
FIRE DOOR

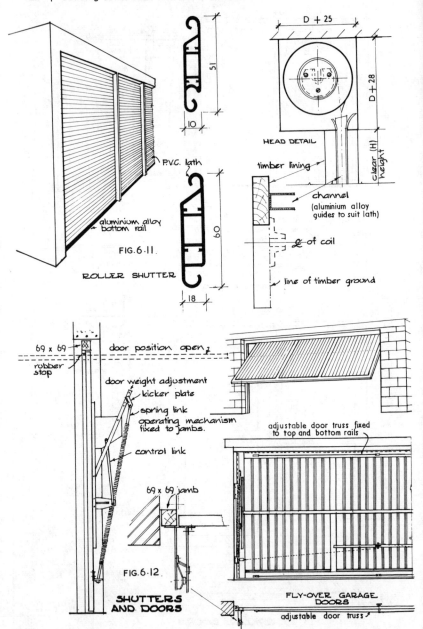

51

10

P.V.C. lath

aluminium alloy
bottom rail

FIG.6.11.

ROLLER SHUTTER

60

18

D + 25

HEAD DETAIL

D + 28

clear (H)

timber lining

channel
(aluminium alloy
guides to suit lath)

¢ of coil

line of timber ground

69 x 69

door position open

rubber
stop

door weight adjustment

kicker plate

spring link
operating mechanism
fixed to jambs.

control link

69 x 69 jamb

FIG.6.12.

SHUTTERS
AND DOORS

adjustable door truss fixed
to top and bottom rails

FLY-OVER GARAGE
DOORS

adjustable door truss

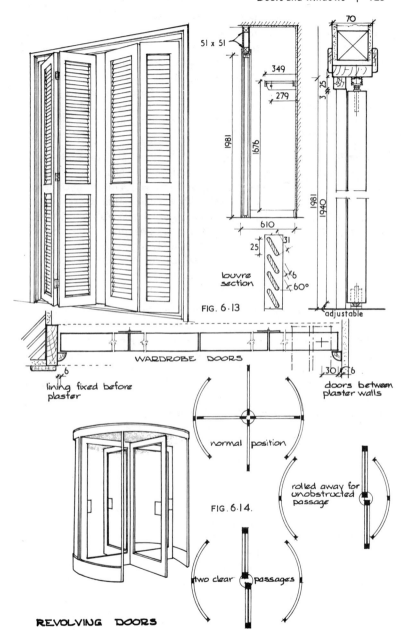

51 x 51

349

279

1981

1676

610

25   31

louvre
section

×6
60°

FIG. 6·13

70

3   25

1981

1940

adjustable

WARDROBE DOORS

lining fixed before
plaster

6

doors between
plaster walls

30   6

normal position

rolled away for
unobstructed
passage

FIG. 6·14.

two clear passages

REVOLVING DOORS

Fly-over doors are extensively used today and are available in both wood and metal. They are used chiefly for garages for one or two cars; over this size separate doors would be needed. Fig. 6.12 shows an elevation of ribbed aluminium, though the section is shown in timber. The latter would be rather heavy for a two-car garage in tropical hardwood, particularly if framed and ledged, though the battens could be slatted to reduce weight and permit ventilation. Structural openings of up to 2100 mm are usual, with a maximum width of 4500 mm in aluminium if trussed as shown.

Flexible doors of rubber or tough plastic in a metal frame are in common use in industries where goods are transported by trolley, fork-lift truck, or wheeled conveyance, through openings which need to be self-closing as a protection against dust, heat, and noise: see Fig. 6.8. The door may be of black rubber panels with clear plastic windows; or wholly of transparent vinyl, as shown. The door springs back into place by means of a horizontal spring return unit, which also acts as a hinge. The doors are for internal or external use. Frames can be made up of tubes or flat steel sections as shown. There are three thicknesses of door, from 8 to 13 mm, the size depending on the opening.

Internal sliding folding-doors are now gaining favour, particularly for wardrobes where space is limited as rooms become smaller. They allow maximum space with minimum projection and are designed to fit into recesses provided by the builder. Dimensions vary but recess openings must be to manufacturers' sizes. The doors are supplied as a complete k.d. set and no floor-track is necessary. They are usually flush and of West African hardwood veneer; but they could be louvred, though adequate blocking of pivots must then be provided to take the extra weight, Fig. 6.13. For louvred doors three hinges are necessary.

WINDOWS

6.5 Introduction

Modern windows are made in a variety of types and materials. They may be side-hung, top- or bottom-hung, tilt-and-turn, louvred, horizontal or vertical sliding, etc. These in turn may be coupled together by means of transoms and mullions to create extensive glass frontages in a variety of materials. They are made in timber, steel, glass, aluminium, plastics, glass fibre, or combinations of these. Double glazing, too, is of value in providing both heat and sound insulation and can be fitted to most types of window without difficulty.

(*continued on page 131*)

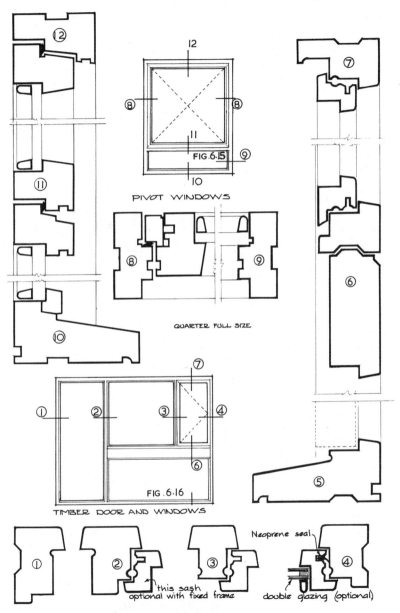

PIVOT WINDOWS

FIG.6.15

QUARTER FULL SIZE

FIG.6.16

TIMBER DOOR AND WINDOWS

this sash optional with fixed frame

Neoprene seal

double glazing (optional)

TIMBER WINDOWS

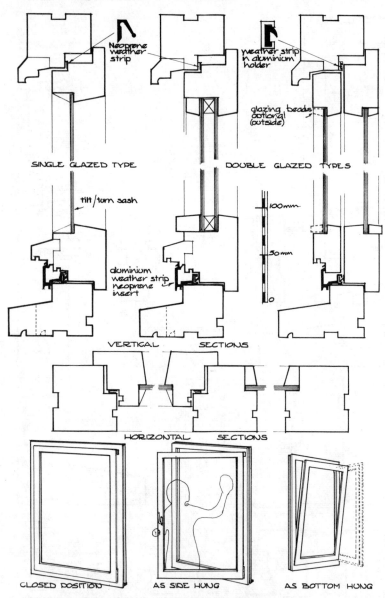

Neoprene
weather
strip

weather strip
in aluminium
holder

glazing beads
optional
(outside)

SINGLE GLAZED TYPE

DOUBLE GLAZED TYPES

tilt/turn sash

100mm.

50mm

aluminium
weather strip
neoprene
insert

0

VERTICAL   SECTIONS

HORIZONTAL   SECTIONS

CLOSED POSITION

AS SIDE HUNG

AS BOTTOM HUNG

TURN AND TILT   TIMBER WINDOWS   FIG. 6.17

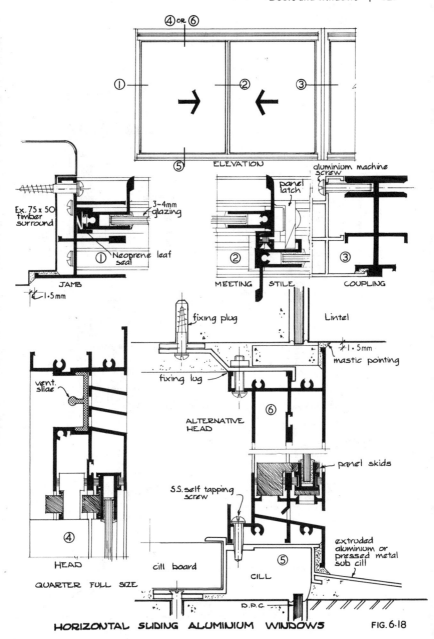

ELEVATION

Ex. 75 x 50 timber surround

3-4mm glazing

① Neoprene leaf seal

JAMB

1.5mm

aluminium machine screw

panel latch

② MEETING STILE

③ COUPLING

fixing plug

Lintel

fixing lug

1.5mm

mastic pointing

ALTERNATIVE HEAD

⑥

vent. slide

④ HEAD

QUARTER FULL SIZE

panel skids

S.S. self tapping screw

⑤ CILL

cill board

extruded aluminium or pressed metal sub cill

D.P.C.

HORIZONTAL SLIDING ALUMINIUM WINDOWS

FIG. 6·18

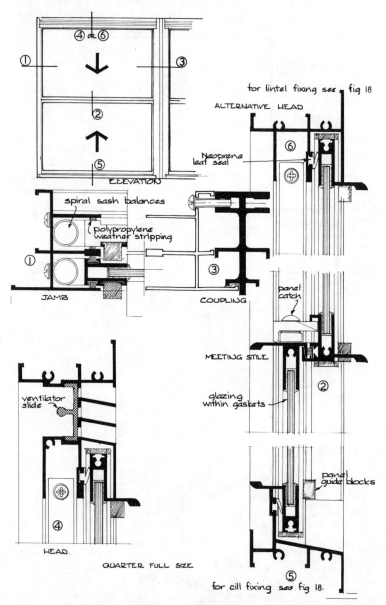

ELEVATION

for lintel fixing see fig 18

ALTERNATIVE HEAD

Neoprene leaf seal

spiral sash balances

polypropylene weather stripping

JAMB

COUPLING

panel catch

MEETING STILE

glazing within gaskets

ventilator slide

HEAD.

QUARTER FULL SIZE

panel guide blocks

for cill fixing see fig 18.

VERTICAL SLIDING ALUMINIUM WINDOWS

FIG.6.19

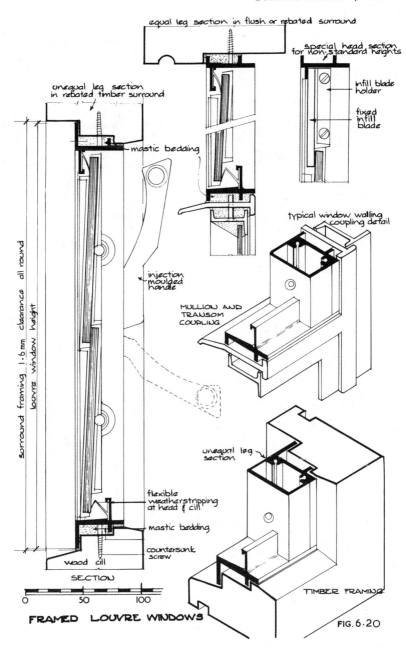

equal leg section in flush or rebated surround

special head section for non-standard heights

infill blade holder

fixed infill blade

unequal leg section in rebated timber surround

mastic bedding

typical window walling coupling detail

injection moulded handle

MULLION AND TRANSOM COUPLING

surround framing 1·5 mm clearance all round

louvre window height

unequal leg section

flexible weatherstripping at head & cill

mastic bedding

countersunk screw

wood cill

SECTION

0   50   100

TIMBER FRAMING

FRAMED LOUVRE WINDOWS

FIG. 6·20

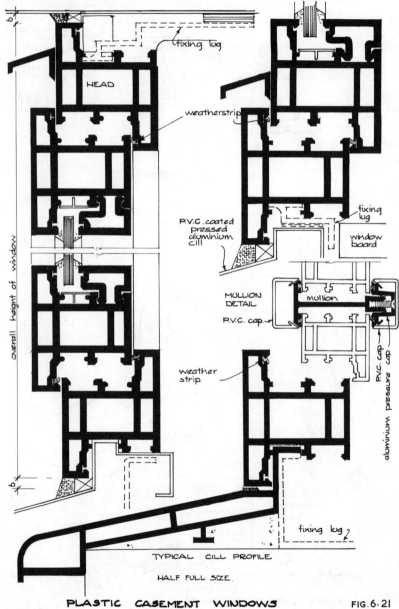

HEAD

fixing lug

weatherstrip

P.V.C. coated
pressed
aluminium
cill

overall height of window

MULLION
DETAIL

mullion

P.V.C. cap.

fixing
lug

window
board

P.V.C. cap

aluminium pressure cap

weather
strip

fixing lug

TYPICAL CILL PROFILE

HALF FULL SIZE.

PLASTIC CASEMENT WINDOWS

FIG. 6·21

Timber windows are still the most widely used. Fig. 6.15 shows a standard type of pivot frame in European redwood, treated with preservative against attack. Fig. 6.16 shows a composite door and window unit. Such units will resist tropical winds and rain when fitted with neoprene sealing strip. Fig. 6.17 shows a more expensive proprietary design, with double-glazing designed to tilt back and turn inwards as shown. This facilitates glass cleaning where outside access is difficult, and also provides maximum ventilation. The window is made up in sizes up to 2400 mm wide and 1500 mm deep.

Aluminium has been used for windows for some years now, and examples have been given in Volume 1. A fairly recent development, however, has been in sliding sashes both horizontal and vertical, Figs. 6.18 and 6.19. With horizontal sliding units each panel is fitted with moulded end-skids which slide on fins within the sill section. With vertical sliders the window is kept in position with spiral sash-balances and the panel held between guide-blocks to prevent rattling. While not providing such a large opening area as other types, they are neat and secure. They can be made up to widths of 1200 mm and heights of up to 1800 mm, but by using couplings any size of opening can be obtained.

Louvred windows are, without doubt, still the most popular type in use today. Since details were given in Volume 1 a number of improvements have been made. Weathertightness has been increased, and special infill blades have been designed to accommodate non-standard heights. Framed louvred windows now comprise a suite of sections to allow simple installation of louvred panels into a variety of surrounds, either as a separate window unit or as a component within a composite frame or curtain walling area. Fig. 6.20 shows the normal arrangement.

Plastic windows. Unplasticized polyvinyl chloride (U.P.V.C.) is now in demand for all types, including horizontal and vertical sliders, particularly in modern housing development. Glass is secured in position with rigid P.V.C. slip-on beads. Double-glazing units, up to 24 mm thick, can also be accommodated. Single units can be coupled together using mullions and transoms, as with other types. The profiles are normally supplied in light grey or white, Fig. 6.21.

Glass-reinforced polyester resin (G.R.P.) is also used in window manufacture. It has excellent weathering properties and is impervious to insect and fungal attack.

7 | Floors

7.1 Introduction

This chapter will deal with various types of floor, omitting those described previously. Owing to the variety of proprietary floors available, only one example of each will be discussed. Preference has been given to those most suited to warm and/or developing countries and to the treatment of the facilities required for erection. The reader is advised to consult the many publications on the subject, some of which are listed at the end of this book. Characteristics such as fire-, heat-, and sound-resistance have been discussed elsewhere, and only construction will be described here.

7.2 Concrete floors

A typical ground layout for a terraced or semi-detached house, designed for repetitive production of precast units or timber walls, is shown in Fig. 7.1. Service inlets include water, electricity, radio or TV relay where applicable, drainage, and other items necessary to the erection of the superstructure. These include dowels, starter-bars, holding-down bolts, floor-stops for erection, stays and checks in walls for concrete thresholds, etc. The starter-bars have been inserted where concrete panel joints occur which have been grooved and recessed for *in situ* fixing. The slab under the loadbearing partitions would be thickened to take the extra weight. With timber walls, holding-down bolts would be inserted at intervals of not more than two metres. Floor noggins could be used as in Volume 1.

Duct covers. With the rapid increase in services of all kinds in recent years, it is advisable in industrial areas and developments to consider the use of ducts or channels to carry pipes or cables. For small workshops and offices covers may consist simply of reinforced slabs such as those in common use on roadway channels where rainfall is high. Where traffic is heavy, however, whether on highways or on factory floors, it is advisable to design rather more substantial covers, especially where traffic is on the increase. An example of a good duct cover is given in Fig. 7.2. Almost any size can be made to this pattern, from 225 to

1200 mm, and built up to any length.

Expansion joints. A wide range of extruded aluminium sections has been designed for surface fixing over expansion joints in large structures, including industrial floors. Examples are given in Fig. 7.2b, where (a) shows compressed rubber cork strip, (b) serrated surface for tile and (c) flexible gasket.

SUSPENDED FLOORS IN CONCRETE. Proprietary floors are manufactured in a fairly wide range, and Figs. 7.7 and 7.8 show typical examples of precast products, both channel and hollow beam sections. They are made in various widths to suit the span and load and also the handling facilities on site. As an example, a hollow beam section, Fig 7.7, 150 mm deep with 38 mm topping added, weighs 285 kg/m^2 or 101 kg per metre run. With an imposed load of 222 kg/m^2 it will span 5·7 m; but exact loading for each span needs further calculation.

Prestressed concrete floors. These have been in use in Europe for some years now, and one of the earliest systems was the Stahlton as illustrated in Fig. 7.9. It consists of factory-made clay or concrete planks containing grooves. The planks are lined up on beds many metres long. Wires are then inserted into the grooves and prestressed; the grooves are then filled with mortar and the planks vibrated. When the mortar has set, the planks are cut to required lengths up to about 10 metres maximum. They are easily handled and erected, as they are light in weight and need no heavy lifting equipment on site. No other reinforcement is necessary, skilled labour is not required, and only light propping is needed during erection and floor-setting time.

Prestressed hollow beams are also available and can be supplied up to 20 metres in length, Fig. 7.9. Prestressed concrete plank sections are also shown; these are light and convenient but require extra reinforcement depending on load and span.

Hollow steel mould floors. These are particularly suitable for flats, hospitals, hotels, etc., where live loads are light; and are designed to carry floors over long spans with few intermediate beams. The moulds can be used many times over and are economical where beams and floors are standardized. A typical cross-section of a building is shown in Fig. 7.10. Spans of beams can be up to 10 metres or more depending on the superimposed load and depth of the beam. The latter may be varied using the same mould by inserting packing pieces as shown. Beams can also be prestressed if desired.

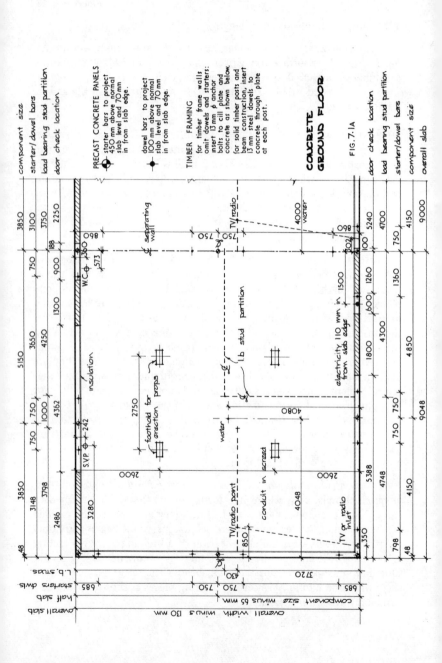

CONCRETE
GROUND FLOOR

FIG. 7.1A

PRECAST CONCRETE PANELS
⊕ starter bars to project 450 mm above normal slab level and 70 mm in from slab edge.

• dowel bars to project 100 mm above normal slab level and 70 mm in from slab edge.

TIMBER FRAMING
for timber frame walls omit dowels and starters: insert 13 mm ø anchor bolts to cill plate and concrete as shown below. (or solid timber posts and beam construction, insert 13 mm steel dowels to concrete through plate at each post.

component size
starter/dowel bars
load bearing stud partition
door check location

door check location
load bearing stud partition
starter/dowel bars
component size
overall slab

L.b. studs
starters dwls
half slab
component size minus 65 mm
overall slab

overall width minus 130 mm

separating wall

TV/radio

4000 water

insulation

footbold for erection props

2750

l.b stud partition

electricity 110 mm in from slab edge

water

conduit in screed

TV/radio point

TV or radio inlet

WC ⊕ 350
573

SVP ⊕ 242

2600

4080

4048

850

98

3850 750 188
3100 900
3750
2250

98
750 750

1500

1260
1360
600
102
100

5240
4700
4150
9000

1800 4300 4850 9048
750 750

5388
4748
4150

3280

5150
3650
4250
1300

3850 750 1000
3148 750
3798 4362
2486

48

430
750 750
3720
685 685

2600

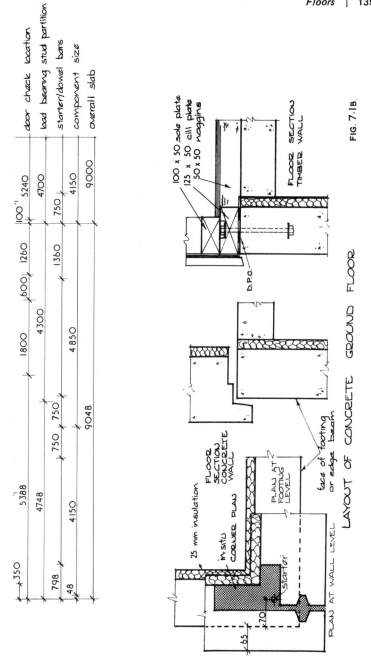

door check location
load bearing stud partition
starter/dowel bars
component size
overall slab

350	5388		1800	600	1260	100'1	5240
	4748		4300		1360		4700
798	750	750	4850			750	4150
48	4150						
	9048						9000

100 x 50 sole plate
125 x 50 cill plate
50 x 50 noggins

FLOOR SECTION TIMBER WALL

D.P.C.

FIG. 7.1B.

25 mm insulation

FLOOR SECTION CONCRETE WALL

in situ CORNER PLAN

face of footing or edge beam

PLAN AT FOOTING LEVEL

starter

70

65

PLAN AT WALL LEVEL

LAYOUT OF CONCRETE GROUND FLOOR

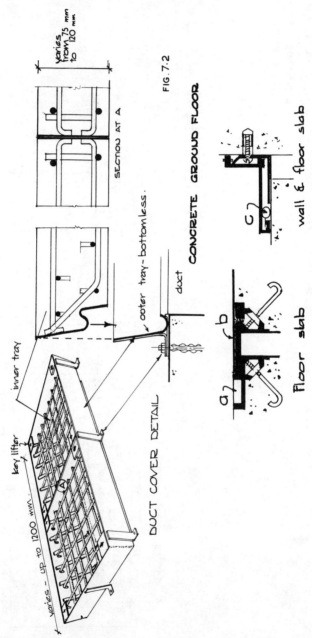

varies from 75 mm to 120 mm

SECTION AT A

FIG. 7.2

CONCRETE GROUND FLOOR

outer tray – bottomless.

inner tray

duct

key lifter

DUCT COVER DETAIL

varies – up to 1200 mm.

wall & floor slab

Floor slab

FIG. 7.2b ALIFAB EXPANSION JOINT

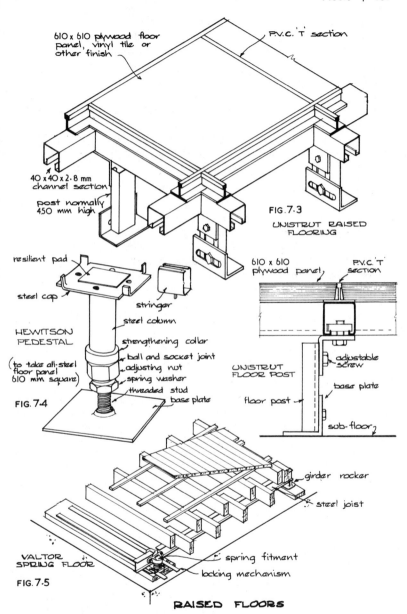

610 x 610 plywood floor panel, vinyl tile or other finish

P.V.C. 'T' section

40 x 40 x 2.8 mm channel section

post normally 450 mm high

FIG. 7.3

UNISTRUT RAISED FLOORING

resilient pad

steel cap

stringer

steel column

HEWITSON PEDESTAL

strengthening collar

ball and socket joint

adjusting nut

spring washer

threaded stud

base plate

(to take all-steel floor panel 610 mm square)

FIG. 7.4

610 x 610 plywood panel

P.V.C. 'T' section

UNISTRUT FLOOR POST

adjustable screw

base plate

floor post

sub-floor

girder rocker

steel joist

VALTOR SPRING FLOOR

spring fitment

locking mechanism

FIG. 7.5

RAISED FLOORS

Lightweight waffle floors. As can be seen from Fig. 7.11 these floors can be constructed by use of light polypropylene moulds, which are supported on bearers while concrete is being poured. The floors are designed to span in two directions. As the moulds are light in weight and stack closely when packed, they can be transported quite economically to all parts of the world. Container services operate in many countries; and they can even be air-freighted. After the floor has set the moulds can be quickly stripped by the use of compressed air supplied by a standard Schraeder nozzle through a boss in the crown, the floor being left supported until ready for striking. Such a floor will meet all normal fire- and sound-resistance requirements.

7.3 Other floors

Raised floors. An alternative means of overcoming service problems where ducts and channels are inadequate is shown in Figs. 7.3 and 7.4. The raised floor is becoming very popular for such buildings as telephone exchanges, information centres, computer rooms, etc., where outlets are needed at almost any point in the floor. The pedestals are quickly secured to the sub-floor by means of cartridge guns, and adjusted to keep the floor level. Floor panels can be removed at will. In tropical countries this has several advantages: air conditioning ducts can be taken under the floor to outlet grilles where needed; the raised floors can replace suspended ceilings, if this is found more convenient; service inlets are often simpler to deal with; and a floor is often better for being raised a little higher than street level.

Spring floors, as illustrated in Fig. 7.5, are usual in community centres, public buildings, and places where assembly halls serve for both meetings and recreational purposes. They are much to be preferred for dancing, and for other purposes can be locked firm when required.

Suspended floors in timber. To revert again to housing; Fig. 7.6 gives a typical plan of an upper floor with details of framing. Joist sizes have been given as an example, and would depend in practice on local by-laws and timber available. It will be noticed that dimensions have been taken from inside walls, omitting insulation; this assists standardization, as insulation thicknesses can vary. Joist spacing has been designed to take plasterboard sheets of 1200 × 2400 mm. As with partitions and walls, boards should be cut only at openings and junctions, and not in mid-span. All noggins for fans, lights, partitions, and trimming must be shown, as must straps and similar items needed to support prefabricated panels, where applicable. Solid blocking to stiffen the joists is to be shown as required by the local authority.

(*continued on page 145*)

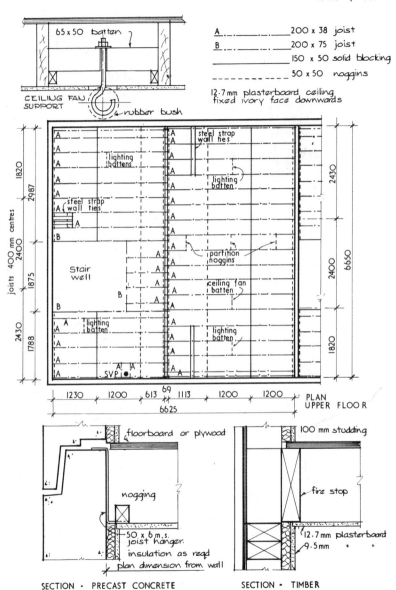

UPPER FLOOR IN TIMBER

FIG. 7·6

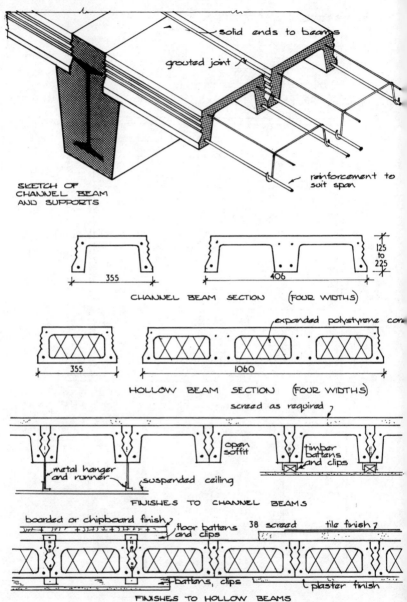

SKETCH OF
CHANNEL BEAM
AND SUPPORTS

solid ends to beams

grouted joint

reinforcement to
suit span

125
to
225

355

406

CHANNEL BEAM SECTION (FOUR WIDTHS)

expanded polystyrene core

355

1060

HOLLOW BEAM SECTION (FOUR WIDTHS)

screed as required

open soffit

metal hanger
and runner

suspended ceiling

timber battens
and clips

FINISHES TO CHANNEL BEAMS

boarded or chipboard finish

floor battens
and clips

38 screed

tile finish

battens, clips

plaster finish

FINISHES TO HOLLOW BEAMS

PRECAST CONCRETE FLOORS

FIG.7·7

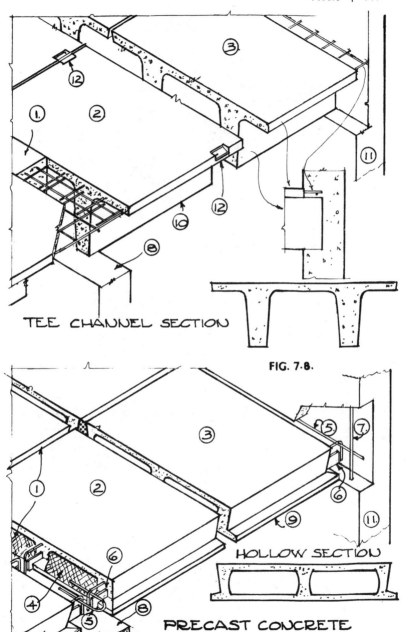

TEE CHANNEL SECTION

FIG. 7·8.

HOLLOW SECTION

PRECAST CONCRETE

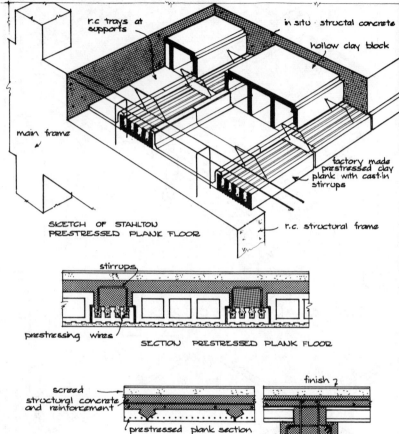

r.c trays at supports

in situ structal concrete

hollow clay block

main frame

factory made prestressed clay plank with cast-in stirrups

SKETCH OF STAHLTON PRESTRESSED PLANK FLOOR

r.c. structural frame

stirrups

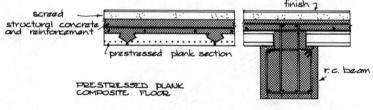

prestressing wires

SECTION PRESTRESSED PLANK FLOOR

screed
structural concrete and reinforcement

prestressed plank section

finish

r.c. beam

PRESTRESSED PLANK COMPOSITE FLOOR

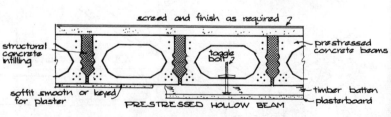

screed and finish as required

structural concrete infilling

toggle bolt

prestressed concrete beams

soffit smooth or keyed for plaster

timber batten
plasterboard

PRESTRESSED HOLLOW BEAM

PRESTRESSED CONCRETE FLOORS

FIG. 7.9

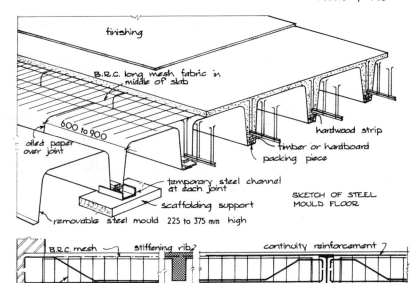

finishing

B.R.C. long mesh fabric in middle of slab

600 to 900

oiled paper over joint

hardwood strip

timber or hardboard packing piece

temporary steel channel at each joint

scaffolding support

removable steel mould 225 to 375 mm high

SKETCH OF STEEL MOULD FLOOR

B.R.C. mesh

stiffening rib

continuity reinforcement

stirrups

steel or concrete beam

LONGITUDINAL SECTION

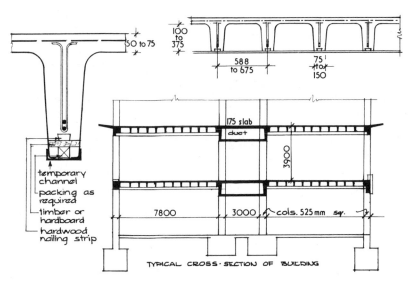

50 to 75

100 to 375

588 to 675

75 to 150

175 slab

duct

3900

temporary channel

packing as required

timber or hardboard

hardwood nailing strip

7800

3000

cols. 525mm sq.

TYPICAL CROSS·SECTION OF BUILDING

HOLLOW STEEL MOULD FLOORS

FIG.7·10

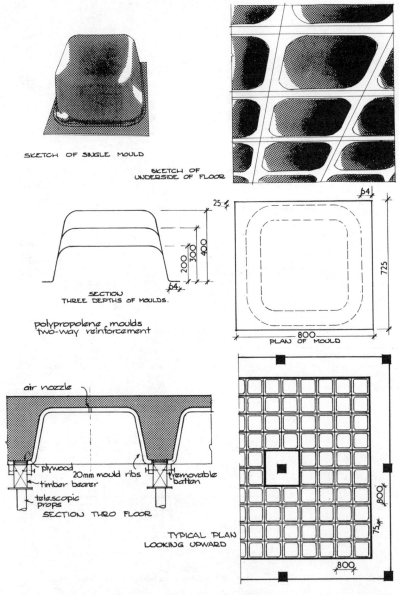

SKETCH OF SINGLE MOULD

SKETCH OF
UNDERSIDE OF FLOOR

25

400

300

200

64

SECTION
THREE DEPTHS OF MOULDS.

polypropolene moulds
two-way reinforcement

64

725

800

PLAN OF MOULD

air nozzle

plywood

20mm mould ribs

removable
batten

timber bearer

telescopic
props

SECTION THRO FLOOR

800

800

75

800

TYPICAL PLAN
LOOKING UPWARD

LIGHTWEIGHT   WAFFLE   FLOORS

FIG. 7·11.

7.4 Machine bases

Where machines and other equipment need to be fixed to floors (or walls or ceilings), a variety of devices exists sufficient to meet most fixing requirements. A few of these are shown in Fig. 7.12. The Lewis or rag bolt (a) is perhaps the most widely used, and is normally held in position by jigs while the concrete is being poured; or it may be grouted in predetermined holes after the floor is laid. The indented bolt (b) serves the same purpose. A more sophisticated method of securing machine bases is by means of a Korker (c). A hole is first drilled and the Korker is dropped in; it is then hammered with a setting tool which spreads the base, holding it firmly in place ready to receive the bolt. A range of sizes is available.

A Rawlplug wedge anchor (d) is another popular device, which is tightened in the hole in the same way as (c). The casing can also have its own cutting edge, enabling it to drill its own hole. It is then removed, and a hard steel cone is dropped into the hole. The casing is then punched tight over the cone, causing it to expand in the hole.

The anchor bolt (e) is a simple device for small-scale work. It uses an ordinary bolt with a special serrated washer placed over the shank; this is followed by a lead collar, which is caulked tight, causing the serrations to expand. It is cheap and effective.

The method (f) consists of concrete inserts, and has many uses. The channel is obtainable in a range of sizes and lengths, and is simply cut and positioned for concreting. Fixing components are supplied to suit the machines, and can be adjusted to fine limits before tightening. The foam-plastic infill is merely to keep the channel clear of concrete, and can be removed with a screwdriver after the concrete is poured. There are a number of other devices made, some designed to reduce vibration. Rubber mountings are sometimes used, and also steel-reinforced neoprene washers. The method will depend on the degree of vibration to be tolerated. C.P. 3: Chapter 111, 1974 gives further guidance on machine mountings.

Large machines subject to vibration, such as universal drilling and boring machines, usually depend on a substantial concrete base to absorb shock by virtue of its mass. A base shown in Fig. 7.12 illustrates such a case. The levelling-screw acts as a pivot, situated in the base of the machine at its centroid. As the cantilevered machine-head is swung through an arc, the holding-down bolts come into play to prevent overturning. The machine on delivery weighs up to 5 tonnes, but can be manoeuvred without a crane by means of rollers, block and tackle, or hydraulic jacks. Fixing accessories are not supplied unless specially ordered, so it may be necessary to have them made. They would

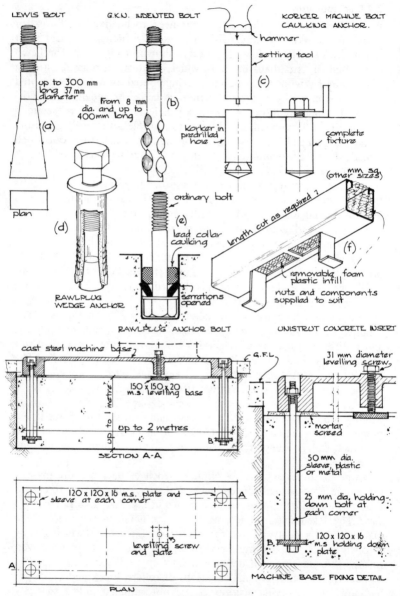

LEWIS BOLT

GKN. INDENTED BOLT

KORKER MACHINE BOLT CAULKING ANCHOR.

up to 300 mm long 37 mm diameter

(a)

From 8 mm dia. and up to 400mm long

(b)

plan

hammer

setting tool

(c)

Korker in predrilled hole

complete fixture

(d)

ordinary bolt

(e)

lead collar caulking

serrations opened

RAWLPLUG WEDGE ANCHOR

RAWLPLUG ANCHOR BOLT

mm sq. (other sizes)

length cut as required

(f)

removable foam plastic infill

nuts and components supplied to suit

UNISTRUT CONCRETE INSERT

cast steel machine base

G.F.L.

150 x 150 x 20 m.s. levelling base

up to 1 metre

up to 2 metres

B

SECTION A-A

120 x 120 x 16 m.s. plate and sleeve at each corner

A

levelling screw and plate

A

PLAN

31 mm diameter levelling screw

mortar screed

50 mm dia. sleeve, plastic or metal

25 mm dia. holding-down bolt at each corner

120 x 120 x 16 m.s. holding down plate

B

MACHINE BASE FIXING DETAIL

FLOOR FIXING DETAILS

FIG. 7.12

consists of:

No. 4 holding down bolts 25 mm dia. or as specified.

No. 4 120 × 120 × 16 mm m.s. base plates drilled with hole slightly larger than holding down bolt.

No. 4 cardboard, plastic or similar sleeves about 50 mm dia. to length shown on drawing.

No. 1 150 × 150 × 19 mm m.s. base plate (not holed) as levelling plate.

Stout plywood template to size of base, with holes slightly larger than bolt diameters drilled at corners in correct bolt positions, and also a square hole slightly larger than the levelling plate at the centroid position.

Concreting material and plant, tools, and levelling instruments would also be required. The normal procedure would then be as follows:

Dig pit to required dimensions.

Concrete up to underside of holding-down bolts at 'B'.

Scoop out handhole in concrete for bolt head.

Place almost dry mortar mix at 'B'. Bed all four plates in position using level and template. Remove carefully leaving imprint and mark for return to same position. Scoop out handhole to take bolthead.

Hold bolt in approximately upright position with head in handhole; mortar in; put washer over shank; replace plate over shank; put on sleeve; and support bolt temporarily in position. Put cotton waste in sleeves.

Fill pit with concrete to level of sleeve tops, i.e. about 12 mm below bottom of machine base. Vibrate if possible and leave to set.

Bed levelling plate in mortar in centroid position with template, and make level to underside of base. Screed corners with cement mortar to same level to form seating. Allow to set and cure. Lower machine into position over bolt holes. Check level with bubble in machine base and screw down. Fill up and make good around floor base.

8 Roofs

8.1 Introduction

In recent years a number of changes have come about in roof construction, not the least important being improvements in the insulating properties of the various coverings. These have been achieved either by the use of sandwich construction, i.e. two layers of sheeting with insulation in between, or simply by using a double layer of roof sheeting. In some cases advantage has been taken of the greater rigidity of the double layers to increase the span between the supports, so reducing the cost. Other improvements have been achieved by increasing the stability of roof-beams and purlins by the use of new shapes and by extra trussing and reinforcement, thus achieving the tested strength with less material. These changes will now be discussed.

8.2 Steelwork in roofs

Owing to heavy transport costs and site erection problems, universal steel beams have not yet been dealt with to any extent, other methods being preferable. As more steelwork is now being used, however, some knowledge of its construction is desirable. The size of the U.B. or steel joist for simply supported beams is usually obtained from makers' handbooks, though care must be taken to use the steelwork economically. A normal type of construction is shown in Fig. 8.1. In this case a parallel U.B. is anchored firmly into reinforced concrete piers, using the rods themselves as anchors. This helps to reduce the size of the beams needed. Alternatively, the beam may be secured by bolting or welding to a steel stanchion, Fig. 8.2c. By cutting the beam longitudinally through the web and reversing the ends a tapered beam may be formed by welding the seam, Fig. 8.2a. For balconies and verandas supported by means of such beams, see Fig. 8.2b.

The use of galvanized mild steel (g.m.s.) purlins of 'Z' section is now quite common, and enables the span between main beams or supports to be increased. Steel tables are available giving safe loads and spans for different sizes of purlins. Special sections are obtainable where required for positions such as eaves beams, Fig. 8.2b. Valley gutters may be laid

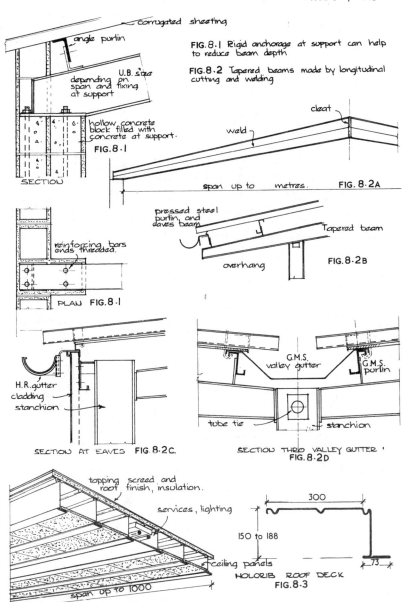

corrugated sheeting

angle purlin

U.B. size

depending on span and fixing at support

hollow concrete block filled with concrete at support.

FIG. 8·1

SECTION

FIG. 8·1 Rigid anchorage at support can help to reduce beam depth

FIG. 8·2 Tapered beams made by longitudinal cutting and welding

cleat

weld

span up to   metres.   FIG. 8·2A

reinforcing bars ends threaded.

PLAN FIG. 8·1

pressed steel purlin, and eaves beam

Tapered beam

overhang

FIG. 8·2B

H.R. gutter
cladding
stanchion

SECTION AT EAVES FIG. 8·2C.

G.M.S. valley gutter

G.M.S. purlin

tube tie

stanchion

SECTION THRO VALLEY GUTTER
FIG. 8·2D

topping screed and roof finish, insulation.

services, lighting

ceiling panels

span up to 1000

HOLORIB ROOF DECK
FIG. 8·3

300

150 to 188

73

STEELWORK IN ROOFS

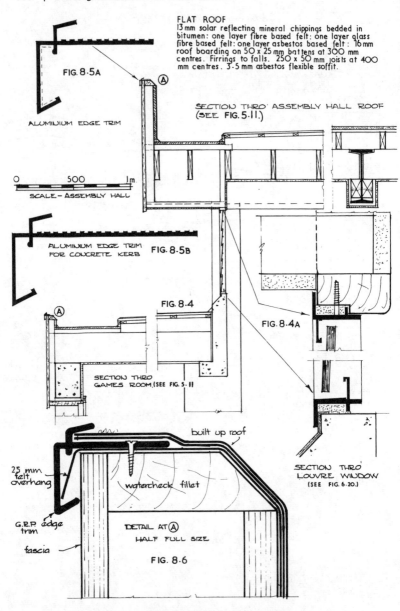

FLAT ROOF
13 mm solar reflecting mineral chippings bedded in
bitumen: one layer fibre based felt: one layer glass
fibre based felt: one layer asbestos based felt : 16 mm
roof boarding on 50 x 25 mm battens at 300 mm
centres. Firrings to falls. 250 x 50 mm joists at 400
mm centres . 3·5 mm asbestos flexible soffit.

FIG. 8·5A

ALUMINIUM EDGE TRIM

0   500   1m

SCALE - ASSEMBLY HALL

SECTION THRO' ASSEMBLY HALL ROOF
(SEE FIG. 5.11.)

ALUMINIUM EDGE TRIM
FOR CONCRETE KERB   FIG. 8·5B

FIG. 8·4

FIG. 8·4A

SECTION THRO'
GAMES ROOM. (SEE FIG. 5.11)

SECTION THRO
LOUVRE WINDOW
(SEE FIG. 6·20.)

built up roof

25 mm
felt
overhang

watercheck fillet

G.R.P. edge
trim

fascia

DETAIL AT (A)
HALF FULL SIZE

FIG. 8·6

FLAT TIMBER ROOFS AND EDGE TRIM

by the use of vertical slotted holes in the purlin or cleat, Fig. 8.2d. Vertical corrugated cladding secured with 'Z' capping held by steel brackets welded to the stanchion is also shown in Fig. 8.2c. A special eaves-filler has been introduced to seal off the roof corrugations against the entrance of lizards and other small animals and insects.

A further type of long-span g.m.s. deck is shown in Fig. 8.3. These sections can span up to 10 metres depending on the size and loading, and are made in lengths up to 13 metres to cover multiple spans. The section can carry acoustic tiles, panels, recessed lighting, services, etc., and can be adapted to curved or sloping surfaces.

8.3 Timber roofs

In areas where tropical timbers are available, the flat roof is widely used. A roof of joists supported by U.B.s and designed for the community hall, Fig. 5.11, is illustrated in Fig. 8.4. The gutter may be laid to falls by the use of gutter bearers. Two types of edge-trim are shown, with a further one to be used with concrete edging. Both aluminium and G.R.P. are in common use and are shown screwed to the water check at 600 mm centres. A typical specification is also given. The clerestory window is detailed as a louvred light, Fig. 6.20.

Trussed timber rafters. With the introduction of steel connector plates and fasteners in trussed rafters and girders, timber roof construction has been revolutionized in most countries. A general selection of trusses is shown in Fig. 8.7. Various plates and fastenings are marketed, the most common for domestic use being the connector shown in Fig. 8.8. For domestic and smaller span buildings the *Fink* or *Fan* truss is the most popular. Trusses can be supplied complete by sawmills, stored on site, and, provided a suitable crane is available, can be hoisted into position in assemblies of up to ten or more, previously linked together at ground level. For simplicity of erection such buildings usually have gable ends as shown, but hipped ends are also sometimes used. These, however, need extra hipped girder trusses and jack rafters, which are more expensive and need skilled labour to erect.

As with floors, Fig. 7.6, provision for service outlets etc. has to be made in advance, and water storage in the roof space has to be allowed for. The size and location of the tanks will vary according to local needs and the requirements of the water authority. More will be said of this in the chapter on services. Assuming the tank in Fig. 8.10 meets requirements, it will have to be firmly supported, with the load spread over four trusses by bearers.

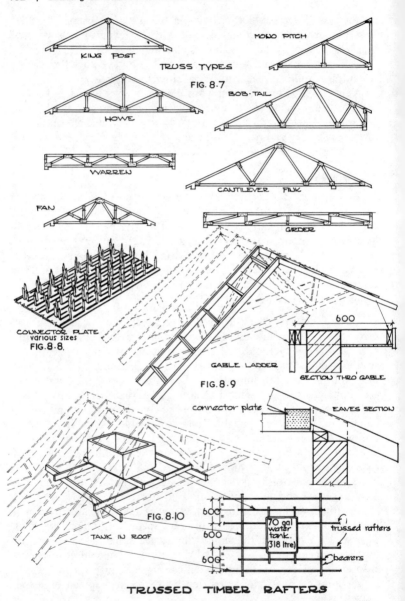

KING POST

MONO PITCH

TRUSS TYPES

FIG. 8·7

BOB·TAIL

HOWE

WARREN

CANTILEVER FINK

FAN

GIRDER

CONNECTOR PLATE
various sizes
FIG. 8·8.

GABLE LADDER

FIG. 8·9

600

SECTION THRO' GABLE

connector plate

EAVES SECTION

FIG. 8·10

TANK IN ROOF

600

600

600

70 gal
water
tank.
(318 ltre)

trussed rafters

bearers

TRUSSED TIMBER RAFTERS

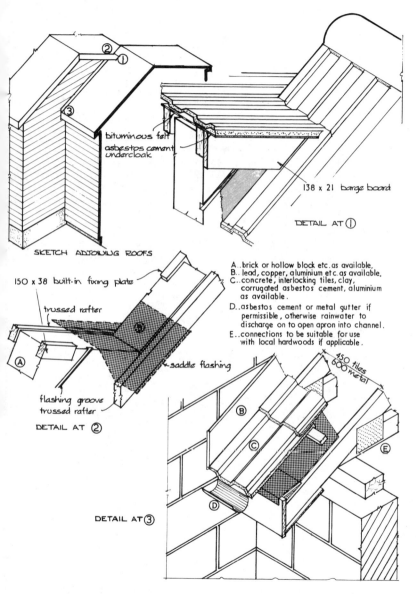

SKETCH ADJOINING ROOFS

bituminous felt
asbestos cement
undercloak

138 x 21 barge board

DETAIL AT ①

150 x 38 built-in fixing plate
trussed rafter
saddle flashing
flashing groove
trussed rafter

DETAIL AT ②

A..brick or hollow block etc. as available.
B.. lead, copper, aluminium etc. as available.
C..concrete, interlocking tiles, clay,
   corrugated asbestos cement, aluminium
   as available.
D..asbestos cement or metal gutter if
   permissible, otherwise rainwater to
   discharge on to open apron into channel.
E..connections to be suitable for use
   with local hardwoods if applicable.

450 tiles
600 metal

DETAIL AT ③

STEPPED AND STAGGERED ROOFS   FIG. 8·11

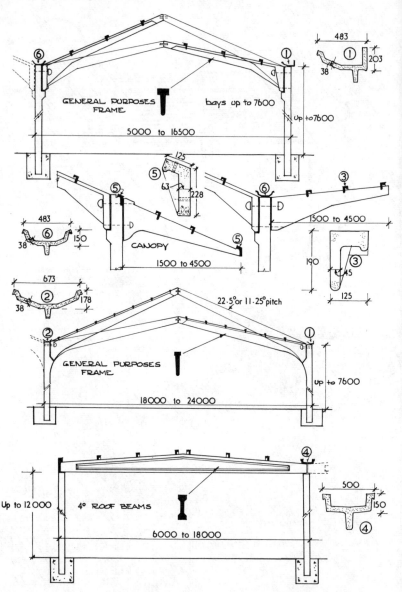

PRECAST CONCRETE FRAMES

FIG. 8·12

STEPPED AND STAGGERED ROOFS. Housing compounds and settlements in warm climates are frequently to be found in hilly or undulating country, especially in mining areas. A usual roofing arrangement in these conditions is shown in Fig. 8.11. The barge board in this case is fixed to the wall. The construction is quite straightforward except for the metal saddle-flashing placed between adjoining houses. This can be beaten, cut, or welded to shape depending on the material available. The roof-covering may be of concrete pantiles, corrugated sheeting, normal tiling, etc. The stepped and staggered condition is the same as that shown in Fig. 4.26.

8.4 Precast concrete frames

Where demand is sufficient, fairly large areas intended for use as stores, warehouses, halls, community centres, workshops, etc., can be covered economically by using precast concrete frames. When factory-produced, they are usually of consistent quality, accurately dimensioned, and quickly assembled on site. Such frames are rust- and vermin-proof and can withstand humid conditions and moisture. They are easily clad, insulated, and provided with natural lighting and ventilation. They are designed in a range of sizes, can be supplied in one or more storeys, and can usually be expanded at will. Provided a demand can be created, existing concrete factories should be capable of being adapted without expensive capital equipment.

A few types of precast concrete frames are shown in Fig. 8.12, and there are several others available. All components, such as eaves, gutters, purlins and accessories, can be supplied together with special units, such as canopy and lean-to sections, bolted on to the standard frame. Details of precast concrete industrial frames can be had from the Building Centre, London.

8.5 Roof decks and coverings

Metal deck roofing. A variety of metal decks are now available, three of which are given in Fig. 8.13. The first is designed as a heavy floor or roof and consists basically of steel ribbed sections, 600 mm wide, laid between supports up to 5 metres span, or 11 metres over two spans. Concrete topping may be added as necessary. A special feature is the provision of a service duct inserted where needed. The floor needs no reinforcement or formwork. A lighter, inverted section is also made which, instead of concrete topping, has insulation board clipped into position as shown. Tables of safe loadings, are available from manufacturers together with details of components and accessories supplied. Provision for openings in the roof is easily made.

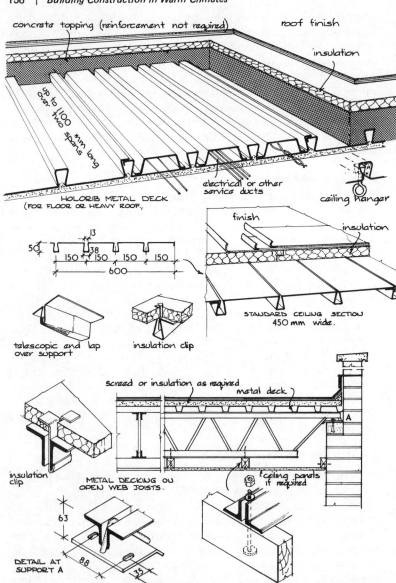

concrete topping (reinforcement not required)

roof finish

insulation

up to 1100 mm long over two spans

HOLORIB METAL DECK
(FOR FLOOR OR HEAVY ROOF.)

electrical or other service ducts

ceiling hanger

finish

insulation

STANDARD CEILING SECTION
450 mm wide.

telescopic end lap over support

insulation clip

screed or insulation as required

metal deck

insulation clip

METAL DECKING ON OPEN WEB JOISTS.

ceiling panels if required

DETAIL AT SUPPORT A

METAL DECK ROOFING

FIG. 8.13

A further type of metal deck is shown supported by open-web steel beams. The deck may be used as before, as permanent formwork, with reinforcement added as required. For roofwork, the metal deck and concrete can be dispensed with and wood-wool or similar insulation clipped direct to the top flange of the open-web joist. A ceiling may be added as shown, if required, though this is often omitted. Wood-wool slabs may be laid either longitudinally or transversely, depending on the beam span, loading, or special conditions. Wood-wool slabs can, in themselves, be used as formwork to concrete, and provide a good key for plasterwork.

The open-web beam dealt with briefly in Volume 1 is a popular form of floor and roof support and could, if desired, replace the U.B. shown in Fig. 8.4. It can be fabricated from high-yield steel in a wide range of sizes from 200 mm to 600 mm deep, and in lengths from 2 to 15 metres. The top boom can be fitted with timber battens to the twin-angled flanges, and covered with roof boarding.

Wood-wool slab decking. Where insulation slabs have to be supported over wide spans, special channels of galvanized steel may be used, Fig. 8.14. In this way spans of up to 4 to 5 metres may be covered. If desired, steel beams of small sizes may be used with the channel or independently, according to the span. The tee sections are normally of 16 s.w.g. of any required length, fixed at 600 mm centres to take 51 mm wood-wool slabs. These are usually 2 metres long, though other lengths are available. Tables of safe loads are supplied by the manufacturers. Lightweight tees are also used to conceal the butt joint of the slabs.

Asbestos roof cladding. In addition to the normal types of cladding referred to in Volumes 1 and 2, asbestos double sheeting is now in general use and suitable for tropical roof insulation. Details are given in Fig. 8.15. The uncompressed insulation material is usually 25 mm glass-fibre mat, though this can be varied to suit local demand. The roof construction also has a high fire-resistance and construction is simple and straightforward, as may be seen from the examples given.

Roof coverings. Though coverings have already been dealt with to some extent, the following have been included because of special properties which make them suitable for warm climates. Fig. 8.18 illustrates the use of *Fural* aluminium sheeting. It has been in use for a number of years now, and is suited to tropical conditions for several reasons. It has high reflectivity; is capable of thermal expansion in a horizontal direction; is quieter in use than normal corrugated aluminium sheeting, which creaks noisily with changes in temperature; and is not punctured

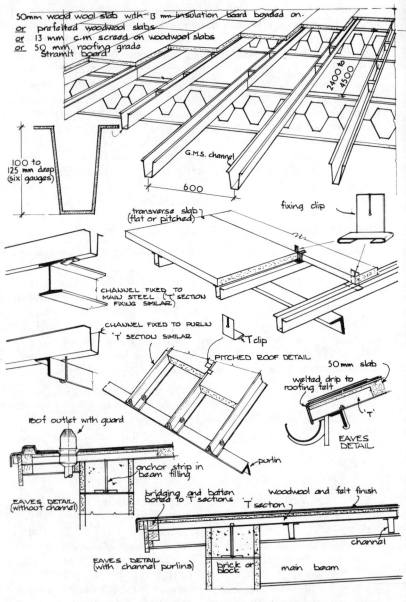

50mm wood wool slab with 13 mm insulation board bonded on.
_or_ prefelted woodwool slabs
_or_ 13 mm c.m screed on woodwool slabs
_or_ 50 mm roofing grade
      Stramit board

2400 to 4500

100 to 125 mm deep (six gauges)

G.M.S. channel

600

transverse slab (flat or pitched)

fixing clip

CHANNEL FIXED TO MAIN STEEL ('T' SECTION FIXING SIMILAR)

CHANNEL FIXED TO PURLIN 'T' SECTION SIMILAR

'T' clip

PITCHED ROOF DETAIL

50 mm slab

welted drip to roofing felt

'T'

EAVES DETAIL

roof outlet with guard

anchor strip in beam filling

purlin

EAVES DETAIL (without channel)

bridging and batten bolted to 'T' sections

'T' section

woodwool and felt finish

'T' section

channel

EAVES DETAIL (with channel purlins)

brick or block

main beam

WOOD WOOL SLAB DECKING   FIG. 8.14

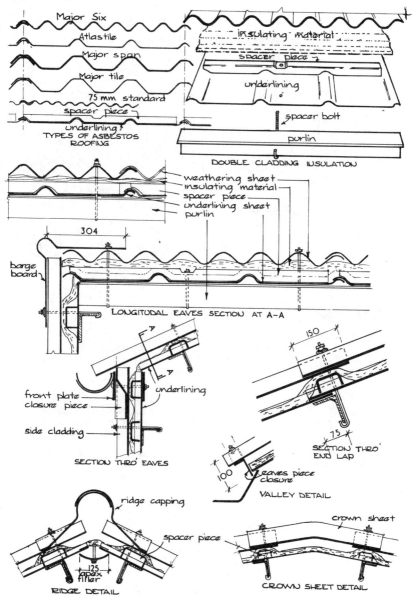

ASBESTOS ROOF CLADDING

FIG. 8·15

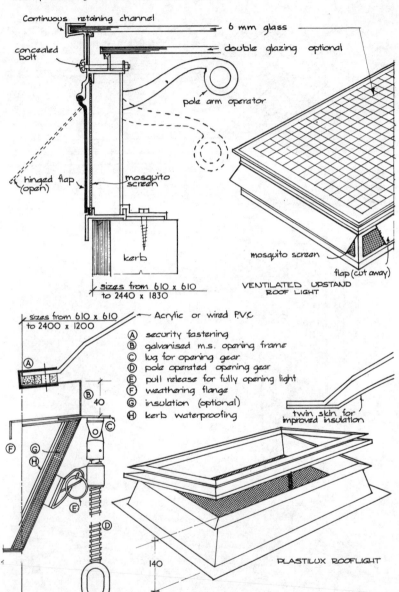

Continuous retaining channel

6 mm glass

concealed bolt

double glazing optional

pole arm operator

hinged flap (open)

mosquito screen

kerb

sizes from 610 x 610 to 2440 x 1830

mosquito screen

flap (cut away)

VENTILATED UPSTAND ROOF LIGHT

sizes from 610 x 610 to 2400 x 1200

Acrylic or wired P.V.C.

Ⓐ security fastening
Ⓑ galvanised m.s. opening frame
Ⓒ lug for opening gear
Ⓓ pole operated opening gear
Ⓔ pull release for fully opening light
Ⓕ weathering flange
Ⓖ insulation (optional)
Ⓗ kerb waterproofing

Ⓐ

Ⓑ 40

Ⓕ

Ⓖ
Ⓗ

Ⓒ

Ⓔ

Ⓓ

twin skin for improved insulation

140

PLASTILUX ROOFLIGHT

ROOF LIGHTS AND VENTILATORS

FIG. 8.16

by nails but clipped firmly in position, so that it will resist high winds and suction. It is fairly easy to fix, and a full range of accessories is available for flashing and trimming. It has a U value of $1 \cdot 51$ W/m² °C. when laid on wood-wool 25 mm thick fixed direct to rafters.

Fig. 8.19 shows asbestos-cement cavity decking which consists of 'U' shaped units simply laid on purlins or other supports and bolted together to form a strong roof and ceiling combined. Ceiling boards may be added if necessary to the battens provided or to suspenders, as shown. With felt roof-covering, and a cavity filling of 25 mm glass-fibre, it has a U value of $0 \cdot 738$.

Fig. 8.20 shows a concrete-tile flat roof where a promenade is required. It acts in much the same way as the concrete basement floor tile shown in Chapter 4, by allowing a flow of water underneath the tile without any tendency to build up deposits of silt or dirt. It also improves solar insulation by reflection from the light-coloured surface, and has extra thermal resistance provided by an air cavity under the tile. Owing to its mass it can also regulate temperature changes. It is claimed that the U value is improved from $0 \cdot 54$ to $0 \cdot 42$ W/m² °C solely as a result of the presence of the cavity.

8.6 Roof lights and ventilators

Roof lighting in warm countries, whether of corrugated plastic sheeting, patent glazing, or lantern light needs some study before any attempt to provide it is made. Not all plastics are capable of withstanding strong sunlight and, under the hazy or overcast skies usual in humid zones, they can deteriorate fairly rapidly. Also, unless some form of insulation is used, they transmit a great deal of heat, and, conversely, lose it rapidly if the temperature drops at night. Double-glazing has now provided some relief and, provided precautions are taken against glare, it can be used to advantage, especially when ventilation is incorporated. As light intensity is usually greater in tropical climes than in the country of manufacture, skylight calculations need to be modified to suit local conditions. Experience is often the best guide in these cases.

Fig. 8.16 shows an example of a ventilated rooflight with provision for both double-glazing and mosquito-screening. It is also secure against illegal entry. Glare can be reduced, if necessary by use of an inner sheet of coloured glass. The flaps should enable the rooflight to withstand extreme conditions of wind and rain, and it should also be fairly dust-proof in sandy regions.

The second example shows an opening light, with domes which can be made of several different materials. These can be of diffused or clear acrylic, wire laminate P.V.C., or fibre-glass. In addition to the effects

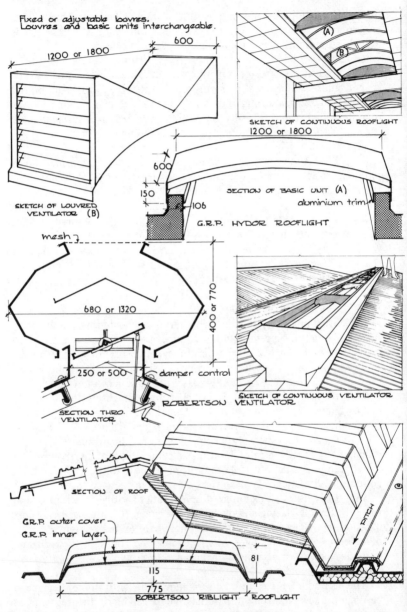

Fixed or adjustable louvres.
Louvres and basic units interchangeable.

600

1200 or 1800

SKETCH OF CONTINUOUS ROOFLIGHT

1200 or 1800

600

SECTION OF BASIC UNIT (A)

aluminium trim

150

106

G.R.P. HYDOR ROOFLIGHT

SKETCH OF LOUVRED
VENTILATOR (B)

mesh

680 or 1320

400 or 770

250 or 500

damper control

SECTION THRO.
VENTILATOR

ROBERTSON VENTILATOR

SKETCH OF CONTINUOUS VENTILATOR
VENTILATOR

SECTION OF ROOF

G.R.P. outer cover
G.R.P. inner layer

PITCH

81

115

775

ROBERTSON 'RIBLIGHT' ROOFLIGHT

ROOF LIGHTS AND VENTILATORS

FIG. 8.17

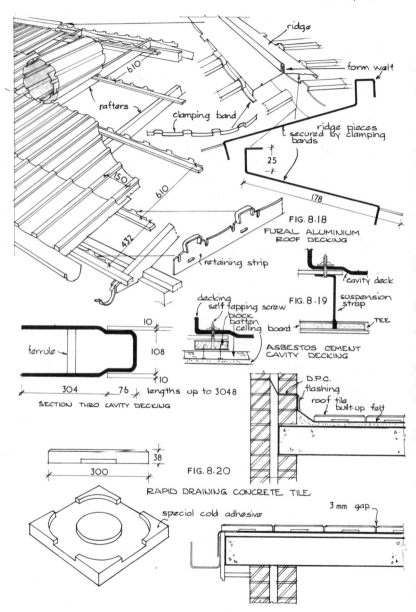

ridge

form wall

ridge pieces
secured by clamping
bands

610

rafters

clamping band

25

178

150

610

FIG. 8.18

FURAL ALUMINIUM
ROOF DECKING

432

retaining strip

cavity deck

decking
self tapping screw
block
batten
ceiling
board

FIG. 8.19

suspension
strap

TEE

ferrule

10

108

10

ASBESTOS CEMENT
CAVITY DECKING

304

76

lengths up to 3048

SECTION THRO CAVITY DECKING

D.P.C.
flashing
roof tile
built-up felt

38

300

FIG. 8.20

RAPID DRAINING CONCRETE TILE

special cold adhesive

3 mm gap

ROOF COVERINGS

of ultra-violet rays on plastics fire precautions may also have to be considered. The use of acrylics in skylights is usually restricted in areas where the risk of fire is high. The shape of the covering, whether flat, domed, or pyramidal, will depend on the material. G.R.P., for instance, can be produced in any shape, but wired P.V.C. can only be used for domes and continuous rooflights.

Where continuous rooflighting is required, G.R.P. lights are simple and effective, especially when used in conjunction with the louvred ventilator shown in Fig. 8.17. The louvres may be fixed or adjustable; units are interchangeable with rooflights to give lighting and ventilation as required. Continuous ventilation can also be provided by use of the Robertson ventilator. This is a well-tried design, available in several sizes, which provides exhaust and air changes to suit estimated wind speed and temperature. This ventilator may also be used with a Rib-light rooflight of double skin G.R.P., which could provide translucent lighting where conditions are suitable. The reduction in the amount of solar heat-gain also results in a better working environment.

9 | Stairs

9.1 Introduction

This subject has also been dealt with in Volume 2, when covering aspects of traditional construction. However, there are other considerations which need to be examined concerning prefabrication, composite stairs, precast work, and metal construction. The staircase is frequently the main feature of design, particularly in public building, and much attention is devoted to its siting as a focal point. In housing, where prefabrication and mass-production is practised, a different approach is required.

9.2 Prefabrication of timber stairs

It is usual on housing sites where different types of houses are being erected by the same organization to standardize the staircase as far as possible. This usually means that the stairs must be independent of loadbearing walls and partitions, particularly where handling problems could be encountered. A typical timber close-tread stair is shown in Fig. 9.1. Elaborate joints and connections, needing craft skills, have been avoided, giving freedom to adjust for slight variations where these occur. The dimensions given are those which enable the stair to conform with most building regulations; these are usually met by the following conditions:

Headroom of not less than 2 metres measured vertically above the pitch line; and a clearance of 1·5 metres minimum measured at right angles to the pitch line.

The sum of the going plus twice the rise to be not less than 550 mm and not more than 700 mm. (In Fig. 9.1 this would be 592 mm.)

The rise must not be more than 220 mm and the going not less than 220 mm.

The pitch must not be more than 42 degrees.

The above applies only to private stairways. Common stairways used by the general public must have a rise of not more than 190 mm and a going of not less than 230 mm. The pitch must not be more than 38 degrees, with not more than 16 risers in any one flight. The balustrade

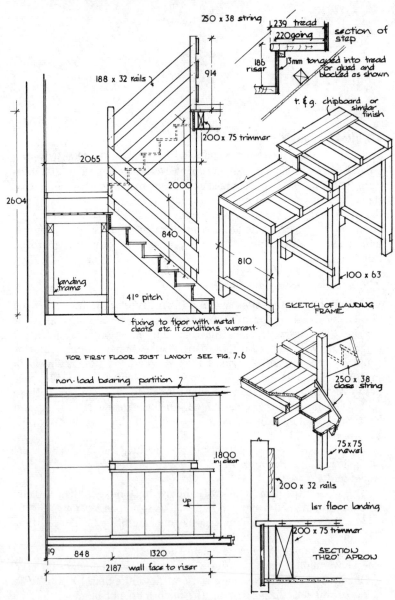

250 x 38 string

239 tread

220 going

section of step

914

188 x 32 rails

186 risar

3mm tongued into tread or glued and blocked as shown

section of step

t. & g. chipboard or similar finish

200 x 75 trimmer

2065

2000

2604

840

810

landing frame

41° pitch

SKETCH OF LANDING FRAME

100 x 63

fixing to floor with metal cleats etc. if conditions warrant.

FOR FIRST FLOOR JOIST LAYOUT SEE FIG. 7.6

non-load bearing partition

250 x 38 close string

1800 in clear

75 x 75 newel

200 x 32 rails

1st floor landing

UP

200 x 75 trimmer

19   848   1320

2187 wall face to riser

SECTION THRO' APRON

PREFABRICATED TIMBER STAIRS

FIG. 9.1

for private stairs should not be less than 840 mm high, as shown, and not less than 900 mm at landings. The landing balustrade for common stairs should not be less than 1·1 metre high. The sizes of the timbers given are for general use, but may be reduced if hardwood is used and local by-laws permit.

Fig. 9.2 shows a simple but ingenious design produced by the Finnish Plywood Development Association, using blockboard and chipboard. It consists basically of a standard flight of seven steps prefabricated as shown. This is connected to a landing by means of a horizontal batten 80 mm deep glued to the side of the landing. The top flight is similar to the bottom but has the added portion shown hatched in section. This is used to secure the top flight to the landing. When a dog-leg stair is needed a half landing is used with the newel left, as shown, but the top flight moved around to form the dog-leg. Should a straight flight be wanted, the landing is discarded and the support batten held by two newels described on the drawing.

9.3 Component and composite stairs

This type of open-tread staircase can be made in a variety of styles and materials. As a stepped timber carriage stair with heavy timber tread (Fig. 9.3) it still enjoys great popularity in public building, department stores, and private residences. An attractive type of carriage made to customers' requirements is that of laminated timber (1a) where selected hardwoods are cut, shaped, and glued to form graceful curves and interesting profiles. African hardwoods are in demand for this form of construction. The timber carriage shown is a small example and large arched spans are quite common. The bearers (2a) are usually made of the same material, to match the carriage. The tread (3a) may either be of one piece of hardwood or may be made up if preferred.

A box section of welded angles, channels, or of rectangular hollow steel or aluminium section is shown (1b), cranked by butt welding with steel brackets to support the treads. A hot rolled steel channel can also be used as a carriage (1c), with a steel flat of 60 × 6 mm section shaped to form steps and adjusted to correct profile with shim plates, (2c). Treads can be of prestressed concrete, hardwood, steel, or aluminium plate chequered as appropriate. A typical hardwood tread, drilled for baluster and bolted to bracket, and blocked and pelleted is shown (3c).

9.4 Precast concrete stairs

Where stair units are repetitive, as in the case of multistorey buildings, high-rise flats, or large office blocks, it is usual to prefabricate them,

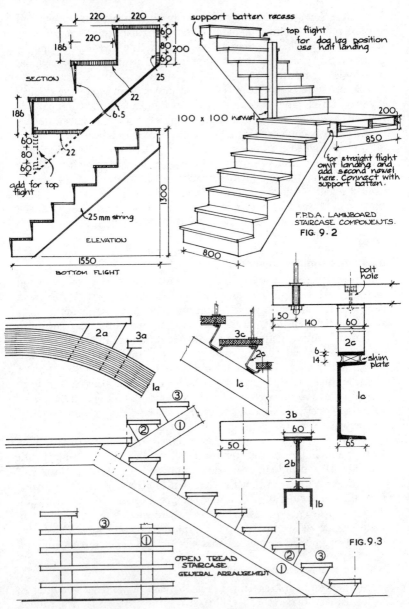

SECTION

SECTION

ELEVATION

BOTTOM FLIGHT

25 mm string

add for top flight

support batten recess

top flight
for dog leg position
use half landing

100 x 100 newel

for straight flight
omit landing and
add second newel
here. Connect with
support batten.

F.P.D.A. LAMINBOARD
STAIRCASE COMPONENTS.
FIG. 9.2

bolt hole

shim plate

OPEN TREAD
STAIRCASE
GENERAL ARRANGEMENT

FIG. 9.3

COMPONENT AND COMPOSITE STAIRS

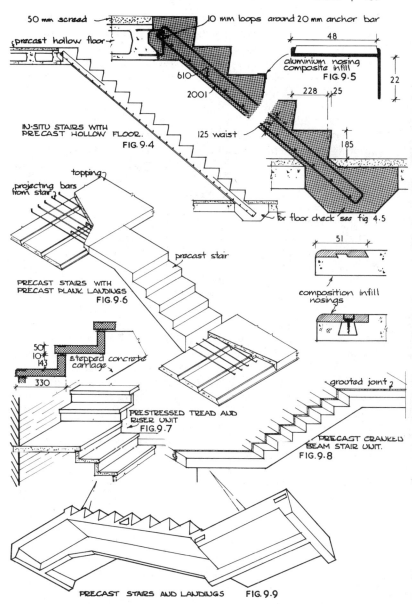

50 mm screed

10 mm loops around 20 mm anchor bar

precast hollow floor

aluminium nosing composite infill
FIG.9.5

48

22

610

2001

IN-SITU STAIRS WITH
PRECAST HOLLOW FLOOR.
FIG.9.4

228

25

125 waist

185

for floor check see fig 4.5

topping

projecting bars from stair

precast stair

PRECAST STAIRS WITH
PRECAST PLANK LANDINGS
FIG.9.6

51

composition infill nosings

stepped concrete carriage

50
10
143

330

PRESTRESSED TREAD AND
RISER UNIT
FIG.9.7

grouted joint

PRECAST CRANKED
BEAM STAIR UNIT.
FIG.9.8

PRECAST STAIRS AND LANDINGS   FIG.9.9

PRECAST CONCRETE STAIRS

deliver them to the site, and set them in position by crane as the work proceeds. Figs. 9.6 to 9.9 show some of the methods used. Precast work allows for better finish, speedier construction, and good quality-control. Preparation and moulds are expensive, however, and a flight such as is shown in Fig. 9.9 would not be produced in numbers of, say, twenty or less.

Where one stair only is needed, as would be the case in the example of the community centre, the flight could be cast *in situ* and anchored to the precast concrete hollow floor as shown, Fig. 9.4. (This has been designed for the floor in Fig. 4.5.) A typical precast stair unit used in conjunction with a Bison precast plank floor is shown in Fig. 9.6. The projecting reinforcement is concreted into the topping of the floor plank to give a strong and economical structure. Fig. 9.8 shows a cranked beam unit, which has several advantages for developing countries. It is easily cast in inexpensive moulds, and can be lifted into position by a small gang of men with a hand winch and without mechanical aid. It is fairly easy to transport and can be built to any width using the same mould. Extra flights can be built up simply by reversing the units, provided that the landings are the same size. The joggled joint is grouted after placing the unit in position.

Fig. 9.7 shows one variation of a precast or prestressed tread and riser which is frequently to be seen in modern construction. The units can be carried by a cut string or carriage of either steel or concrete, shaped as in Fig. 9.3, or built into a brick wall, in which case the riser height could be adjusted to suit the brick courses. Sometimes the combined tread and riser is inverted, and the riser splayed, with a nosing or infilling of non-slip composition inserted at the time of casting. Three types of nosing are shown in Fig. 9.5; there are many others. Aluminium extruded section with a composition infill is popular; but the nosing can be secured to the step if preferred. The screw-down type shown has the advantage of being capable of renewal after wear.

9.5 Metal strings and balustrades

The types of stairs shown in Figs. 9.10 and 9.11 would be rather elaborate for a normal community hall or centre. However, in most capital cities such designs are to be found in public buildings, embassies, administrative offices, and good department stores. Stainless steel or bronze over a hardwood core, Fig. 9.10a, is not a new process, having been used in good-quality fixtures and fittings for a number of years. The design can, of course, be varied to produce different types of balustrading, stair-treads or strings. A central carriage can also be added for stairs over, say, 2 metres wide.

Fig. 9.12 shows a cheaper form of R.H.S. stainless or pressed steel which can be made to measure and supplied k.d. with all accessories and fixing instructions. It is a simple and durable structure, and when fabricated can be handled for installation by a very small gang. Tread-coverings have a non-slip composition front edge, and the remainder is covered with a flooring to suit the customer's choice. The normal width between stringers is between 750 and 1500 mm.

9.6 Metal stairs and fire escapes

There are a number of options open to designers of metal staircases, depending on function and local by-laws. The latter can vary from stringent regulations based on those of developed countries, to very few, sometimes with no implementation at all. Special precautions have to be taken with, say, refineries and storage depots in oil-producing countries; but the examples given in Figs. 9.13 to 9.15 should meet most needs.

A straightforward design for a single flight, Fig. 9.13, shows an independent central-spine stair supported at landing level by a cantilever steel joist. The stair is an all-welded structure of either m.s. or aluminium with not more than 16 risers to the flight. The simplest type of spine is a single rolled m.s. channel or box section made up of two angles or channels welded together, as in columns, Volume 1. The size would depend on the dimensions of the stairs and the purpose for which they are intended. Box sections of this nature would need to be protected internally by dipping or painting before welding, especially in humid or corrosive areas. The treads could be of chequer steel plate or aluminium, welded to a light metal frame. The balustrade is shown with R.H.S. newels, intermediate supports, and handrails with 20 mm solid square balusters as shown.

Spiral stairs. As a simple and effective form of stair the circular design shown in Fig. 9.14 has many advantages. It consists basically of two main components, the tread and the central core. The latter is made up of units of riser height, each fixed firmly to the one below. This is done with high-tension bolts inserted through a hand-hole in the core and tightened from above with a socket wrench. The treads are either embossed or have an applied non-slip finish. If used internally they may be carpet-recessed or of solid hardwood in mahogany or iroko. If used externally as a fire escape a galvanized hot-dip finish is usually applied.

External stairs with landing. A typical arrangement for a straight-flight metal stair with more than 16 risers — thus requiring a landing —

Metal strings and balustrades

Fig. 9.10 Stainless-steel open-tread stairs

A	Oval handrail 18 s.w.g. S.S. over timber core
B	38 × 31 × 6 mm channel core rail
C	38 × 25 mm baluster
D	10 mm toughened glazing to balusters
E	String of 16 s.w.g. S.S. cladding over 10 s.w.g. steel rectangular sections
F	Stair treads and landing of 10 s.w.g. S.S. sheet
G	Tread bolted/welded to string infilled with cement screed and terrazzo, hardwood etc.

Fig. 9.11 S.S. balustrade and string cladding

H	S.S. balustrade
J	12 s.w.g. m.s. continuous core
K	Timber ground
L	18 s.w.g. S.S. end cover
M	Lamin/chipboard
N	50 × 6 mm m.s. straps
P	Concrete step
Q	16 s.w.g. bronze sheet
R	18 s.w.g. S.S. end cover
S	Plaster soffit
T	38 × 38 mm handrail
Z	38 × 10 mm knee rail

Fig. 9.12 Aluminium alloy staircase unit

U	76 × 44 mm extruded handrail
V	38 × 38 mm extruded balusters
W	44 × 38 mm newel
X	127 × 45 mm strings
Y	280 × 32 mm treads with rubber composition inset

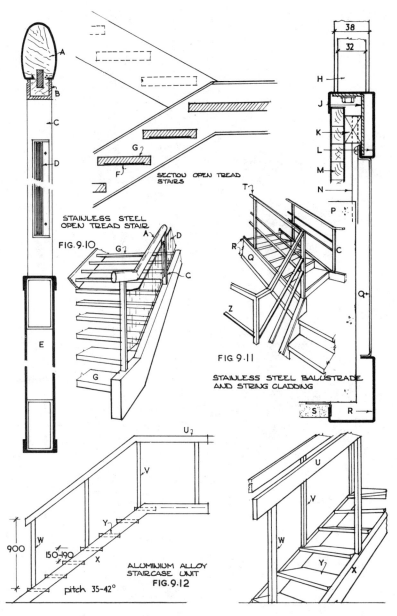

38

32

H

J

K

L

M

N

P

SECTION OPEN TREAD STAIRS

STAINLESS STEEL OPEN TREAD STAIR

FIG. 9.10

A

B

C

D

E

G

F

G

A   D

G

C

C

T

R

Q

Z

Q

FIG. 9.11

STAINLESS STEEL BALUSTRADE AND STRING CLADDING

S   R

U

V

Y

X

W

900

150-190

pitch 35-42°

ALUMINIUM ALLOY STAIRCASE UNIT

FIG. 9.12

U

V

W

Y

X

METAL STRINGS AND BALUSTRADES

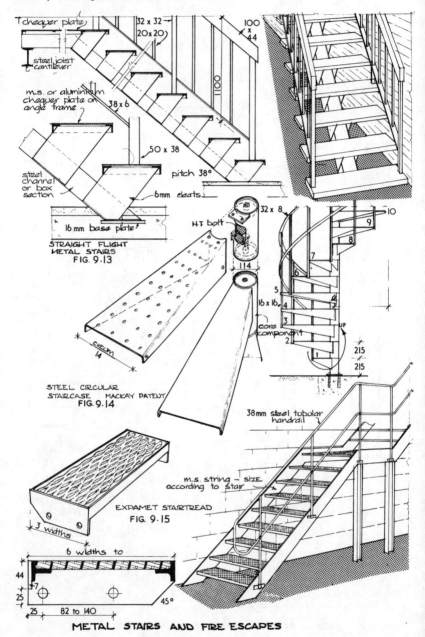

chequer plate

steel joist cantilever

32 x 32
20 x 20

100 x 44

100

m.s. or aluminium chequer plate or angle frame

38 x 6

50 x 38

pitch 38°

steel channel or box section

6mm cleats

16 mm base plate

STRAIGHT FLIGHT METAL STAIRS
FIG. 9.13

H.T. bolt

32 x 8

114

16 x 16

core component

10

9

8

7

5
4
3
2
1

UP

215
215

circum 14

STEEL CIRCULAR STAIRCASE   MACKAY PATENT
FIG. 9.14

38mm steel tubular handrail

m.s. string – size according to star

3 widths

EXPAMET STAIRTREAD
FIG. 9.15

6 widths to

44

25

25   82 to 140

45°

METAL STAIRS AND FIRE ESCAPES

is given in Fig. 9.15. The landing can be put at any convenient level, but is usually at mid-height. With a stair of under 1200 mm wide the construction is usually 6 mm strings and unsupported chequer plate treads of the same thickness welded to the strings, though this would depend on use and location. Treads could be strengthened if required with small steel angles or tees across the width of the stairs. Metal stairs should be located in a shaded position or protected by a canopy, which can also act as a shelter from heavy rain.

For specialized industrialized work in factories and workshops, treads may be made up of expanded metal in m.s. or tee frames. These are supplied in a number of sizes by the manufacturer, and welded or bolted to strings to form stairs of the required dimensions. The steel columns are designed as reqiured. Handrails for such stairs are usually of galvanized steel tube, shaped and welded, and secured to strings by steel collars. Instead of welded tube connections, ball socket connectors are available, which are much neater. These are simply slipped over the end of the tube at each connection and tightened with 'Kee Klamps' to give a firm joint.

Stairways of this nature frequently give access to gangways, catwalks, or walkways constructed in the same way as the stair itself. The normal method is to clamp sheets of X.P.M. of suitable gauge in frames of light m.s. angle up to 2 m long, which are then secured to the main structure. Such frames are strong, light, non-slip, and permit the free air passage which is essential in places where hot processes are used.

10 | Surface coverings and finishings

10.1 Renderings and plastering

A brief introduction to this subject has been given in previous volumes, but with the development of tropical countries and the need for changes from traditional practices a wider approach to the subject is now necessary. New methods and materials will be introduced or improvements discussed as appropriate.

Mesh. Expanded metal is widely used in plastering and rendering, and is now made in galvanized metal as well as in forms coated to resist damp- or salt-laden atmospheres. Improvement has also been made in the provision of metal stops where different finishings occur in the same wall-face, Fig. 10.4. These also avoid disfiguring cracks due to difference in movement between the various backgrounds and materials. Expanded metal, however, should be rendered on both sides, preferably in three coats when used independently of other backing.

Mesh has other uses apart from lathing; in tropical lands it is in demand as partition screens and grilles, and when supplied in aluminium, copper, or stainless steel it can be decorative as well as protective. It is made in a range of mesh sizes and gauges in anodized, galvanized, plastic-covered, or painted finish. Screens can also be made of chain link, or woven, or wire-netted, or of wirework or metal lattice. Mesh is a very versatile material.

FINISHINGS. A widely-used external and often internal finish to walls is that of cement and sand, usually applied as two-coat work, and finished with a wooden float or trowel. Other finishes are gradually being introduced; one is that of pebble-dash, where small stones up to 5 mm diameter are thrown on to freshly-applied mortar. The pebbles adhere better if the rendering contains lime or plasticizer — in small quantities so as not to impair the quality of the mix. Roughcast, a mixture of wet mortar and sometimes fine pebbles flung on to the wall, is also used, and later finished with cement paint, emulsion, or distemper. It is also applied as a first coat on concrete as a key for the rendering.

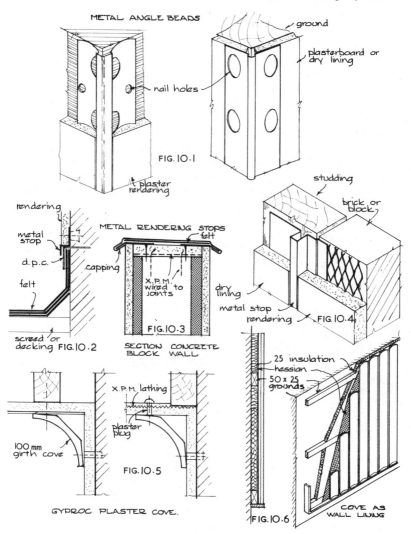

METAL ANGLE BEADS

ground

plasterboard or dry lining

nail holes

FIG. 10·1

plaster rendering

rendering

metal stop

d.p.c.

felt

capping

screed or decking FIG.10·2

METAL RENDERING STOPS

felt

X.P.M. wired to joints

FIG.10·3

SECTION CONCRETE BLOCK WALL

studding

brick or block

dry lining

metal stop rendering

FIG 10.4

X.P.M. lathing

plaster plug

100 mm girth cove

FIG. 10·5

GYPROC PLASTER COVE.

25 insulation

hessian

50 x 25 grounds

FIG. 10·6

COVE AS WALL LINING

RENDERING AND PLASTERING

Scraped finishes are now becoming general. These are formed with a board studded with nails, a float covered with expanded metal, or a steel straightedge. The treatment is applied when the rendering is firm but not hard. Textured finishes with patterned surfaces are also usual in some areas, but only where this is a traditional craft, as it is skilled work. It consists of treating the freshly-applied coat with various tools to produce ribbed stucco, stippled stucco, fan texture, etc. Machine spatter finish known as Tyrolean has been mentioned in Volume 1. Where renderings finish against flashings, copings, and the like, metal stops give a clean, efficient finish, as shown in Figs. 10.2 and 10.3. Pigments may be added to colour the mix if desired, but tend to fade if exposed to strong sunlight. For protection against tropical sun and rain, eaves and canopy cover is desirable, though not always essential when plain rendering is used with a good key.

Additives or plasticizers for cement and concrete are now in use in a variety of brands. As workability admixtures, wetting, dispersing, and water-reducing agents these chemicals are quite efficacious when added in small quantities. For rendering in wet areas a water-repellent and workability admixture is usually desirable. The former prevents the absorption of rain, but generally cannot be used in situations where pressure may occur, as in basements or tanks. Workability is not as important as waterproofing; increasing the quantity of cement in a mix may make it more waterproof and workable but could be more expensive than an admixture. Such products are well-advertised in building literature.

Plastering. Whilst tropical countries normally follow their own traditions, local practices are gradually being modified by the use of imported materials. The introduction of Portland cement has caused internal plastering to be applied in the same way as external rendering, though trowelled smooth instead of floated. On first-class work the mason is sometimes trained to use the plasterer's rectangular steel float for this purpose. This is also used to apply Gypsum C neat plaster, as mentioned in Volume 2, but the practice is not universal. However, the application of undercoat and finish on backgrounds as described is becoming more widespread in large towns.

Special plasters are sometimes imported for finishing coats to meet specific needs, such as improved insulation, hardness, or texture, but are not yet normal practice. Such aids as external angle beads at corners, however, are in use for *in situ* plastering or for finishing coat to plasterboard or dry lining, Fig. 10.1.

FIBROUS PLASTER. Some years ago, when mouldings and carvings were more of a feature than they are today, plasterwork for internal decoration was often precast in workshops, to be fixed on site. Some of this work was very elaborate and may be seen on older public buildings, theatres, and hotels. It is still carried out today, though to a lesser extent. Some firms, however, do produce standard mouldings such as covings, cornices, beading, architraves, and surrounds to be cut to position on site, fixed with galvanized nails, and made good. Fig. 10.5 shows a simple example of coving around a ceiling. This not only serves as a neat ornament but provides a straight-line finish to wall or ceiling decoration, especially where these are in different colours and materials. A further use for such coving can be seen in Fig. 10.6, where it is formed into wall panels mounted on battens, with a backing of insulation to keep out excessive heat and sound. Decorative plaster-work can be a highly skilled art, and is practised extensively in some lands.

10.2 Concrete finishes

Cast stone, sometimes referred to as artificial or reconstructed stone, has been in use in different forms for many years. First used as precast stone, its appearance changed with the introduction of white Portland cement, and later even more with the use of the actual stone being simulated, broken to use as aggregate. Cast stone may be homogeneous throughout or consist of a facing to structural concrete. The facing should be at least 19 mm thick, and cast at the same time as the block. Better quality stone is manufactured by hammer-compacting the semi-dry facing mix in the mould and adding a core as soon as possible afterwards. Only the minimum amount of water consistent with adequate compaction is used.

Cast stone can be moulded to fairly intricate detail, but undercutting features should be avoided. It may also be carved to a required finish if desired. It is equal in strength to natural stone, though durability will depend on conditions of exposure. Normally concrete and stonework weather better in tropical conditions than in colder situations, though scour in sandy regions can cause erosion. Atmospheric acids in industrial areas can cause some surface deterioration, though this is not widespread. Cast stone can be economical where a number of units can be cast from the same mould. Fixing is carried out as for natural stone, using cramps, joggles, and dowels. 'U' shaped cramps are embedded in chases to hold copings, cornices, and projecting courses, though thinner facing stones sometimes have plain non-ferrous metal strips inserted in the units.

Cast stone for sills is usually fine concrete with aggregate not more than 13 mm nominal maximum size. Sills and similar members are not

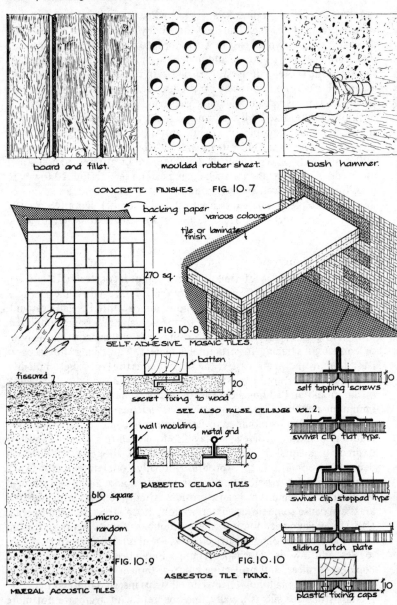

board and fillet.

moulded rubber sheet.

bush hammer.

CONCRETE FINISHES   FIG. 10.7

backing paper

various colours

tile or laminate finish

270 sq.

FIG. 10.8
SELF-ADHESIVE MOSAIC TILES.

fissured

batten

secret fixing to wood

20

SEE ALSO FALSE CEILINGS VOL. 2.

self tapping screws

10

wall moulding   metal grid

20

swivel clip flat type.

610 square

RABBETED CEILING TILES

swivel clip stepped type

micro. random

sliding latch plate

FIG. 10.9

FIG. 10.10

ASBESTOS TILE FIXING.

plastic fixing caps

10

MINERAL ACOUSTIC TILES

CONCRETE AND TILE FINISHES

more than one metre long, though longer lengths can be cast by adding galvanized reinforcement. The mortar for cast stone should not be stronger than the stone itself. If possible crushed stone should be used as fine aggregate, giving a 1:2:9 cement/lime/fine aggregate mix; or else cement/lime/sand may be used, reserving the stone dust mix for pointing.

Concrete cladding as discussed in Chapter 5 is widely used today, often where repetitive units can be used, as in housing developments, large office blocks, flats, and administrative buildings. An example of the ribbed cladding now universally popular for commercial enterprises or large-scale construction is shown in Fig. 5.9. The majority of these claddings are manufactured by system builders and precast concrete firms, who provide a variety of finishes; these are in effect cast stone, though not usually made to resemble natural stone, but rather graded with different aggregates and cements either to stock specification or to architects' requirements.

Sometimes sections are cast in stock units, such as concrete weather-boarding made in various coloured cements. The units can be supplied in 600 mm lengths for convenient handling and are about 27 mm thick, giving a 100 mm lap. Some firms manufacture concrete blocks whose face has a weatherboard profile, giving a shiplapped effect to the completed wall. For further finishes see Volume 1 and Chapter 5.

Formwork finishes. In warm regions, *in situ* concrete walls are constructed using hardwood panels as shown in Volume 1 and supported by props, bolts, and proprietary fixings. The finished surface as it leaves the formwork, usually of hardwood or plywood, is not generally satisfactory owing to lack of grain, variation of aggregate and grading, blemishes or rough patches, and oil or laterite stains.

To improve the finished effect, the concrete formwork may be panelled, ribbed, overlapped, stepped, shiplapped, or lined, to produce a smooth, textured, or patterned finish. One such lining showing a moulded rubber sheet finish is given in Fig. 10.7. Another way is to rib the face, or to bush-hammer, point-tool, or chisel the face by hand. Care must be taken not to reduce the cover of the reinforcement to less than about 13 mm. This means providing an original cover of between 38 and 50 mm.

A cheap and effective formwork can be had by using a corrugated-iron sheet; but this should be fairly new and well-positioned, or else the concrete should be hammered later. The cement film may also be removed by the use of a wire brush after the concrete has set but not hardened. Sometimes retarding paints are applied to the formwork before the concrete is placed – though uneven boarding could prevent uniform distribution of the retarder, with patchy results if it is applied carelessly. Most integral finishes on concrete, especially when exposed aggregate is used, are much superior when used with precast concrete factory products rather than with those cast *in situ*. Precast units are normally preshrunk and, with proper site jointing, give a uniform, crack- and craze-free finish. This is dependent upon the water/cement ratio, uniformity of batching, mixing, even compaction, and freedom from laterite and rust stains. Rigid moulds are constructed, to produce panels of a size set within fine limits; and this leads to speed and economy, at least where repetition justifies the cost of the moulds.

10.3 Tiling and ceramics

Wall tiling ranges from best-quality ceramic and glass mosaic to cheap plastic substitutes with self-adhesive backing. However, mass-production of good-quality thin-glazed tiles and special adhesives now enables walls to be covered by masons with little or no traditional skills in this work. The tiles are made in a variety of colours and patterns, in single units or in mosaic panels. They are used as a decorative, durable, and easily-cleaned surface for walls, table-tops and slabs.

Wall tiles should be fixed with proprietary adhesives, which usually have greater holding power than cement and sand mortar. For those with a keyed back a thicker bed adhesive is used. Tiles may be obtained in white or in a wide range of colours, and may be of high glaze, eggshell, or matt finish. For household purposes modern ceramic tiles usually 108 × 108 × 6 mm thick laid square bedded in adhesive and grouted are used. The tiles normally have cushioned edges, i.e. slight radius and spacer lugs, though edges can be square if preferred. Various accessories, such as cappings, border strips, hooks, soap dishes, and toilet-roll holders, are available in addition to those shown in Volume 2. With wall tiling it is always advisable to follow the manu-facturers' instructions regarding adhesives, method of application, and grout finish between joints.

Thermal expansion is not necessarily a serious problem, provided that the wall is not exposed to the hot sun or adjacent to boilers or refrigerated areas. It is advisable to leave a small expansion gap at internal angles, and not to grout up joints with inflexible material such as neat cement. Tiled walls over 5 m long should have expansion joints, especially if the walls are subject to sharp changes of temperature. For better-quality work, a considerable range of patterned ceramics is made, as shown in Volume 2.

Marble, glass, and ceramic tiles are also supplied as small tiles or tesserae. The mosaics mentioned in Volume 2 are now made up into panels of self-adhesive tiles 270 mm square. After removing the backing, the tiles are simply pressed on to the surface and the joints grouted. They are made in a variety of colours and can be cut and arranged to form patterns as desired, Fig. 10.8.

Precast marble tiles may be made, where material is available, from pieces placed together in tile size. Some tiles are formed from mortar of white cement coloured to match the natural marble and marble chippings. This mixture is then compressed, vibrated, and, when hard, polished. The same result can also be obtained by placing large lumps of marble in a mould. The spaces between them are then filled as above; and, after vibrating and setting, the blocks are sawn to size.

Marble tesserae, to be used for wall and floor finishes, may also be imported loose in 50 kg bags. On walls the natural rough texture is quite pleasing and much in demand where tiles are traditional. The chippings are normally mounted in reverse on stiff paper and applied as described. If marble is not available, or too expensive, materials such as pebbles, glass, slate, etc., can be used if they are durable and aesthetically suitable.

10.4 Thin surface finishings, internal

Sheet materials are extensively used today for internal finishing; the main ones being asbestos sheet, wood-based board, glass, marble, plasterboard, insulation board, and laminates.

Asbestos sheet, which may also be used externally, is supplied either as fully-compressed or semi-compressed wall board or insulation board. Fully-compressed boards usually have two smooth surfaces and are used in vulnerable places such as worktops, bath panels, and window sills. They can receive a mineral-glazed finish in standard colour ranges. Semi-compressed flat sheets are natural grey with a smooth face and textured back, and are used for infill panels, bath panels, and shelving. They will accept protective finishes such as chlorinated rubber-based paint; but oil-paints require an alkali primer. Asbestos insulation boards consist of asbestos fibre, usually with cement or lime binders. They are used for fire-protection purposes around structural steelwork, partition linings, flue casing, and fire-check door linings; and they are also useful in damp or humid situations. They are supplied in lengths from 1800 to 3000 mm and in widths from 900 to 1200 mm.

Fibre building boards are sheet materials which usually have a thickness exceeding 1·5 mm and are manufactured from wood or vegetable-fibre

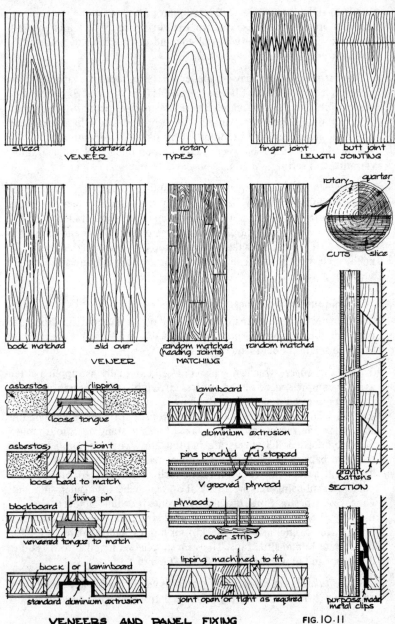

sliced   quartered   rotary   finger joint   butt joint
VENEER   TYPES   LENGTH JOINTING

book matched   slid over   random matched   random matched
VENEER   (heading joints)
                MATCHING

rotary   quarter
CUTS   slice

asbestos   clipping
loose tongue

asbestos   joint
loose bead to match

blockboard   fixing pin
veneered tongue to match

block or laminboard
standard aluminium extrusion

laminboard
aluminium extrusion

pins punched and stopped
V grooved plywood

plywood
cover strip

lipping machined to fit
joint open or tight as required

gravity battens
SECTION

purpose made metal clips

VENEERS AND PANEL FIXING   FIG. 10.11

material. They are classified as hardboards, medium boards, and insulation boards. Standard hardboards usually have a density of 800 kg/m^2. They are used for wall and ceiling linings, partitions, flush doors, floor coverings, etc., and are also supplied perforated, moulded, or pre-decorated with paint, plastic veneer, or printed wood grain.

Special boards are available, such as boards tempered to increase their strength (either perforated, plain or pre-decorated), or boards for cladding, sheathing, formwork lining, or flooring. Low-density or duo-faced boards with smooth surfaces on both sides are also made. Boards can be used laminated to core material, such as particle boards or cellular paper core; fibre insulated soft board is also in wide demand, with a density of 350 kg/m^2. Applied finishes include paint, paper, metal foil, cloth, P.V.C., etc.

Bitumen-impregnated insulation board is used to improve moisture resistance. Acoustic boards and tiles are also used as low-density boards, drilled or grooved to increase sound-absorption, Fig. 10.9. Flame-retardant board, i.e. fibre-board treated to reduce the spread of flame by impregnation with a flame-retarding chemical or paint, is manufactured; or the board can be cladded with asbestos or foil.

Particle board is the general term for wood chipboard, flax board or hemp board. It can be a standard chipboard, a fine-surfaced board, a primed, painted, polyester-surfaced board, a veneered, melamine-faced particle board, a vinyl-faced board, prefelted board, and so on. Flax or hemp board is manufactured from the residue of the plant after it has been processed to produce linen or hemp.

Woodwool slabs are made from chemically-impregnated wood-fibres and cement. They are intended for non-loadbearing application, such as partitions, ceilings, wall linings, permanent formwork, and insulation. Fixing and jointing accessories such as nails, clips, taping, etc. can be obtained from the manufacturers.

Wood laminates such as plywood, blockboard, and laminboard are extensively used for internal finishings and partition linings. Plywood consists of veneers, assembled by gluing and crossing alternate veneers to improve strength properties and minimize movement. Some plywoods have an equal number of veneers, with the two central veneers running parallel. Blockboard consists of a core made from strips up to 25 mm wide placed together, with or without glue between each strip, to form a slab sandwiched and glued between one or more outer veneers, with the grain veneer at right angles to the grain of the core strip.

Laminboard has a core built up from thin strips, about 1·5 mm to

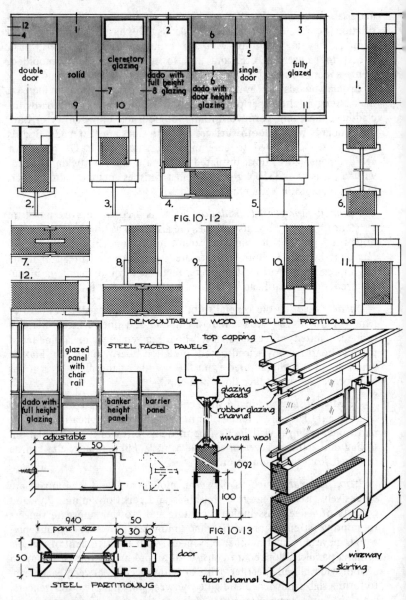

double door

solid

clerestory glazing

2

dado with full height glazing

6

dado with door height glazing

5

single door

3

fully glazed

1.

2.

3.

4.

5.

6.

FIG. 10·12

7.

12.

8.

9.

10.

11.

DEMOUNTABLE WOOD PANELLED PARTITIONING

glazed panel with chair rail

STEEL FACED PANELS

top capping

dado with full height glazing

banker height panel

barrier panel

adjustable

50

glazing beads

rubber glazing channel

mineral wool

1092

100

FIG. 10·13

940 panel size

50

10 30 10

door

50

STEEL PARTITIONING

wireway

skirting

floor channel

CORED STEEL AND WOOD FACED PARTITIONING

7 mm wide, glued together face to face to form a strip like blockboard; but it is heavier than blockboard. Face grading, bonding, and construction of plywood manufactured from tropical hardwoods have to comply with codes and standards laid down by various bodies, e.g. British Standards. Many boards made of tropical hardwoods which are exported to Europe follow this or other grading standards. Blockboard and laminboard are assessed for stability and rigidity.

Glass is a (usually) transparent solid, whose principal constituents are silica, soda, and lime, which are fused together at a very high temperature. Internally it may be used in a variety of ways and is now in great demand for internal decoration. It may be transparent, translucent, wired, or mirrored, and designed for insulation, solar control, or safety. The variety of uses and finishes are endless. With solar radiance, the amount of light transmission and reflection, and of solar-radiant heat transmitted and absorbed can be accurately assessed for many types of glazing, including roughcast, patterned, or toughened glass.

Plasterboards are also available in considerable variety. A well-known type for standard covering is Gyproc wall board which has one ivory face for direct decoration and one grey surface to receive gympsum plaster. The latter is not widely used where craftsmen are in short supply, though some firms and organizations train their own. The board is intended for internal lining, primarily for direct decoration. It is available with taped, square, or bevelled edges. Gyproc plank is a narrow-width plasterboard and is thicker than other types. It is used for dry wall systems where a high standard of fire-protection or sound-resistance is required without plastering. It is supplied with two grey faces for plastering, or, if preferred, with one ivory surface for wall board. A coloured P.V.C. film called Deksheen bonded to the face and returned around the longitudinal edges, is also available. It is used as a lining for walls of public buildings, hospitals, schools, hotels, etc. It can be supplied in three standard colours with a textured finish.

Other boards consist of a laminate of plasterboard and polyurethane foam (Purlboard). The standard board consists of 9·5 mm plasterboard and 13 mm polyurethane foam, with a moisture-barrier backing paper and a polythene vapour-barrier between the laminates. It has a tapered edge and an ivory face for decoration, is intended for internal linings with insulation combined, and is used particularly for lining concrete beams and lintels or solid walls. It should be noted that when using plasterboard all four edges should be supported by noggins to hold it in position.

Insulation slabs are now being used to an increasing extent because of the need for better control of interior temperature and sound. They are made either from minerals, asbestos fibre, ceramic fibres, celloluse, rock wool, slag wool, etc., or from plastics – polystyrene, expanded P.V.C., foamed polyurethane – or organic materials, such as fibreboard, wood-wool-slabs, etc. There are many insulation slabs which are composed of two different substances: one common type consists of resinous binders used as a matrix for mineral fibres; and another consists of slabs which have been bonded or laminated to other material, such as building paper, foils, or plastic shells.

Plastic laminates are also used for durable and decorative finishes to interior surfaces. They can be of melamine laminated or similar sheets bonded to suitable backing framed in panels as already described. They can be supplied in sheets from 1525 × 760 mm to 3660 × 1525 mm and in thicknesses from 0·5 to 6·35 mm. The finish can be glossy, matt, or textured as required. They have a good resistance to staining but should not be placed in contact with hot surfaces or cooking utensils. The colour fastness to light is quite good but contact with light through glazed openings should be avoided. All laminates are capable of being sawn, drilled, milled, or tapped without splitting if the operation is properly carried out.

Adhesives. A wide selection of proprietary brands is produced for bonding most products, and adhesives can also be compounded for specific needs. P.V.C. faces are usually fixed to walls with rubber-based compounds, as are rigid decorative melamine laminates. Hardboard, plywood, and similar materials may be bonded to interior plasterwork with P.V.A. based compounds. The subject is a wide one and much literature has been published on it.

Veneered panel and wood-faced partitions. As panels and partitions probably incorporate the greatest single use for surface coverings and finishings, further examples of applications are given in Figs. 10.9 to 10.13. Types of tiles and veneers are shown together with methods of fixings panels to walls and ceilings. The various partition units may be supplied direct from stock for normal heights with provision made for slight adjustment to suit floors and ceilings where necessary.

10.5 Thin surface finishings, external

Overlap tiles which may be described under this heading as cladding consist of material other than plain tiles, e.g. asbestos, slate, shingle and concrete, some of which are mainly roofing products.

Shingles are traditional where timber is plentiful, though they frequently deteriorate due to lack of preservative treatment. One type

of shingle or overlap tile of asbestos cement, which needs no such protection, is shown in Fig. 10.16a (see page 200). It is made in five
different colours and easily fixed without skilled labour. A further type
of cladding is shown in Fig. 10.16b, consisting of hardboard siding held
in place with metal mounting strips and supplied, made ready to fix, in
several colours.

Overlap sheets are also available in a range of materials, many of
which have already been discussed. These include aluminium, G.R.P.,
plastics, steel, weatherboard, wood, copper, etc. Though plastics are
extensively used in colder climates it is worth repeating that some tend
to degrade when exposed to sunlight and warmth. Galvanized sheet will
last up to 30 years; and aluminium is excellent, though it can corrode
when in contact with cement, or in industrial areas.

Some types of cladding incorporate integral insulation material, or a
double skin hollow section, Fig. 8.15. Sections are often designed to fix
directly to an external wall, in order to give weather protection and
provide a decorative appearance. Some types have simple lap joints, and
are fixed with profile either horizontal or vertical, Figs. 10.14 & 15.
They can also be used for cladding columns, canopies, fascias, etc.

Claddings can be rolled, pressed, extruded, moulded, or machined;
and all materials except steel can be used self-finished. Most manufacturers supply accessories, including starting and finishing pieces, edge
trims, and internal corner pieces. Some provide make-up pieces to
enable cladding to be used in non-modular space. Many different
profiles and shapes are produced, and trade literature will provide
details of the standard lengths and of the maximum lengths obtainable.
(The term 'length' refers to the dimension parallel to the profile.) In
some cases the design of the joint permits adjustment, within limits.
Adjustable accessories at corners and edges compensate for inaccuracies
in fixing and for the use of non-modular sizes. Jointing and fixing
systems in the tropics must allow for thermal movement.

Maintenance. Products should be assessed for their liability to
impact damage and ease of repair and replacement; consider, in
particular, whether individual sections can be replaced without having
to remove adjacent sections.

Blinds. Many types of sun blinds, shop blinds, awnings, pleated
rollers, Venetian blinds and vertical vanes are now in use in tropical
lands. They can be in cotton, canvas, synthetic material, aluminium,
paper, steel, P.V.C., or wood. Fabrics may be polyurethane-proofed or
vinyl-impregnated. Other materials may be stove-enamelled, anodized,
galvanized, polished, painted, or sheradized. Blinds serve many purposes,

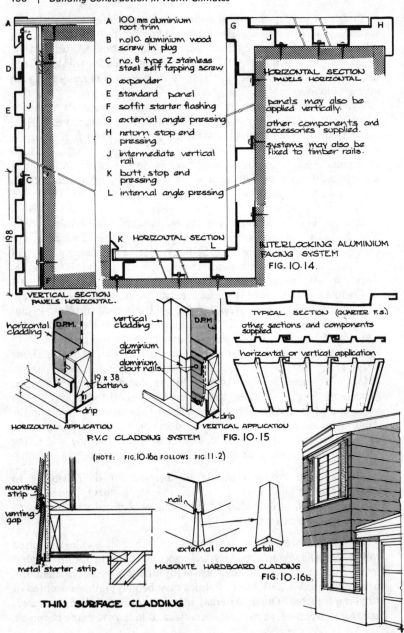

A 100 mm aluminium roof trim
B no.10 aluminium wood screw in plug
C no. 8 type Z stainless steel self tapping screw
D expander
E standard panel
F soffit starter flashing
G external angle pressing
H return stop end pressing
J intermediate vertical rail
K butt stop end pressing
L internal angle pressing

VERTICAL SECTION
PANELS HORIZONTAL.

HORIZONTAL SECTION
PANELS HORIZONTAL

panels may also be applied vertically.

other components and accessories supplied.

systems may also be fixed to timber rails.

HORIZONTAL SECTION

INTERLOCKING ALUMINIUM FACING SYSTEM
FIG. 10.14.

horizontal cladding
D.P.M.
19 x 38 battens
drip
HORIZONTAL APPLICATION

vertical cladding
D.P.M.
aluminium cleat
aluminium clout nails
drip
VERTICAL APPLICATION

P.V.C. CLADDING SYSTEM   FIG. 10.15

TYPICAL SECTION (QUARTER F.S.)
other sections and components supplied
horizontal or vertical application

(NOTE: FIG.10.16a FOLLOWS FIG 11.2)

mounting strip
venting gap
metal starter strip

nail
external corner detail

MASONITE HARDBOARD CLADDING
FIG. 10.16b

THIN SURFACE CLADDING

including black-out, dim-out, light control, acting as simple barriers to vision. External Venetian blinds may be used in hot countries to give extra protection from the sun's rays before they reach glazed areas. Such external blinds are very suitable for air-conditioned buildings, where solar heat gain must be minimized. The raising or lowering, or tilting of the slats, either singly or in groups, is controlled from inside, either by hand-rod or mechanically. When made of aluminium the blinds can withstand fairly severe climatic conditions. Other types are made for internal use and for double-glazing.

Stone. The practice of using natural stone in thin slabs as a facing material has recently become more general; it is used like marble or granite, but in rather thicker slabs. Bonding is often ignored, as vertical joints are now preferred. Stonework is supported at each level by stainless-steel or bronze angles, using methods adapted from those shown in Chapter 5.

10.6 Floor finishes

Cement-based screeds. The thickness of a screed depends on the base on which it is laid. When put on *in situ* concrete within three hours of placing, complete bonding is obtained. With such monolithic construction the screed need only be between 12 and 19 mm thick. If the base has set and hardened, then it must be thoroughly cleaned, hacked, and grouted immediately before the screed is placed. With separate construction of this kind the screed should be a minimum of 40 mm thick. With monolithic construction the size of the bays for screeds can be quite large, up to 30 m² where the base is 150 m thick. For other construction the bay should not be greater than 15 m². Gradual drying is necessary to prevent cracking and curling.

The wide use of admixtures in from 1:3 to 1:4/5 sand/cement screeds has greatly improved the quality of the floors. P.V.A. and acrylic resin emulsion, added in the proportion of approximately 15 per cent of the cement is sufficient where the laying has been properly carried out. To level up uneven floors prior to the laying of wood blocks, parquetry panels, cork, linoleum, P.V.C. tiles, etc., proprietary latex screeding compounds are available. These, when mixed according to instructions, give strong adhesion, coupled with flexibility, to existing forms of floor, including wood, concrete, stone, iron, and tiles. To improve the quality of industrial floors, toppings of both monolithic and separate construction can have chemical admixtures added to suit the nature of the floor and the degree of wear it may sustain. Where it has to sustain abrasion or impact loading, particularly in trucking lanes, such treatment can be invaluable. There are a number of admixtures to choose from, such as chemical plasticizers, liquid surface-hardeners, or pulverized iron aggregates. Steel-clad flags and anchor plates are also used where traffic is heavy.

Flexible P.V.C. flooring can now be supplied in either sheet or tile form, and in various sizes, qualities, and thicknesses. Some are resistant to wheel trucking, some can be laid on old floors and others, for use in hospitals, have anti-bacterial properties. A few have interlayers of foam to make the floor quiet and non-slip, and others have high resistance to indentation. Several have electrically-welded joints. Tile sizes vary from 250 to 600 mm square, but are usually 300 mm square. Sheets are normally 1·5 m wide. For hard wear, an anti-slip deck tread of moulded asbestos resin material is used, for example in swimming pools, passenger vehicles, etc. It is supplied in sheets 2130 × 915 × 2·3 mm thick, but can be cut to the standard tile size if preferred. It can be bonded to aluminium, steel, or wood, using neoprene-type contact adhesive or epoxy-type adhesive.

Ribbed or studded flooring, as illustrated in Fig. 10.17, is now in general use in transport terminals, railway stations, concourses, airports, etc. The ribbed surface gives an interesting finish, which is non-slip and makes footmarks less conspicuous. It is not suitable for garages or workshops where it could come into contact with oil or grease. The tile size is 500 mm square and 10 mm thick. It may also be used as ribbed stair nosing. The tiles are laid direct on a 1:3 cement/sand screed 38 mm thick a day or two after placing, while it is still green. The tiles are first laid on a fresh screed and carefully lined up to shape. They are then turned face downwards with cement mortar 1:1 trowelled into the dovetails. A wetter mix is then buttered on to the screed and the tiles pressed firmly into place, and the surface is then cleaned off with sawdust and a wire brush.

Clay tiles for floors provide a cool, durable, and decorative finish. Floor tiles are usually vitrified, as glazed ceramic tiles are suitable only for walls and light floor traffic. Sizes are usually from 100 × 100 × 9·5 mm, to 200 × 100 × 9·5 mm. To prevent arching and ridging as a result of shrinkage, thermal movement or creep, tiles should be bedded in a 1:3 cement/sand mortar 20 mm thick which has been separated from the base by building paper, felt, or a polythene sheet. Direct bedding may be used on concrete or screeds not containing water-repellent mixtures. Wetted tiles are usually bedded on 1:3 mortar screed, 15 mm thick for tiles 10 mm thick, and up to 20 mm thick for thicker tiles. Another method is to use a semi-dry mix as described in Volume 1. A typical example of a good-quality tiled floor is given in Fig. 10.19. These

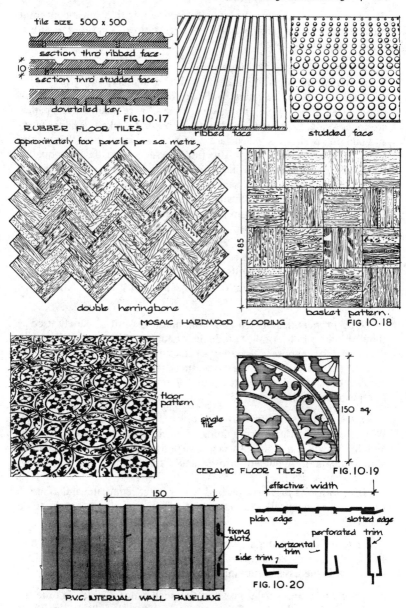

tile size 500 x 500

section thro' ribbed face.

section thro' studded face.

dovetailed key.

FIG. 10.17

RUBBER FLOOR TILES

ribbed face

studded face

approximately four panels per sq. metre.

double herringbone

basket pattern.

485

MOSAIC HARDWOOD FLOORING

FIG 10.18

floor pattern

single tile

150 sq.

CERAMIC FLOOR TILES.

FIG. 10.19

150

effective width

plain edge

slotted edge

fixing slots

perforated trim

horizontal trim

side trim

FIG. 10.20

P.V.C. INTERNAL WALL PANELLING

FLOOR AND WALL FINISHINGS

tiles are best bedded by the semi-dry process, with a movement-joint provided around the perimeter and an expansion-strip inserted as shown for marble in Volume 2.

Hardwood mosaic flooring. Wood-block flooring is used in areas where hardwood is readily obtainable. Unfortunately, owing to inferior workmanship, poor seasoning, and poor bedding materials, wood-block floors often give trouble after a short time. Where wood mosaic panels are made under controlled conditions, either felt-backed or paper-faced and laid on a well-prepared surface with bitumen emulsion adhesive, the result is quite satisfactory. The timber used is usually good-quality teak, oak, or West African hardwood laid in panels, as shown in Fig. 10.18. Borders are supplied separately, and should be laid with a small movement-gap under the skirting and around pillars and obstructions. This, together with a dry gap between each panel as laid, will provide sufficient flexibility to prevent lifting. The panels should be laid according to suppliers' instructions if these are different from those given above.

10.7 Painting and wall coverings

Paints have been described in Volume 2, and the list is now extended to include the following:

Wood primers, lead-based and aluminium, are not as widely used as they should be in timber-producing countries. They are especially necessary for exposed secondary-species hardwoods when these are used structurally as balconies, handrails, and external stairs. Acrylic emulsion-based and traditional wood primers are suitable for lighter non-resinous species.

Metal primers contain red lead, red oxide, and calcium plumbate. Wash- or etch- primers containing zinc chromate resin are used on aluminium, zinc, and other surfaces. Metallic paints containing flakes of aluminium and copper are also available for priming, rust inhibiting, and decorative work.

Alkali-resisting primers are used with plaster, concrete, and asbestos to resist the chemical reaction caused by free alkali present in these materials.

Surface fillers for metal, wood, and plaster are now in common use. For wood and metal, semi-liquid material is used, and several applications may be necessary, rubbing down between each coat.

Water fillers are usually powders to which water is added in order to produce a fine paste for filling cracks in internal walls and ceilings.

Water-thinned distempers are non-washable, but are used extensively in low-cost housing and where labour is not a problem.

Water-thinned emulsions are tough, elastic, and widely used over concrete block, brickwork, and plaster. The water evaporates leaving the synthetic resins to form a coating on the surface. Several qualities are obtainable of which the alkyd resin types are probably the best.

Water-thinned cement paints made from Portland cement and obtainable in a variety of shades, can be applied to brick, concrete, or blockwork. They have a coarse texture suitable for roughcast and uneven surfaces.

Bituminous paints, except those containing aluminium, are normally dark-coloured, and when used in situations exposed to sunlight, give protection against disintegration. They have excellent water-repellent properties and are resistant to corrosion. They are useful for exterior steel work, for coating buried pipes, or for use in heavy industrial areas, but can only be overpainted with ordinary paint after a suitable sealer has been applied.

Oleo-resinous paints are used for high-gloss hard surfaces and contain natural and synthetic resins. Oil-modified alkyd resins also give full gloss and are very durable. Acrylic resins are also popular, because they hold colour well and are durable.

APPLICATION. The correct application of paint is still a problem in areas where unskilled labour and lack of trained supervision are major drawbacks. The need for cleanliness, preparation of surfaces, mixing, and application is often insufficiently understood. New plaster and renderings should be left to dry out during rainy seasons and in humid areas. Efflorescence should be sponged off with clean water and left to dry. Emulsion paint in two or three coats is best where immediate decoration is necessary. Porous flat oil paint in two coats may be applied to dry surfaces. Metalwork is usually treated with paint for protection. A primer and two coats of paint is the minimum in contaminated areas and correct surface preparation and application is essential.

Wood is not always primed as it should be when joinery is made on the site. Though good-quality hardwoods will survive, others deteriorate rapidly when the end-grain is exposed in humid conditions. Some hardwoods require special primers and advice should be sought in cases of doubt. Wood-based boards are usually filled with surface fillers and primed. They may then be decorated with two or three coats of oil or two coats of emulsion paint. Fibre-board can be treated with either emulsion or distemper. Wood-wool slabs may be treated with two coats of paint with a flat oil finish or emulsion.

Application of paints and distempers by roller or spray, though increasing, is still not general practice. Rollers need little skill and produce a good surface, especially with emulsion, though they need careful cleaning after use. Spray painting is rarely seen, though when used for multicoloured coatings they produce a pleasing dappled or flocked finish.

WALL COVERINGS AND FINISHINGS. A considerable range of papers and fabrics, in both natural and synthetic materials are made as wall coverings, though they are not yet in general use in warm countries. Paper can be made as flock, hand printed, lining, machine printed, photomural or wood chip. Natural materials include cork, glasscloth, felt, jute, linen, silk, wool, and other fabrics. Synthetic materials include foamed vinyl, glass-fibre, metallized vinyl film, nylon, polystyrene, etc.

Where wall coverings have been used they seemed to have stood up well in dry and semi-humid climates and also during rainy seasons, though experience in damp areas is still limited. One type of U.P.V.C. internal decorative panel which should be satisfactory in most zones is shown in Fig. 10.20. Panels are available in lengths up to 2 or 3 metres, 150 mm wide, and each interlock when fixed. They may be glued, stapled, or nailed to backings of plasterboard, blockboard, or normal plaster renderings as appropriate, and can be supplied in a range of colours. Simple trim sections can be used around openings or as finishing to panel edges.

11 | Services

DRAINAGE AND WASTE DISPOSAL

11.1 Rainwater systems

Rainwater systems are now available in aluminium alloy, asbestos cement, cast iron, mild steel, and P.V.C., and also in copper, stainless steel, and lead. Roof outlets are made in cast iron, gun metal, spun steel, H.D. polythene, copper, and P.V.C. Purpose-made gutters are also obtainable in precast concrete, and gutters and r.w.p.s are produced in a range of fittings to form complete systems. Roof outlets are also designed to be used in conjunction with rainwater- or soil-systems to provide surface drainage from roofs, balconies, canopies, and paved areas. They are supplied in different profiles and can also be obtained with fascia and soffit assembly complete.

P.V.C. systems for houses and smaller buildings are employed mainly for internal use and should not be used externally, unless shaded from the sun. Industrial buildings usually use asbestos cement systems. The gutter must be of adequate size to suit the rainfall of the appropriate area. It is not possible to make generalizations about such a range of countries, but, as a rough guide, some roof calculations allow for a fall of 75 mm per hour. Short storms of 150 mm per hour may have to be allowed for and here overflowing for brief periods is usually tolerated. One formula for calculating flow load is $1.25 \times$ actual roof area in m^2 to give litres per minute, but this can be modified to suit the locality. Falls, sizes of pipes, and radii of bends can affect the result. P.V.C. gutters can be snap-fit jointed using factory seals; alternatively joints may be solvent welded. Asbestos gutters can be joined with dry compression strips and bolted in the normal way. Some types of connectors are shown in Volume 2. It should be noted that U.P.V.C. is not suitable for hot-water wastes.

11.2 Drainage

Soil pipes. A typical single-stack soil system in P.V.C. is shown in Volume 2. It is suitable only for low-level housing development where it is used almost exclusively, Fig. 11.1. For multistorey buildings a one-pipe system is used where all soil and waste discharge into a common pipe, and all branch wastes use a single main ventilating pipe. Rainwater connects to the same drain at the bottom of the system, provided a combined system of drainage is in operation. This of course depends on location and annual rainfall. The traps should have a 75 mm seal as shown in Volume 2. A modified one-pipe system is now in general use, as shown in Fig. 11.1. It differs from the one-pipe system in that the branch ventilating pipe is taken only from the w.c.s. There are certain minimum diameters and radii specified in the design of single-branch fittings and the reader is advised to consult local by-laws before proceeding.

Soil drainage treatment. Small communities and private institutions sometimes have a packaged installation to treat sewage of a domestic character. The septic tank shown in Volume 2, though widely used in tropical countries, does not always treat sewage to the standard required, both with regard to siting and the quality of the effluent produced. Self-contained units may be obtained which are generally manufactured from steel, and suitably treated for corrosion. Sewage installations made up from factory-produced components also have fibre-glass or concrete tanks assembled from modular units. The plant can be installed above or below ground, but foundation work, pipework, and electricity supply are usually provided independently. One such type of package is shown in Fig. 11.2. It includes tanks with reinforced concrete sections, filters, and distributors in sizes for 4 to 150 persons. The raw sewage enters chamber No. 1 where suspended matter settles out; the effluent then overflows into chamber No. 2. After further settlement it passes into a well-ventilated filter-bed where any remaining organic matter is broken down by bacteriological action before being discharged into a soakaway or stream.

Drainage below ground. Underground drainage systems can have cast iron, clay, pitch fibre, or P.V.C. pipes and fittings, with G.R.P. or P.V.C. inspection chambers. Plastic pipes intended for underground drainage are generally brown in colour for ease of identification. They can be of solvent weld or push-fit joints, the latter being available as a loose coupling for use with plain-ended pipes or as integral sockets. For most domestic work, pipes of 100 mm and 150 mm are used, with a gradient of 1:80 if 100 mm pipes are used, when serving not less than five housing units, and 1:150 gradient for 150 mm pipes when serving not less than ten units. The 100 mm pipe which serves only the first unit in any terrace should have a fall of not less than 1:70.

Damage to external drainage caused by settlement is very common, largely because of the traditional practice of laying pipes with rigid

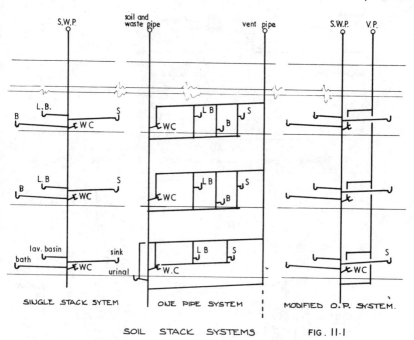

SINGLE STACK SYTEM

ONE PIPE SYSTEM

MODIFIED O.P. SYSTEM.

SOIL STACK SYSTEMS

FIG. 11.1

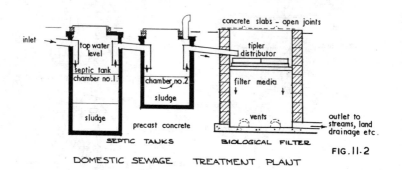

SEPTIC TANKS

BIOLOGICAL FILTER

FIG. 11.2

DOMESTIC SEWAGE TREATMENT PLANT

SOIL DRAINAGE

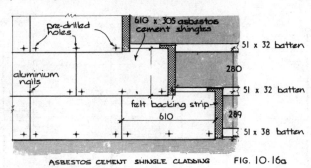

pre-drilled holes
610 x 305 asbestos cement shingles
51 x 32 batten
280
51 x 32 batten
aluminium nails
felt backing strip
610
289
51 x 38 batten

ASBESTOS CEMENT SHINGLE CLADDING   FIG. 10.16a

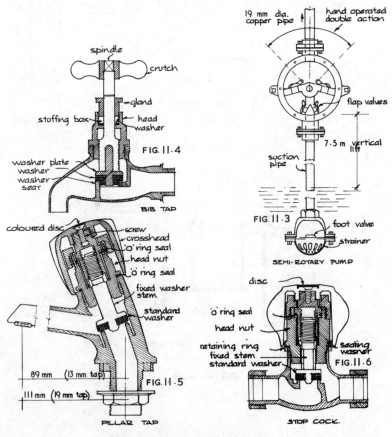

spindle
crutch
gland
stuffing box
head washer
FIG. 11.4
washer plate
washer
washer seat

BIB TAP

19 mm dia. copper pipe
hand operated double action
flap valves
7.5 m vertical lift
suction pipe
FIG. 11.3
foot valve
strainer

SEMI-ROTARY PUMP

coloured disc
screw
crosshead
'o' ring seal
head nut
'o' ring seal
fixed washer stem
standard washer

89 mm (13 mm tap)
111 mm (19 mm tap)
FIG. 11.5

PILLAR TAP

disc
'o' ring seal
head nut
retaining ring
fixed stem
standard washer
sealing washer
FIG. 11.6

STOP COCK.

WATER SUPPLY FITTINGS.

mortar joints on an *in situ* bed. Modern practice is to use flexible joints and to lay in a base of compacted sand and gravel. Pitch fibre pipes as shown in Volume 2 are cheap and satisfactory but where they are liable to deformation from overfill they should be covered with compacted granular fill before backfilling.

Gullies and gratings. With the growth of state-controlled housing projects over the years, carefully-prepared lists and schedules of sanitary and other fittings in clay and cast iron have been produced to standardize production. Specimens of these are given in Drg. No. DCED/G/1, produced by the Department of the Environment, England. In order to illustrate their use, reference has been made to them in the drainage layout plan, Fig. 2.9. Further reference to manholes will be made in the next chapter.

11.3 Refuse and rubbish

Refuse collection and disposal is sometimes a major problem in large towns. Some authorities provide bins or sacks to expedite clearance and reduce nuisance, but this practice is by no means widespread in tropical countries. In some housing blocks over three storeys high, vertical ventilated chutes are provided, passing from floor to floor with openings to connect with hoppers where required. The bottom of the chute discharges into a chamber with refuse containers emptied by the local authority when necessary. Sometimes the tenant himself brings his refuse to the container direct, which simplifies disposal. Unfortunately the practice of placing rubbish in open pens at street corners, which attracts scavengers, buzzards, gulls, and vermin, is still prevalent, though more authorities are now using incineration.

The general types of waste for burning fall into three categories:
Rubbish is waste paper, cardboard, textiles, rags, wood scraps, either domestic, commercial or industrial, or sawdust in timber towns.
Refuse is a combination of the above and wet vegetable and meat waste.
Animal and vegetable waste usually comes from cafes, restaurants, hotels, and markets. In areas where oil is produced, oil-burning incineration is advantageous; units are available in all sizes, from small wall-units for sanitary-towel disposal to municipally-run crematoriums. Often colleges, hostels and municipal or community buildings have their own incinerators. They can be fired by oil, wood chips, fine dust, propane or butane gas, or coal. Some have natural draught firing designed for general destruction of paper, cardboard, and combustible material. Incineration is the most effective method of sterilizing organic refuse and reducing its bulk, but the cost of installation and firing should first be investigated as the amount of waste for disposal may not warrant the cost.

202 | *Building Construction in Warm Climates*

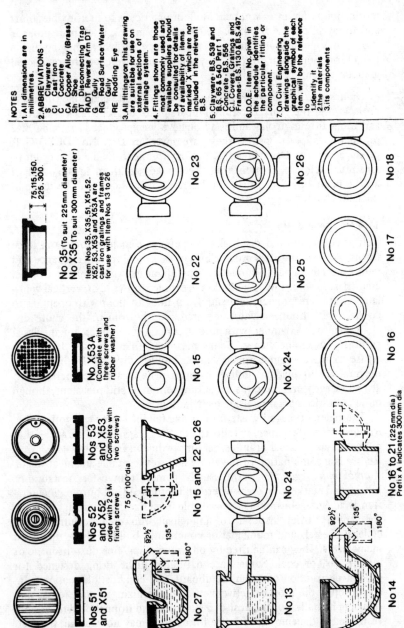

NOTES

1. All dimensions are in millimetres.

2. ABBREVIATIONS –
   - S   Clayware
   - CI   Cast Iron
   - C   Concrete
   - CA   Copper Alloy (Brass)
   - Sh   Shoe
   - DT   Disconnecting Trap
   - RAD Reverse Arm DT
   - G   Gully
   - RG   Road Surface Water Gully
   - RE   Rodding Eye

3. All fittings on this drawing are suitable for use on external sections of drainage system.

4. Fittings shown are those most commonly used and available suppliers should be consulted for details of availability of items marked X which are not included in the relevant B.S.

5. Clayware:-B.S. 539 and B.S. 65 & 540 Part 1. Concrete:-B.S. 556. C.I. Covers, Gratings and Frames:-B.S.1130 & B.S.497.

6. D.O.E. Item No. given in the schedule identifies the particular fitting or component.

7. On Civil Engineering drawings alongside the graphical symbol of each item, will be the reference to
   1. identify it
   2. the materials
   3. its components

C(90·5)

DCED/G/1-2 FEB 1975

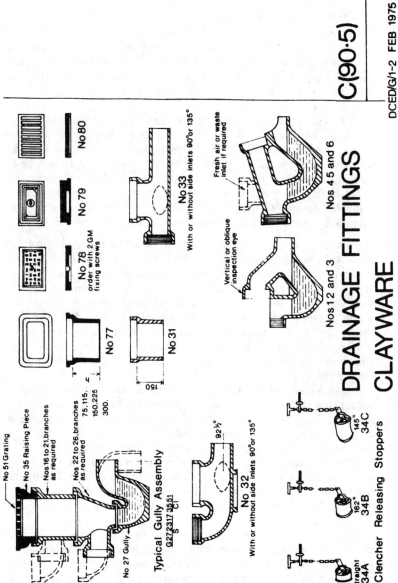

DRAINAGE FITTINGS
CLAYWARE

No 80

No 79

No 78
order with 2 GM fixing screws

No 77

No 31

No 33
With or without side inlets 90°or 135°

Nos 1 2 and 3
Vertical or oblique inspection eye

Nos 4 5 and 6
Fresh air or waste inlet if required

No 51 Grating

No 35 Raising Piece

Nos 16 to 21,branches as required

Nos 22 to 26,branches as required
75.115.
150.225
300.

Typical Gully Assembly
G 27 23 17 3 5 51
S CI

No 27 Gully

No 32
With or without side inlets 90°or 135°
92½°

Straight
34A

162°
34B

145°
34C

Clencher Releasing Stoppers

PLUMBING, SANITARY FITTINGS

11.4 Hot and cold water supply

Many towns and villages in rural areas depend entirely for their water supply on lakes, rivers, shallow wells, and boreholes. Others create catchment areas or dam suitable rivers and streams to make small reservoirs. Frequently no mechanical means of raising water is employed, with the exception of jack pumps. Supplies are often irregular and hardship sometimes ensues.

Wells and boreholes are the most frequently-used method of obtaining water. Local custom in well digging is almost always followed, the digger using the traditional hoe with baskets of soil hauled up by helpers. Digging is usually done in the dry season and the well lined with concrete block, brick, or even plastic bags filled with dry concrete. After digging, the well is covered with a large concrete slab containing a manhole through which the water is raised by means of bucket and rope. Wells should always be sited away from cesspits and septic tanks.

Pumps. In new developments boreholes would be dug using hand drills and mechanical augers, and the holes would be lined with steel tube. A wide variety of pumps exist, though where power is not available the traditional lift or jack pump is the most popular. This will raise water to a height of 7·5 m. Lift and force pumps are also in use where greater height is wanted. The semi-rotary pump, Fig. 11.3, is a sturdy and reliable piece of apparatus in use throughout the world. A 19 mm suction pipe will raise water to a height of 7·5 m after priming, and should deliver up to 200 gallons per hour if the delivery pipe is fairly short.

Where a continuous supply of water is available, as from a spring or stream, the hydraulic ram is used. It is entirely automatic and acts by forcing a large volume of water of low velocity into a compression chamber, which in turn raises a smaller volume to a higher level. It is widely-used in India and elsewhere where suitable conditions exist.

Windmills are in fairly wide use where prevailing winds make them worth-while, though hand labour is preferred if available. Electricity is now more widespread than hitherto and used to drive small centrifugal pumps of various kinds. A typical submersible pump is shown in Volume 2.

Where lakes and reservoirs are used for supplying urban areas, they are usually controlled by water authorities under the basic legislation of water acts and by-laws. They are concerned with such matters as purity, contamination of main water, provision of drinking water, quality of materials, stop valves, draw-off taps, ball valves, storage systems, and all matters concerned with new water undertakings. The source, examination, and treatment of water is outside the scope of this book, though the question of filtration and sterilization could involve the reader at some time. The work of the architect and engineer usually starts with connections to the water mains, which in towns usually means the provision of a stop-cock to the cold water service pipe as shown in Fig. 11.14.

Where supplies are intermittent, as is often the case, intermediate pumping systems are frequently installed to feed water towers of various capacities which are coupled direct to the main supply. The towers, which can be many thousands of gallons in capacity, are normally made up of Braithwaite panels, each over one metre square, sealed at all edges, and braced internally. The height depends on the site altitude and the head required. It is usual to fill the tank at off-peak times or during the night. In areas of growing water-shortage it is the increasing practice for householders to install extra storage tanks near their houses and fill them from hosepipes when supplies are available. Isolated or detached residences often have permanent tanks of concrete or blocks designed on the lines shown in Fig. 11.15. The tank is filled either by hand, pump, or by rainwater from the roof. The example given shows filter chambers with a semi-rotary pump to fill the roof storage tank. Often the tank is for plain storage, the water, when used for drinking, being passed through standard household candle filters and perhaps boiled and cooled before use.

Water fittings. Pipe fittings have been discussed in Volume 2. These include capillary, compression, and solvent welded joints. Fittings for P.V.C. pipes have spigot and socket joints and a range of adaptors and fittings are available to enable pipes to be connected to taps, ball valves, and other metal threaded fittings. Valves, stop valves, and stop cocks are used to regulate isolated supplies and are of four basic types: screw-down stop valve, gate valve, plug valve, and diaphragm. They may be activated manually by screwing, or be self-acting, as with a ball or float valve. The screw-down valve, Fig. 11.6, controls the water along the pipe line. It is very positive but its resistance to flow is high. The gate valve enables the valve to be lifted vertically clear of the flow and so offers a lower resistance than the screw-down. It is not suitable, however, for regulating service and a screw-down is essential where the main service enters the building. The ordinary bib tap, Fig. 11.4, is the most widely used for general purposes and is normally fixed to stand-pipes or walls. The pillar tap is usually fixed to the bath, basin, or sink direct and is now made in a variety of shapes and sizes. Some are self-closing, others non-splash, while others are spray taps designed to save water when hand washing.

Hot water supply. In most warm climates hot water is not in great demand except in hospitals, kitchens, canteens, and some industrial premises. In some places showers and baths are taken direct from tap water. Modern buildings, however, usually have electric immersion heaters, Fig. 11.14. Various sizes are available to suit the ambient temperature and number of users. As a general guide, a tank with a nominal capacity of 140 litres with a 3 kW heater should provide several baths when the cold water temperature is already about 28 degrees C. A typical hot water distribution diagram for domestic use is shown in Fig. 11.14. Cold water is fed from the roof tank to the base of the cylinder and the hot water drawn from above the cylinder, as shown. An expansion pipe is taken over the top of the c.w. tank with sufficient height to prevent the hot water discharging into the tank. The remainder of the diagram is self explanatory. It should be noted that the hot water supply to the cloakroom wash basin, Fig. 11.10, may be omitted where the pipe is fairly long, as its use is only intermittent. A small separate unit heater is sometimes installed in such cases as these.

11.5 Sanitary fittings

Water closets. A wide selection of w.c.s and cisterns is now obtainable, ranging from the straightforward wash-down to the syphonic and bidet types shown in Volume 2. The wash-down is still the most widely used as it is less liable to become blocked, though syphonic types are more silent and direct in action.

Wash basins. These are also supplied in a variety of colours and materials, including enamelled iron, plastics, and steel. They may be corner, pedestal, or wall basins or vanitory basins set in cabinet or counter top with metal trim. Small vertical hand basins are also made which project only 125 mm from the wall which is recessed to make the maximum use of space. Makers' catalogues can supply all details.

Baths. Like basins, baths can be supplied in a range of colours to match bathroom suites. The most widely-used materials are porcelain and enamelled or cast acrylic sheet (Perspex). The latter is popular when supplied from abroad, owing to its lightness and ease of fixing. It will withstand detergents, disinfectants, and boiling water and can be supplied with twin hand-grips and a slip-resistant base.

Urinals. Those shown in Volume 2 are still widely used and can be supplied in fireclay or vitreous china. Slab urinals, however, are becoming more widely used and can be obtained either in fireclay or stainless steel in various sizes and with fitments to suit individual tastes.

A fairly typical bathroom and cloakroom layout is shown in Fig. 11.13. As with kitchen layout, all details must be shown including position of towel rails, mirrors, shelves, door hooks and other accessories. Proprietary fittings must be measured from the makers' catalogues to ensure that sufficient space is available, but the dimensions need not be given in elevation. Heights of wall-basins as distinct from pedestal-basins must be given. Precise information must also be supplied to manufacturers as to catalogue numbers, space available, thickness of partitions or ducts for ranges with concealed plumbing, handing of fitments, position of bath panels, height of windows as well as details of traps, flushpipes, taps, etc. It is also necessary to ensure that non-loadbearing partitions will carry any fitments attached to them. A copy of an elevation drawing such as Fig. 11.13 would help the supplier considerably.

Cloakroom lockers. Community centres, gymnasiums, stadiums, and other such buildings usually have a sanitary annex incorporated, particularly when they are open to the public. These normally contain single or twin lockers or locker combinations, basket hangers, clothing lockers or protected shelving. Cloakroom equipment is available in a considerable range of cupboards, cubicles, and cabinets, usually made of metal though white melamine is now becoming popular.

HEATING, VENTILATING, LIGHTING

11.6 Heating and ventilating

Gas supplies in warm countries depend almost entirely on liquified petroleum distributed, by and large, by the major oil companies or similar organizations. The two principal types are propane and butane. Propane has a higher rate of vaporization from its liquid form making it suitable for storage in large outdoor tanks. Butane is normally supplied in cylinders. Propane can also be supplied in small quantities if desired but is not to be stored internally. The gases are produced initially at refineries where crude oil is refined for petrol. The gas is both colourless and odourless but a stenching agent is added for safety reasons.

Distribution is effected by tanker and cylinder vehicles to depots throughout the regions. A customer on bulk supply would have one or more storage tanks on his premises which are topped up by road tanker

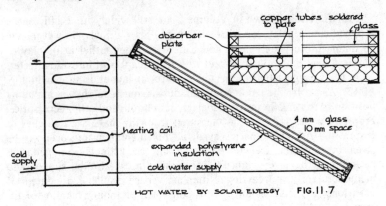

HOT WATER BY SOLAR ENERGY   FIG. 11·7

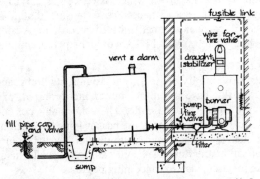

OIL TANK INSTALLATION   FIG. 11·8

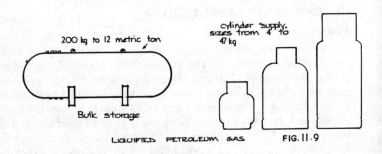

LIQUIFIED PETROLEUM GAS   FIG. 11·9

FUEL STORAGE.

as necessary. Cylinders are in greatest use and are available from a net-work of centres which either provide regular delivery or supply customers direct. The gas is available in cylinder sizes from $8\frac{1}{2}$ lb. (3·8 kg) to 104 lb (47·2 kg). Bulk tanks are supplied to customers on rental and come as complete units including supports, vapour outlets, and pressure-relief valves. Propane for bulk users is stored in these tanks which are available in sizes from 200 kg to 12 tonnes. The usual site requirement is an open-air compound with a level concrete base and access for bulk tankers. Minimum distances from buildings are laid down as are distances between vessels, where more than one is required. Information is avail-able from the supplier who normally installs the vessels.

Cylinders are normally of butane for domestic use and may be installed inside or outside the building, Figs. 11.9 and 11.10. Propane should always be outside. For canteens and such like, where bulk storage is not desirable, propane can be supplied in cylinders and connected in batteries to a common supply pipe. Metered gas for housing estates, both permanent and temporary, can be installed to provide water heating and cooking where demand is sufficiently great. The scheme may be administered by the authority responsible for the estate or by the supplier himself if he considers it viable. It is of course desirable to have alternative sources of power if possible. Modern cooking appliances may now consist of a gas hob with up to four burners set into a kitchen worktop and with a separate electric oven and eye grill for those who can afford it.

Cylinder gas has uses other than those already mentioned. It may be used for domestic or industrial refrigeration, operating factory vehicles and equipment, or site construction including cutting and welding. It is a very versatile substance.

Warm air heaters are sometimes used in tropical areas though their use is not widespread. Problems of seasonal change in temperature are usually met by use of extra clothing, though, where cold nights are frequent, electric heaters, either wall type or portable, can be used. Tubular skirting heaters are sometimes fitted in airing cupboards where humidity is troublesome, but are not always effective. Modern portable gas lamps are now quite popular which can also be used as heaters in an emergency.

Solar heating. In the past few years there has been a surge of interest in energy sources other than those supplied by conventional fuels. The developed world, particularly U.S.A. and Japan, is spending large sums on experiments which may ultimately benefit other countries by providing power for lighting, heating, and industrial use, especially as the cost of oil-based fuels is now comparatively high. Solar energy for powering water heaters, distillation plants, stills, and driers is one such source now being investigated and should ultimately do much to ease the energy problem where fuel is scarce.

The most straightforward application of solar energy in warm countries is to water heating which is shown diagrammatically in Fig. 11.7. More sophisticated plants can be installed which provide ample supplies for domestic use. The rate of flow would obviously depend on the latitude, hours of sunshine, the angle of solar collector, the size of storage tank and whether the solar energy reaching the absorber plate is for direct use or for storage. It is also possible to provide warm water by this system for, say, swimming pools in cooler zones. A number of firms throughout the world are supplying complete solar panels ready for installation. Details can be supplied by the Building Centre, London.

Fuel storage. Energy for industrial and domestic use is currently based on solid fuels or oil, depending on local resources. The type of storage will depend on the fuel available. Some tropical countries depend on wood and wood-products for which storage is simple. Some produce coal or oil, while others, Nigeria for example, produce all three as well as gas. A typical small oil tank installation for industrial use is shown in Fig. 11.8. A tank for domestic use would vary in size between 250 and 650 gallons, though oil is usually confined to industrial areas. Burners are thermostatically controlled and tank installation is normally undertaken by the oil company as it is subject to fairly strict regulations.

11.7 Mechanical ventilation

It has now become standard practice in multistorey buildings to install mechanical ventilation systems particularly in w.c.s and toilets. These are simple to operate and consist only of extract ducts as fresh air is replaced by normal circulation within the building. The system must be provided with a duplicate fan at roof level, however, in case of breakdown. The main duct rises to roof level and connections are made by vertical shunt ducts i.e. short separate branches which help to reduce noise, fumes, and smoke in the event of fire. The pipe is normally of plastic and quite small in diameter and the installation is designed to cope with an air change of $21 \, \text{m}^3/\text{hr}$ for w.c.s only, greater where bathrooms are combined. Other types of mechanical air extraction are mentioned in Volume 2.

Air-conditioning. This subject has been dealt with in Volume 2 as far as is necessary here. Although the principles have not altered greatly, new ranges of cabinet models and styles have now reached the market, particularly in smaller units where the demand is still very high.

11.8 Power and lighting

Light fittings, now termed *luminaires*, cover all aspects of building including commercial, industrial, hospital, and educational. Light distribution may be direct, semi-direct, general diffusing, or indirect. Basic types of indirect lighting are (a) incandescent, which houses a bulb or short tube which emits light by means of tungsten filament heated to a high temperature or (b) discharge lamps which are of two types (i) where *low pressure* mercury discharge is contained within a glass and emits U.V. radiation. A mixture of phosphorus inside the tube convert this into visible radiation. Various colours as well as white can be produced. (ii) *high intensity* where discharge is produced in a small quartz arc containing a mercury or sodium vapour at high temperature.

Filament lamps and high-intensity discharge lamps are not greatly affected by temperature, though the luminaire must be designed to dissipate heat. Fluorescent tubes produce lower temperatures but their light output varies considerably with the ambient temperature. It is therefore important that the luminaire is designed and installed so as to allow heat from the lamp and control gear to escape. Surface mounted fittings should not be fixed direct to combustible materials and in any case should be installed about 4 mm clear of the surface.

Emergency lighting. Buildings providing sleeping accommodation, treatment or care, entertainment, recreation, or education to which the public have access, may have to be supplied with emergency lighting systems to satisfy fire regulations. This usually covers illumination to all stairways and exit routes, including direction signs, and is normally powered by electricity. It should be permanently installed and come into operation immediately on mains failure. The system should be capable of maintaining illumination for up to three hours. In rooms where large numbers of persons assemble, emergency lighting must be kept on at all times when the room is in use, if there is insufficient natural light for escape. The power supply lighting must not be used to supply any other equipment and should come on automatically within ten seconds of failure. The power supply may be a central battery, individual batteries, or generator normally powered by diesel engine. Official technical memoranda exist which give useful information on generating equipment and should be obtainable from the local authority.

KITCHEN FITTINGS

11.9 Domestic kitchen layouts

Industrial kitchen and canteen layout has been discussed in Volume 2 and attention will now be paid to domestic kitchen layout in fairly

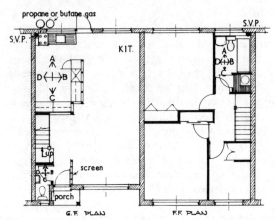

propane or butane gas

S.V.P.

S.V.P.

KIT.

A
D ⟨┤├⟩ B
C

A
D ⟨┤├⟩ B
C

A
D⟨┤├⟩B
C

up

screen

porch

G.F. PLAN

F.F. PLAN

LAYOUT OF TYPICAL TERRACED HOUSE

FIG. 11.10

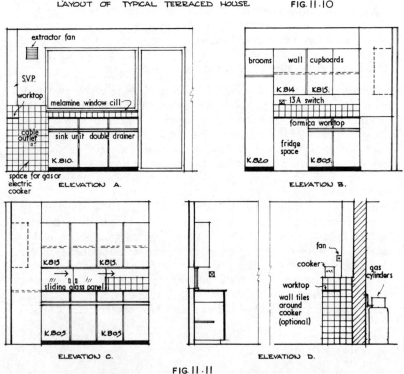

extractor fan

S.V.P.

worktop

cable outlet

melamine window cill

sink unit double drainer

K.810.

space for gas or electric cooker

ELEVATION A.

brooms | wall | cupboards

K.814. | K.815.

13A switch

formica worktop

fridge space

K.820 | K.805.

ELEVATION B.

K.815 | K.815.

sliding glass panel

K.805 | K.805

ELEVATION C.

fan

cooker

worktop

wall tiles around cooker (optional)

gas cylinders

ELEVATION D.

FIG. 11.11

DOMESTIC KITCHEN LAYOUT.

conventional housing design. Kitchen fittings, usually proprietary brands, are normally 900 mm high, 600 mm wide and lengths can be made up of double units or a combination of single units. Manufacturers can produce plinth filler units of 100, 200 or 300 mm increments which can be combined with base units to fill every 100 mm multiple of length. Tall units can be supplied, as a rule, in three heights, 1500, 1950 and 2250 mm, though other non-standard sizes of kitchen units can be obtained, often ready made. Frequently the sink unit is made 20 mm higher than the surrounding worktop, i.e. 920 mm high.

In setting out kitchens several matters require attention. The position of the cooker is important; it may have to be placed near the external wall if supplied with gas where cylinders are put outside and connected by piping taken through the wall. Also cookers should not be placed in a cross draught where the flame could blow out. Where electric cookers are to be installed, care should be taken to check the power supply for correct voltage, reliability of current supply and fluctuation of voltage. Some householders prefer to have gas cookers with standby electric rings to serve as extra hobs or use in an emergency.

The arrangement of fittings is a matter of choice. Some people prefer a separate larder while others opt for such extra fitments as wall cupboards or floor units. The dimensions of the kitchens should be decided with regard to the dimensions of the fitments to be installed so that they may fit exactly. It is necessary to do this when the building is being planned in order that all services can be located in wall or floor at the foundation stage. Proprietary appliances such as cookers, freezers, refrigerators, should be ordered early and checked not only for size but for handing of doors, and space needed for thickness of doors, etc. opening against the wall.

All electric points, switches, and outlets should be drawn in elevation as shown in Fig. 11.11, not as conventional symbols but to actual scale to ensure that they will fit in the space provided and not foul doors, sliding panels, etc. Electric cooker points should not be placed over the cooker itself and should preferably have a cable built into a conduit behind the cooker as shown. Spare cable must be allowed, to enable the cooker to be removed for cleaning.

Wall fitments should have adequate support from walls or loadbearing partitions. Window heights should be checked to ensure that at least one course of tiles can be placed between the sink or the worktop and the window-sill. Overall widths of doors and windows, including architraves, are also important to note if later cutting and fitting is to be avoided. Some suppliers provide sheets of squared paper in their catalogues to enable the architect or householder to plan his layout, par-

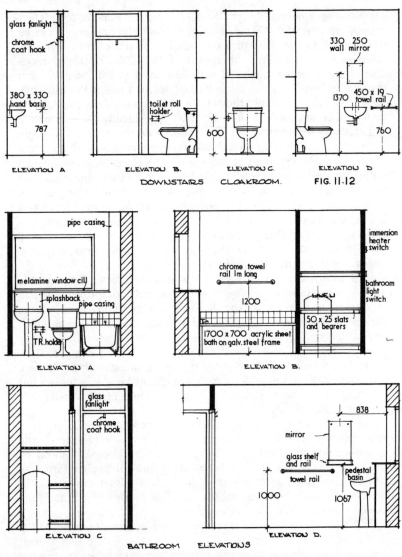

ELEVATION A

ELEVATION B.

ELEVATION C.

ELEVATION D

DOWNSTAIRS CLOAKROOM.

FIG. 11·12

ELEVATION A

ELEVATION B.

ELEVATION C

ELEVATION D.

BATHROOM ELEVATIONS

DOMESTIC SANITARY LAYOUT

FIG. 11·13

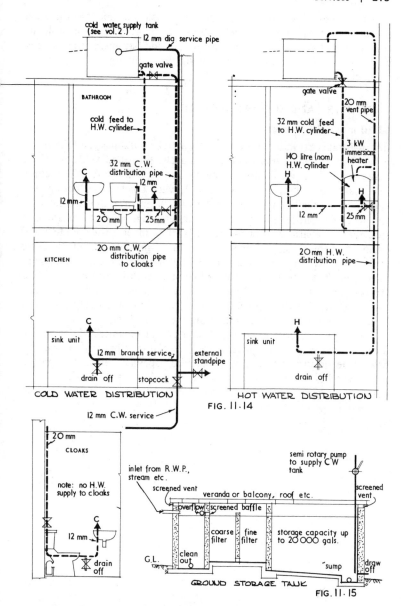

cold water supply tank
(see vol. 2.)
12 mm dia service pipe

gate valve

BATHROOM

cold feed to
H.W. cylinder

32 mm C.W.
distribution pipe
12 mm

C

12 mm

C

20 mm

25mm

20 mm C.W.
distribution pipe
to cloaks

KITCHEN

C

sink unit

12 mm branch service

external
standpipe

drain off

stopcock

COLD WATER DISTRIBUTION

12 mm C.W. service

gate valve

20 mm
vent pipe

32 mm cold feed
to H.W. cylinder

3 kW
immersion
heater

140 litre (nom)
H.W. cylinder

H

H

12 mm

25 mm

20 mm H.W.
distribution pipe

H

sink unit

drain off

HOT WATER DISTRIBUTION

FIG. 11·14

20 mm

CLOAKS

note: no H.W.
supply to cloaks

C

12 mm

drain
off

inlet from R.W.P.,
stream etc.

screened vent

overflow

screened baffle

coarse
filter

fine
filter

veranda or balcony, roof etc.

semi rotary pump
to supply CW
tank

screened
vent

storage capacity up
to 20 000 gals.

G.L.

clean
out

sump

draw
off

GROUND STORAGE TANK

FIG. 11·15

DOMESTIC WATER SUPPLY

ticularly his worktop and fill-in units. It is also convenient at this stage to consider any extras required such as strip lighting, both at wall and ceiling level, two-way switches, curtain rails and battens, extractor fans, waste bins and extra points for ironing, food mixing, electric kettles, rings, etc.

TRANSPORT AND SPECIAL SERVICES

11.10 Internal transport

Internal transport in industrial and commercial buildings can take forms other than trucks, trollies and mechanically-propelled vehicles. These may include lifts, escalators, travelling cradles, conveyors, and ducting. Lifts have been discussed in Volume 2, while conveyors, ducting, and cradles are usually specialist trades which involve the architect or engineer very little except at the design stage. Escalators are also specialist installations, though, like lifts, they involve the architect and engineer more closely. Being speedy, safe, and convenient they are popular in department stores and public and commercial buildings in most countries. They can move people at a rate of between 4000 and 10 700 per hour up an incline of usually between 30 and 35 degrees. They are fitted with emergency brakes, stop buttons, safety covers to handrails, and often have flush lighting at skirting level.

Escalators are usually provided as a complete unit. Builder's work includes access pits at ground level, escalator supports and beams, balustrade or fireproof enclosures at floor openings, electric wiring to main switches, external finishes to sides and soffit of escalator, provision of cranage to effect entry to building installation, and preparation and making good.

12 | External works

12.1 Road drainage

Drainage in external works, with the exception of soakaways, would normally refer to systems outside the boundary or building line of individual properties or private estates, and would come under the control of the local authorities.

Soakaways. These may be of two kinds, one consisting of precast concrete rings, similar to those shown in Fig. 12.1, but perforated, or as described in A–A, Drg. DCED/G/7 and surrounded by hardcore. The notes are self-explanatory.

Manholes. Spun-concrete manholes, Fig. 12.1, are now much in demand but could be expensive unless manufactured locally. They are fairly easy to install, particularly in public works or for deep manholes. An example of a shallow manhole, which is suitable for estate and housing work, is also given. With the advent of plastics, however, domestic inspection chambers of G.R.P. and U.P.V.C. can be supplied ready for use. With push-fit adaptors and flexible joints they are light to transport and easy to assemble, Figs. 12.2 and 12.3.

12.2 Precast paving

Precast concrete and brick paving, Fig. 12.4, are extensively used for patios, paths, pedestrian precincts, and open areas. A variety of choice exists, though flagstones are still the most popular. They are made in a range of colours and surface textures, but the supply would depend on the local demand. Surface features can be created by moulding flags on glass fibre or rubber mats; fluted, squared, rib-faced and other finishes are also cast. Chapter 10 shows a moulded rubber finish used for flags.

Accessories to paving are also available. Standard kerbs, size 127 by 254 mm, are usually made only in natural colour: planting kerbs, tree grilles, gulley gratings and frames, and cycle blocks are all manufactured to match pavings, as are single or double drop footpath crossings, transition kerbs, corner angles, etc. Slabs of 50 mm are adequate for footways and pedestrian acess but 63 mm slabs are necessary for vehicular

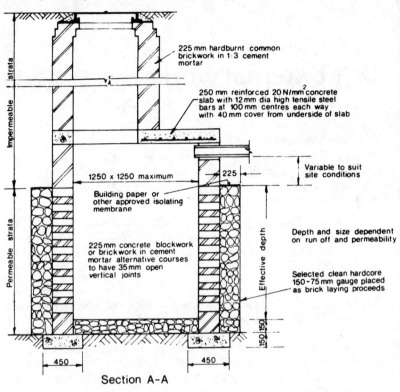

225 mm hardburnt common brickwork in 1:3 cement mortar

250 mm reinforced 20 N/mm² concrete slab with 12 mm dia high tensile steel bars at 100 mm centres each way with 40 mm cover from underside of slab

Variable to suit site conditions

1250 x 1250 maximum

225

Building paper or other approved isolating membrane

Depth and size dependent on run off and permeability

225 mm concrete blockwork or brickwork in cement mortar alternative courses to have 35 mm open vertical joints

Effective depth

Selected clean hardcore 150-75 mm gauge placed as brick laying proceeds

strata

Impermeable

Permeable strata

150 150

450

450

Section A-A

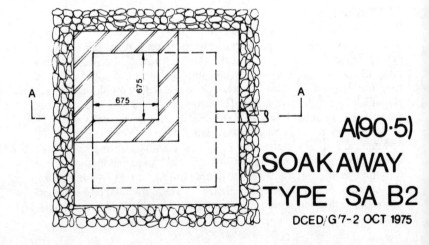

A

A

675

675

A(90·5)
SOAKAWAY
TYPE SA B2

DCED/G'7-2 OCT 1975

traffic. A minimum width of 2 metres is recommended for access ways in residential areas.

Slabs are normally laid on a formation of granular fill which should be compacted if possible with a 2 tonne roller and weed-killer applied, especially in humid areas. Cohesive soils should be blinded with granular material 75 mm thick. Slabs for pedestrian traffic should be bedded in 25 mm of sand or on mortar dots of 1:1:6 cement/lime/sand, each about fist-sized, placed at each corner and at the centre of the slab. For heavy traffic, slabs should be laid in continuous mortar, close-jointed with dry or slurried cement/sand, 1:3 well brushed into the joints.

12.3 Fencing

External boundaries are usually fenced, the type depending on the security, appearance, or use for which it is needed. There is a wide range: some manufacturers supply complete systems including accessories according to requirements. The main categories set out below are based largely on British Standard No. 1722 from which further information can be obtained.

Chain link fencing consists of interlinked galvanized or plastic-coated plain or galvanized wire. Posts may be of steel angle, tube, or hollow sections, timber, or concrete either plain or plastic-coated. Heights vary and can be up to 11 metres with special design. A chain link fence is shown in Volume 2.

Woven wire fencing is commonly used for agricultural purposes. It is available in a number of standard group sizes and heights and can be supplied plastic-coated.

Strained wire is provided in a range of standard heights for fences consisting of rows of plain wire. Steel angle, timber or concrete posts are commonly used.

Cleft bamboo or riven pale fencing consists of split pale joined by strands of galvanized wire. The fencing can be made up into rolls where pales are grown, or assembled on site if preferred. It is normally used for temporary fencing.

Bamboo or straight branch spiles are also driven into the ground at about 300 mm centres and connected by two rows of wire.

Close boarded fencing commonly consists of featheredged boarding fixed back to arris rails and timber or concrete posts. Capping and gravel boards are usually added and replaced as required. Preservative

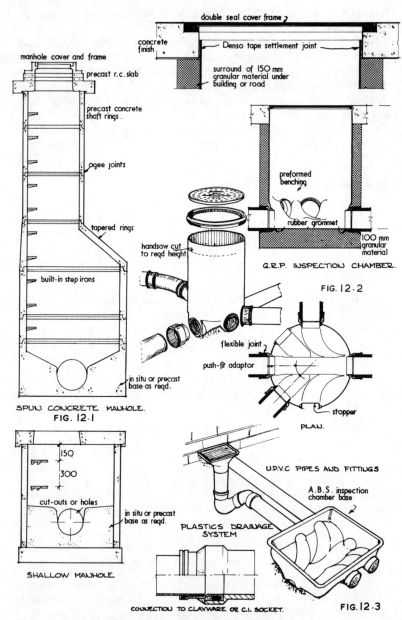

double seal cover frame

concrete finish

Denso tape settlement joint

surround of 150 mm granular material under building or road

manhole cover and frame

precast r.c. slab

precast concrete shaft rings

ogee joints

tapered rings

built-in step irons

handsaw cut to reqd height

preformed benching

rubber grommet

100 mm granular material

G.R.P. INSPECTION CHAMBER.

FIG. 12.2

flexible joint

push-fit adaptor

stopper

PLAN.

in situ or precast base as reqd.

SPUN CONCRETE MANHOLE.
FIG. 12.1

150

300

cut-outs or holes

in situ or precast base as reqd.

SHALLOW MANHOLE

U.P.V.C PIPES AND FITTINGS

A.B.S. inspection chamber base

PLASTICS DRAINAGE SYSTEM

CONNECTION TO CLAYWARE OR C.I. SOCKET.

FIG. 12.3

DRAINAGE BELOW GROUND

prolongs the life of such fencing, especially in humid climates.

Wooden palisade fencing consists of cleft, sawn, or planed vertical timber fixed as shown in Volume 2. It may be applied to one side only if preferred.

Wooden post and rail fencing consists of well-spaced vertical members with horizontal rails mortised to make them strong enough to resist pressure of cattle in compounds. They are also used on busy roads.

Wattle hurdles are used in agricultural work. They are woven from single rods or groups of pliable reeds around stouter posts which are driven into the ground. Hurdles can also be woven from split bamboo or rushes.

Mild steel continuous bar fencing consists of round or flat sections for rails and any suitable sections as posts.

Mild steel unclimbable fences consist of round or square vertical bars secured to flat horizontal rails and fixed to suitable steel posts usually at 2 metre centres. The tops of the bars are usually pointed or curved for extra security. The fence is supplied in panels which may be rigid or may be self-adjusting to sloping ground. Stays may be used for extra strength if required.

Anti-intruder chain link fences are also widely used. They are shown in Volume 2.

Woven wood fencing, consisting of thin interwoven wooden slats in frames panels which are fixed between suitable posts, is popular where timber is available. It is pleasing to the eye and provides privacy, but is not strong in high winds.

Steel palisade and security fences are usually erected around power stations, industrial sites, and high-risk establishments. They have corrugated metal vertical pales with spiked tops and are difficult to climb.

Precast concrete fences are made in all shapes and sizes, often by the local manufacturer, who tends to produce his own designs. The basic types are post and panel, which present a solid appearance, Fig. 12.6, or concrete palisade, Fig. 12.5. Barbed-wire tops may be added for security if required, as shown in Volume 2.

Plastics are widely used for fencing, mainly of the post and rail variety, and are used for demarcating boundaries around, say, housing plots and open areas. Plastic-coated metal or concrete standards are sometimes used with coated wire to give a pleasing appearance. They

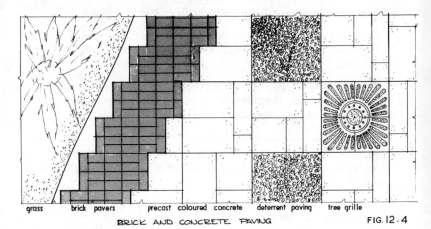

grass   brick pavers   precast coloured concrete   deterrent paving   tree grille

BRICK AND CONCRETE PAVING

FIG.12.4

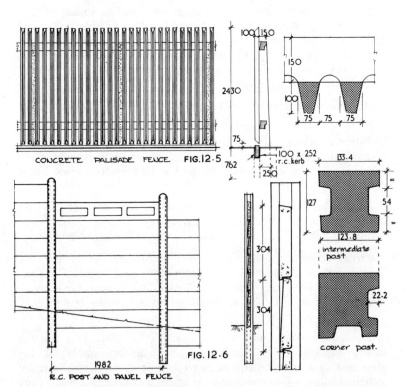

CONCRETE PALISADE FENCE   FIG.12.5

R.C. POST AND PANEL FENCE   FIG.12.6

intermediate post

corner post.

EXTERNAL WORKS. PAVING & FENCING

are not much used in the tropics.

Live fences have been discussed in Volume 2.

Walls and boundaries including screen walls have been dealt with in Volumes 1 and 2 and Chapter 5.

12.4 Hard surfaces, street furniture, parking

The design of hard standings, estate roads, car parks, and areas which take a limited amount of vehicular traffic usually come within the province of the architect or engineer. Two types of carriageway are in general use: the concrete carriageway, where the concrete itself is the wearing surface, and the flexible-surfaced carriageway, usually with a surface of coated bitumen macadam, either on a concrete base or on one that is flexible throughout. A great deal of information has been published on the design of slab thicknesses, weight of reinforcement, spacing of joints relating to anticipated intensity of traffic, nature of subgrade, and also on unreinforced concrete and flexible carriageways with a variety of road bases and sub-bases. This may be found in the British Standard specifications and road notes published by the Road Research Laboratory. The Cement and Concrete Association also offers a free technical advisory service.

Though hot rolled and mastic asphalt is used extensively where heavy traffic may be expected, the usual material for estate and similar work is coated macadam. This is a road material consisting of graded mineral aggregate coated with tar or bitumen, and here the strength of the compacted road base depends on the interlocking of the aggregate. It is always necessary to have a suitable base, depending on the soil, traffic, and type of material used for the pavement. Flexible surfaces are usually of two-course construction unless the existing surface is adequate in shape and strength. Dense Tar Surfacings (D.T.S.) are wearing surfaces of aggregate, filler, and road tar, which provide a close-textured impervious surface.

The types of bituminous material now being employed in the construction and surfacing of light roads, paved areas, hard standings, and sports arenas, together with laying methods, are listed in the publications of the Asphalt and Coated Macadam Association. Alternatively a number of countries now have their own research stations and laboratories from which information may be obtained direct.

STREET FURNITURE. Street and landscape fixtures cover an extensive range of items and accessories including lighting columns, amenity lighting, bollards, litter bins, bus shelters, covered ways, outdoor seats, poster display units, footbridges, guard rails, crash barriers, planting boxes, flagstaffs, clothes-posts, etc.

Lighting columns are made of steel either sheet, tubular, or galvanized in a variety of shapes and sizes, and in precast or prestressed concrete, aluminium etc. They may have raising or lowering gear for lamps, be tapered, round, or octagonal with built-in access for switch gear and fitted with a variety of bracket arms or lanterns. An example of a 5 metre aluminium column is shown in Fig. 12.16.

Amenity lighting can be supplied as low-level posts from 220 to 2800 mm high in aluminium, stainless-steel, or concrete, also as horizontal or vertical mounting on walls. Fittings can be of clear glass, or else consist of opal bowls with graphite or prismatic glass, white plastic, etc. An example of an illuminated bollard is given in Fig. 12.10.

Bollards can be supplied as amenity lighting, internally illuminated with 8 to 20 watt fluorescent tubing for traffic direction, or non-illuminated in precast concrete in a wide selection of finishes. They can be of exposed aggregate, cast stone, or plain concrete all coloured if required, Fig. 12.7.

Litter bins are popular in pedestrian precincts. They can be free-standing, post-mounted, or wall-hung, and are available in a range of sizes. They can be of precast concrete of different textures and colours, Fig. 12.8, or of stove-enamelled galvanized steel, fibre-glass or G.R.P. with separate steel containers, Fig. 12.9. Some are set in teak or iroko slats on galvanized steel posts and rings.

Bus shelters can be either of cantilevered or enclosed type are are made in precast concrete, Fig. 12.12, or steel, with R.H.S. cantilever pillars, aluminium trough decking, and galvanized steel window-frames with toughened glass, Fig. 12.11. They can be supplied in standard units or in multiples with end units added separately.

Outdoor seats are a popular feature in parks and precincts, and are made in a range of styles and materials. Precast concrete bases of exposed aggregate are frequently used with hardwood slats as seats, often of teak or iroko, Fig. 12.14. Some have 50 X 50 X 10 s.w.g. m.s. galvanized tubes with capped ends finished with teak or iroko seats and backs. Some frames are black stove-enamelled, others have nylon-coated metal stands with oiled iroko seats, and others are made of G.R.P. Some consist of benches attached to tables in the Canadian style while others are made of U.P.V.C. plates on die cast aluminium supports. A combination of styles may also be chosen if preferred.

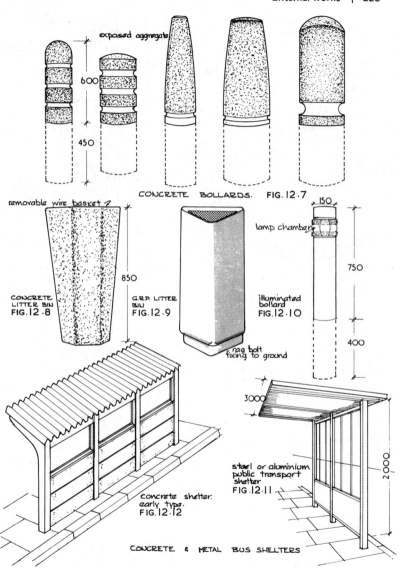

exposed aggregate

600

450

CONCRETE BOLLARDS. FIG. 12·7

removable wire basket

850

CONCRETE
LITTER BIN
FIG. 12·8

G.R.P. LITTER
BIN
FIG. 12·9

illuminated
bollard
FIG. 12·10

rag bolt
fixing to ground

150

lamp chamber

750

400

concrete shelter.
early type.
FIG. 12·12

3000

steel or aluminium
public transport
shelter
FIG. 12·11

2000

CONCRETE & METAL BUS SHELTERS

EXTERNAL STREET FURNITURE

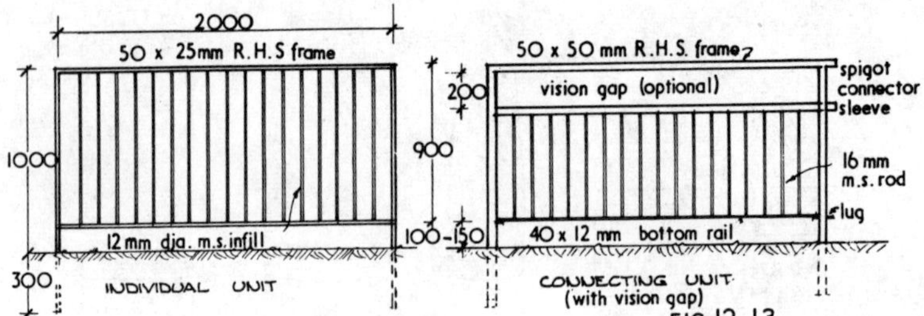

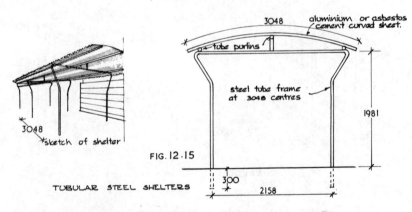

2000

50 x 25mm R.H.S frame

1000

12 mm dia. m.s. infill

300

INDIVIDUAL UNIT

50 x 50 mm R.H.S frame

spigot connector sleeve

vision gap (optional)

200

900

16 mm m.s. rod

lug

100-150

40 x 12 mm bottom rail

CONNECTING UNIT
(with vision gap)

FIG. 12.13

FRAMED PEDESTRIAN GUARDRAILS.

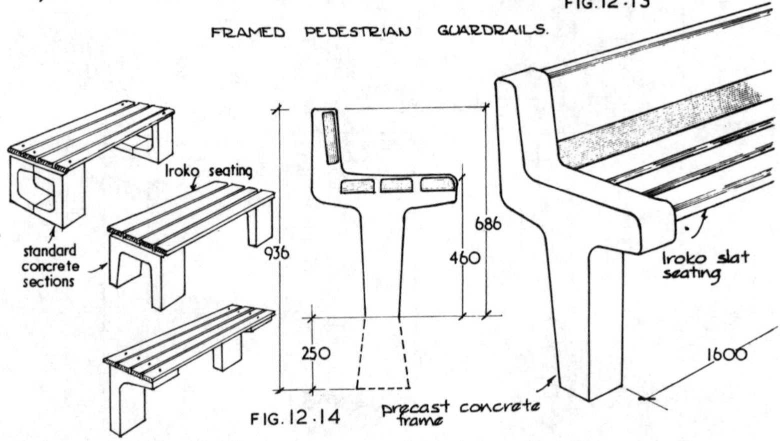

Iroko seating

standard concrete sections

936

686

460

250

precast concrete frame

Iroko slat seating

1600

FIG. 12.14

PATIO SEATS

3048

aluminium or asbestos cement curved sheet.

tube purlins

steel tube frame at 3048 centres

1981

3048
sketch of shelter

FIG. 12.15

300

2158

TUBULAR STEEL SHELTERS

EXTERNAL GUARDRAILS, SEATS, SHELTERS.

Poster display units. These are becoming fairly common in squares and pedestrian areas as well as in indoor foyers of public buildings. They are made in a range of illuminated and non-illuminated units to accommodate all sizes of poster. A typical style is a box section metal frame with a hinged door of clear acrylic sheet. Stainless-steel and aluminium are also popular. They can be wall- or post-mounted, or combined to perform a dual function of telephone box and poster panel. They can also be made to rotate as a poster drum of circular or polygonal shape.

Footbridges over culverts, streams, or recessed areas may be purchased ready-made or made to measure. They are often constructed of structural hollow sections, precast concrete, aluminium alloy, or of well-seasoned hardwood of good quality, preferably treated with preservative. They are usually secured to purpose-made concrete abutments.

Guard rails may be bought as patent aluminium-welded balustrading in approximately 2 metre sections ready for site fixing, or in aluminium alloy panels 2 metres by one metre high, Fig. 12.13. They can be curved or raked as required and are made in mill finish, satin waxed, or silver bronze anodized. Some makes are in continuously-moulded nylon or polypropylene chain in red and white. Fixing posts are sometimes nylon-covered designed for drop-fit into ground-sockets in concrete or they may be free standing.

Crash barriers for road safety can be designed in many types to suit the class of road, bridge, or hazard for which they are required. They may be tensioned, untensioned, or open-box built up of R.H.S. to local authority or other requirements. A range of motorway bridge-railing and security fencing is available, the type depending on impact requirements. Standard fixings of barriers, mainly holding-down bolts for attachment to kerbs, designed for speedy replacement in the event of damage, are usually mandatory. Stout galvanized corrugated-iron barriers secured to steel or hardwood posts are also required at curves, corners, car parks, and danger points.

Vehicle pedestrian parapets as shown in Fig. 12.20 are designed for bridges such as those used, say, at entrances to car parks over deep flood-water culverts or on retaining walls. The parapet shown, type P2, will resist an impact of a 1·5 tonne vehicle at 80 Km/h (50 mph) at a 20 degree angle of incidence. Other systems have been designed for motorways, pedestrian parapets, footbridges, and bridges over railways. The sections are of extruded aluminium alloy and meet the requirements of the Department of the Environment Technical Memorandum No. BE5.

They are produced by High Duty Alloys Ltd. from whom further information can be obtained.

Planting boxes. Square, polygonal, circular and other curved shapes of plant containers are now often seen in pedestrian areas and open precincts, Fig. 12.18. They are available in benches and planters in various combinations of cast stone and plain and exposed concrete, with a brushed finish in various colours. They are also made in iroko, or in G.R.P. with smooth finish.

Flagstaffs, of which there are a number of types, are usually to be found in open squares, spaces, parade grounds, forecourts or prominent buildings, and also on walls and roofs. They can be of timber though fibreglass is now widely used. Poles are usually supplied by manufacturers complete with cap, and truck or finial, which not only gives protection but provides a protective housing for the pulley used for the halyard line which raises and lowers the flag. Though conventional guys and rigging are still in use where displays are traditional, more sophisticated methods of support are now favoured, except in areas of high stress and exposed situations.

The normal method of securing flagstaffs to the base is by means of tabernacles, Fig. 12.17, which consist of two built-in upright planks or steel angles to which the pole is secured. In the humid tropics galvanized steel for tabernacles is more suitable than timber. Tubular aluminium alloy poles are also used as flagstaffs and fixed by being dropped over spigots bedded in the ground. Poles may also have hinged flanges welded to the base which enables them to be lowered horizontally for servicing. Flag-poles are also secured to the front walls of buildings by brackets. All fittings must be strong enough to resist stresses caused by the environment.

PARKING EQUIPMENT. Parking equipment is now in use in most cities and used for heavy and light commercial vehicles, cars, cycles, market trucks, and miscellaneous barrows, etc. Bicycle parking can be a problem in colleges, schools, and factories where this is sometimes the only means of transport. The simplest form of bicycle-stand in a precast concrete block approximately 600 X 300 X 100 mm deep with a cast-in recess to take the front wheel running diagonally across the face of the block. Manufacturers also produce all-steel cycle parks, both covered and open, in many standard types. Outdoor models usually have roofs of galvanized corrugated steel, aluminium, or asbestos supported by a steel tubular frame, Fig. 12.15. Such shelters are in use as market stalls, covered ways, and storage areas, and racks can be inserted for bicycles

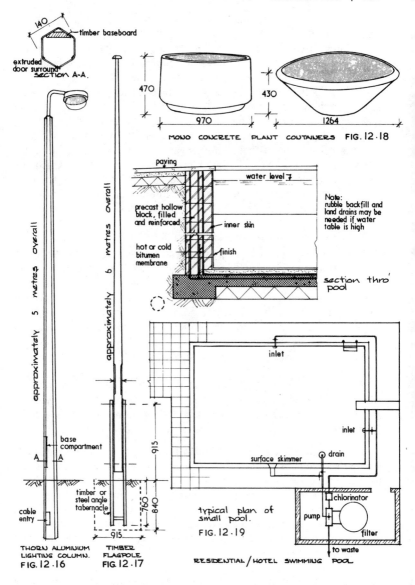

140

timber baseboard

extruded door surround
SECTION A-A.

470

970

MONO CONCRETE PLANT CONTAINERS FIG. 12·18

430

1264

paving

water level

precast hollow block, filled and reinforced

inner skin

hot or cold bitumen membrane

finish

Note:
rubble backfill and land drains may be needed if water table is high

section thro' pool

approximately 5 metres overall

approximately 6 metres overall

base compartment

A   A

cable entry

timber or steel angle tabernacle

915

760

840

915

THORN ALUMINIUM LIGHTING COLUMN.
FIG. 12·16

TIMBER FLAGPOLE
FIG. 12·17

inlet

inlet

surface skimmer

drain

chlorinator

pump

filter

to waste

typical plan of small pool.
FIG. 12·19

RESIDENTIAL / HOTEL SWIMMING POOL

LIGHTING COLUMN, FLAGSTAFF, PLANT CONTAINER, SWIMMING POOL

as required. They are usually in units of four and are easily bolted together on site.

Parking for private cars is usually the biggest problem, both on and off the street. Tree-covered bays are the most popular, though any cover is usually in demand. Car-ports are widely used as described in Volume 1. Individual space control can be obtained by the use of lockable bollards which can be folded or hinged to lie flat on the ground or unlocked and removed completely. Entrance and exit barriers to both private and public car parks are widely used. They consist mainly of rising arms operated either manually or automatically by coin-in-the-slot or season pass key. Collapsible plates at road level, which are vehicle activated, are also in use for one-way traffic. When touched by wheels from the right direction they collapse flat on the ground, springing back up again once the wheels have passed over.

12.5 Soft landscaping, play areas

Soft landscaping can include turfing, seeding, planting, grass cutting and earth finish, the construction of tennis courts, playing fields, golf courses, running or cycle tracks or the landscaping of gardens. Lawns in public and private places can be turfed, but where seeding is simple and soil available this can be cheaper and produce very good results, though some soils need careful preparation. Advice on fertilizers and soil cultivation is necessary in some localities. There are a number of excellent grasses which grow green and thick and also stand up to the machete.

The choice of plants calls for local knowledge. Temperature, rainfall, shade, latitude, and location all demand consideration. Local authorities and specialist firms can advise on large-scale planting, live-fencing, landscaping, and reclamation schemes for different areas.

Tennis courts are often in demand in colleges, recreational centres, and sports clubs. Though coloured and textured macadam or similar hard surfaces are used, perhaps the most popular is that of compact self-binding gravel or hoggin, clay, laterite, etc. Regular watering and rolling maintains an excellent surface. Stabilizing with lime or cement (see Volume 2) produces a harder and more durable surface, especially when drained and graded. Plain concrete is not usually satisfactory and grass courts are difficult to maintain.

PLAY AREAS. Land designated as a play area usually needs to be cleared, drained, levelled, fertilized, and seeded to become suitable for ball games and athletics. Where modern earth-moving equipment can be

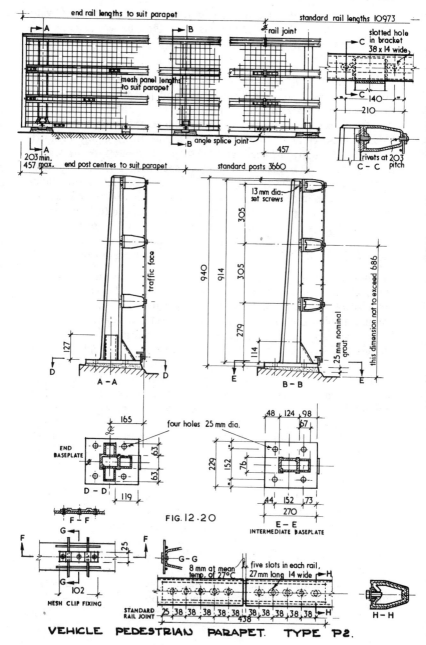

end rail lengths to suit parapet

standard rail lengths 10973

slotted hole in bracket 38 x 14 wide

mesh panel lengths to suit parapet

rail joint

angle splice joint

457

203 min.
457 max.

end post centres to suit parapet

standard posts 3660

rivets at 203 pitch
C – C

210
140

13 mm dia. set screws

305

940
914

305

traffic face

279

114

25 mm nominal grout

this dimension not to exceed 686

127

D

A – A

D

E

B – B

E

END BASEPLATE

four holes 25 mm dia.

165

63
63

D – D

119

48 124 98
67

229
152
76

44 152 73
270

E – E
INTERMEDIATE BASEPLATE

F – F

FIG. 12·20

MESH CLIP FIXING

G

25

102

G

G – G

8 mm at mean temp. of 27°C

five slots in each rail, 27 mm long 14 wide

H

STANDARD RAIL JOINT

25 38 38 38 38 38 38 38 38 38
438

H

H – H

VEHICLE PEDESTRIAN PARAPET. TYPE P2.

used such tasks can be completed quickly, provided adequate water supply is to hand. Surface conditions vary widely between wet and dry seasons and, though rainfall figures may be available, local knowledge is a better guide.

Playground equipment. Most public parks, schools and playgrounds now have at least some facilities. Parks and playgrounds can include slides, rocking boats, rotating equipment, fun frames, etc. or schools could have rope equipment, steel tubular apparatus such as horizontal bars, ladders, climbing frames, etc. Specialist firms supply complete ranges to suit individual needs and the scope of products is considerable. Some suppliers specialize in concrete shapes such as tunnels, bridges, stepping stones, planks, wall cubes, concrete trees, and shaped saddles. Plastics are also in demand for polyhedron units in several colours made from high-density polyurethane which can be arranged to form tunnels, slides, and mushroom seats, and also for artificial grass for playing surfaces. A comprehensive list of standards cover such equipment and the reader is advised to study these before setting up playgrounds.

Finishes: sports surfaces. As an alternative to conventional surfaces for track and field athletics, an increasing number of all-weather surfaces are available. These prove beneficial in countries where natural surfaces are unsuitable or change considerably between wet and dry seasons. Porous surfaces include artificial turf of synthetic fibres, while 'impervious' surfaces are used in hard all-weather areas and are usually wet-poured on site. These comprise natural aggregates, synthetic binders, and bituminous or rubber-based materials. Most suppliers have literature on sub-stratum preparation. Surfaces are usually laid on smooth level concrete or coated macadam and/or stabilized crushed stone aggregate. These surfaces are usually applied to athletic tracks, cricket wickets, golf driving tees, tennis courts, and indoor flooring for ball games.

12.6 Swimming pools

Swimming pools are important in tropical countries and construction techniques range from traditional *in situ* building to installation of prefabricated systems delivered complete to site. Public pools are provided by some local authorities, though these are not widespread, particularly in coastal towns where bathing is possible. Pools, however, are installed by commercial and industrial organizations and are often of straightforward construction. Hotels in large towns frequently have pools set apart for the use of guests and visitors and, although facilities may be limited, visual treatment is often important. Modern schools and colleges sometimes have to cater for teaching swimming and diving, the latter needing a separate recess or pit roped off for safety reasons.

In situ cast reinforced concrete gives freedom of construction though supervision needs to be strict to ensure that the principles of sound mix design are adhered to, with correct water/cement ratio and workability maintained. Cracking may be controlled by incorporating contraction joints at intervals of about 7·5 m. Curing of concrete must be controlled and thorough. Floors and walls of large pools must be cast in alternate bays to control shrinkage and day joints properly arranged and constructed to coincide with contraction joints where possible, or water bars inserted.

Some pools are constructed by 'guniting' i.e. spraying liquid concrete through a nozzle on to formwork lined with hessian. Reinforcement is then added and further spraying applied until the pool is complete. It gives a satisfactory finish but cannot take tiles. Concrete blocks are normally used for pools of straightforward design. Hollow blocks are built up and reinforcement added, followed by infill. A waterproof membrane is then added and an inner wall of blocks built, Fig. 12.19.

Glass reinforced plastic (G.R.P.) pools are prefabricated and arrive on site in sections. These are flanged and welded or bonded with neoprene gaskets. The pool is set on a dry mortar mix floor in a pit and backfilled concrete added as the pool is filled with water.

Aluminium pools are usually fabricated in double skin sections and welded together on site. The exterior must be painted with bituminous paint before concrete backfill is added.

Flexible liners of suitable plastics are sometimes used inside a pit of brick, blockwork, or concrete. Some manufacturers produce wall systems, prefabricated in concrete, steel, or timber, for use both in-ground and on-ground.

Finishes. Renderings applied as backings to tiles, mosaic, or terrazzo are usual, though the rendering itself may be trowelled smooth to take paint. Tiles of all types are used as a finish, together with accessories such as overflow channels, steps, and bullnosed surrounds to match. Paint for rendering should be of chlorinated rubber, co-polymer resin, or epoxy resin. The surrounds themselves should be finished in non-slip textured materials with a fall of 1:24 away from the pool to facilitate washing down. Pool markings may be made on walls and floors by means of coloured tiles or paint. Accessories such as ladders, diving boards and stages, starting blocks, swimming lanes, and special markings will need consideration as far as they apply.

Permeability. Pools must be designed to prevent the escape of water, both to preserve the surroundings and to eliminate waste. Sometimes it pays to design a cavity drainage around the pool rather than make it totally impermeable. Flexible liners are impermeable and are usually used in conjunction with concrete whether *in situ* or in blockwork. G.R.P. and aluminium are suitable, provided they are properly welded or sealed at the seams. Waterproofing of renderings with additives is not usually satisfactory.

Treatment of water is necessary in almost all cases, either by filtering and dosing or by changing water at regular intervals. The latter method is usual in coastal towns where sea water is used and seaweed and sand can be removed at the same time. In small pools sterilization is achieved by adding chemicals direct to the water but in large pools the process is mechanical and automatic. Water from the pool is passed through a strainer, dosed with a coagulant, filtered, then dosed with alkali and chlorine before being returned to the pool, Fig. 12.19. In tropical pools control of algae is important. Normally this is done by the chlorine treatment but where the process is intermittent, algacides may have to be used. In addition tiles need to be cleaned at regular intervals either by hand brooms or underwater electrical equipment.

12.7 Garages

Though traditional garages of brick and concrete block with framed, ledged, and braced doors are normally built, some authorities and organizations find it more convenient to acquire proprietary systems complete. These may be supplied in a standard range of sizes in single or double units or in battery form, in concrete, steel, timber, or asbestos sheet claddings. Some manufacturers, where the scope of the scheme justifies it, will provide a complete service including site clearance, access roads, and concrete bases, and will supply and erect the garages complete with finishings.

Proprietary garages are usually supplied with doors, windows, glazing, and drainage if necessary. Doors may be of timber, though glide-over doors in aluminium, galvanized steel or timber are commonplace, Fig. 6.12. Optional extras include extra doors, pedestrian doors, windows, and partitions. The majority of garage systems have concrete walls of various finishes and colours, though walls of timber, or asbestos cement, and steel frames with asbestos or steel sheeting are also sold.

13 | Inspection and maintenance

13.1 Introduction

The inspection of existing properties covers a number of functions. It may be done prior to buying or selling a property; as a periodic check for repair and maintenance; to investigate one specific aspect, such as subsidence or dry-rot; or with the object of alteration, extension, or change of use.

There is also, of course, the inspection of work in progress, which is usually done by the architect or engineer. Under the terms of agreement between the architect and building owner, it is the duty of the former to inspect the work from time to time to ensure that it is being carried out according to contract. Also, under the Defects Liability Clause, the contractor can be held responsible for making good any faults occurring within a period after the building has been completed, usually six months.

Some organizations such as hospitals, universities, or private industrial concerns employ resident architects or engineers who are responsible for all aspects of construction, including repair and maintenance. Public bodies and local authorities have departments which deal exclusively with land economy, surveying, estate management, public works, and new building. Apart from classical structures of early origin and possibly of religious or military significance, or indigenous constructions, or traditional pre-war building of good quality, much government property is now subject to survey for repair, alteration, or extension. This applies also to early post-war development carried out mainly by European contractors when the need for new building was urgent.

SCOPE OF MAINTENANCE AND INSPECTION. The object of many inspections is to locate the cause and effect of structural defects; these may be due to imperfect design, method, or materials, hostile elements, acts of God, movement, settlement, shrinkage, or failure of component parts, including joints and connections. Inspection is also concerned with prevention as well as diagnosis and treatment. It is not always easy to detect defects while work is in progress as faults tend to get covered up between visits. It is for this reason that the architect or engineer cannot be held responsible under the normal conditions of engagement for failure by the contractor to complete work satisfactorily. But routine inspection of work both existing and partly completed, is the exception rather than the rule and it is usually the case that only when things go wrong do inspections receive priority. The weathering of buildings in warm countries means that high standards of maintenance are required.

13.2 Building surveys

There is, or should be, in every office responsible for inspections, a check list drawn up to meet the needs of the particular department. A report concerned with take-over or purchase would differ from one dealing with defects liability at a final inspection. A surveyor concerned with the former would probably require something on the following lines, which could be adapted to suit other needs:

Preliminary. Check quantity of equipment needed, also its suitability and accuracy. Make necessary arrangements for travelling, etc.

Legal: general. Restrictive covenants as to usage and sale; rights of way; access; names and addresses of legal practitioners and agents; legal charges; rateable value.

Legal: building. Town planning requirements; restrictions and intentions; existing use and development; by-laws; ordinance block plans; submission requirements and forms; existing records; building lines and boundaries; road and development charges; restrictions under any industrial acts (e.g. Clean Air); existing rights and restrictions.

Local information. Existing or anticipated undesirable features in vicinity e.g. sewage-works, factories, quarries, rivers, areas subject to flooding; age of property; local facilities; shops; schools; availability of labour.

Underground hazards. Mineral workings; tunnels; springs; ground movement; erosion; earthquakes; tremors; heaving soil; level of water table.

Adjoining properties. General character; condition and usage; conflicting interests; evidence of subsidence.

Actual site. Orientation; local climate; sea; smoke; fumes; trees; rock outcrops; ponds; soil; subsoil; trial holes. Fences, paths, gates, boundaries. Retaining walls, ownership and condition. Paths, roads, pavings. Levels and sections in relation to benchmark.

Drainage. Surface water, records, levels. Sewage disposal, manholes, gullies, interceptors, traps, drains. Local authorities, connections and charges.

Buildings and outbuildings. Existing plans, construction details, floor levels.

Roofs. Pitch, construction, access, condition, light, materials. Tanks, vents, r.w.p.s, outlets and projections in roof. Fire-hazard; evidence of vermin, corrosion.

Walls and partitions. Construction; foundations; finishes; condition; internal and external renderings and pointing. Ceilings; laylights.

Floors. Construction; condition under boards; sleeper walls; d.p.c.s; surface concrete; vents.

Finishings. Room schedule of finishes including ceilings, walls, windows, sill heights, floor decorations and ironmongery. Stairs; cupboards; fixtures. Existing light and service points. External painting, gutters and pipes; flashings.

Water supply. Adequacy of supply, type and source; piping systems, stop-cocks, cisterns, taps, traps, vents, waste, emergency arrangements, spare tanks, towers, wells, springs; storage, tanks, pumps, rams.

Hot water. Storage, type of heating; fuel and storage.

Electricity. Overhead or underground supply, location of poles, wires, cables, meters, switchgear, phasing, voltage, amperage, current. Types of wiring, conduits, earthing, distribution boards, light points and amperage, fans, cooker points, transformers, special fittings, power point, test report from electrical engineer. Advice on future loading.

Gas. Storage tanks and cylinders, position and size, location of supply, number of appliances. Storage tank test by suppliers.

Other services. Phone, overhead wires, position of poles, lightning conductors, special services. Stand-by generators, compressors, etc.

Structural defects. Dry rot and causes; damp; recent and urgent repairs, settlement cracks, misshapen openings, skirting and window cracks and gaps, bulges in walls, sagging floors. Beams and lintels, deflection, slope, cracks, bearing. Columns and posts, general condition. Evidence of rodents, lizards, cockroaches, beetles, insects.

Special fittings. Educational and visual aids; religious, recreational, industrial accessories and equipment. Fire equipment.

Insurance. Name of company; type of policy.

General. Names of reliable builders; local materials available; source of supply.

13.3 Foundations and walls

Site explorations and types of foundations have been discussed in Chapter 4. Further brief notes are necessary here to deal with inspection and maintenance. The main cause of foundation failure is movement; some of this is inevitable. It may be due to settlement caused by normal consolidation after disturbance, excavation or compression of backfill, or change in soil conditions. Changes in loading and wall heights also create non-uniform pressure on soil, leading to differential settlement and cracking. Soils subject to expansion and contraction caused by changes in the moisture content can also disturb the foundation. Such variations are not usually serious and must be expected. As explained earlier, a foundation on expansive clay may be made stable by increasing its depth or by using piles.

Defects due to overloading could occur in an existing building as a result of increasing its height, or as a result of inserting new openings thus converting uniform loads to point loads on columns via the lintels. Building on porches and annexes can also lead to distortion by creating eccentric loading. Where no adequate bond exists between the main structure and the extension, cracking frequently occurs at the junction of the wall.

Disturbance of soil beneath the footing by natural causes is a common cause of trouble. This may be due to tree roots extracting moisture from the soil or simply growing under the foundations. It could also be caused by decay, shrinkage due to tree felling, underground streams, subsidence in mining areas, slip due to sloping sites or for other reasons. Stepped footings incorrectly built or left out, fractured water-pipes or other buried pipes or gullies, disturbance by digging in adjacent areas, and excessive use of pumping to lower the water table are other causes. Where earthquakes, tremors, and heaving soils are possible hazards, special precautions may have to be taken. This is usually the task of the structural engineer.

Made-up ground is usually suspect, especially as a base for heavy buildings. Where good soil has been used for filling, deposited in layers and consolidated mechanically, it should be stable, though piling may be advisable. Some firms specialize in strengthening walls subject to subsidence or excessive settlement by inserting hollow concrete stools at ground level as described in the section headed 'Underpinning'. A short description of bored piles is given in the same chapter. The reason for getting an engineer to inspect the site before operations commence, as described in Chapter 2 will now be appreciated. Inspection has been found cheaper than correcting faults later.

Walls. Inspection of the wall of an existing building can reveal whether the foundation or the superstructure itself is the cause of the defect. The wall can fail through lack of lateral support, insufficient thickness, too high a slenderness ratio, cracking due to movement, inadequate design, bonding or detailing at junctions, windows, or doors. Failure can also occur through faulty materials, gable ends being out of plumb, lack of wind bracing in the roof, roof thrust at the eaves, imperfect pointing, loose parapets, broken copings and dowels, caused by thermal movement and wind, longitudinal expansion, creep, gable kneelers being broken causing sliding of copings. Renderings could be flaked off, arches and lintels sheared, d.p.c.s missing or defective, gratings and air bricks blocked, tile hanging broken, or materials and workmanship faulty. Some locally-made blocks can be of very poor quality indeed.

A wall can also fail through vibration from heavy traffic or through erosion due to storm, atmosphere, or scour. Overloading of floors by increased storage can cause excessive deflection of beams and joists, leading to eccentric thrust on walls, which may be transferred to the foundation. Where the end of a timber beam in a loadbearing wall perishes, the wall itself may fail. Corroded metal in a wall can also cause damage. The surveyor alone may have to decide whether the cost of repair would exceed that of rebuilding. Replacement of defective materials and workmanship using modern techniques could prove more acceptable. Where this can be concealed within the structure, it is usually done, e.g. with the use of galvanized shoes and fastenings, ties, etc.

To shore or buttress a wall leaning out beyond the middle third could be dangerous — desirable though it may be to preserve the facade. Each case must be decided individually. An example of a repair to a central support is shown in Fig. 13.2. This has been taken from the shoring example used in Volume 2, and is self explanatory. A simple method of tying in walls is also shown, with ties running both parallel and at right angles to the wall. A typical flying shore erected between two buildings is shown in Fig. 13.3.

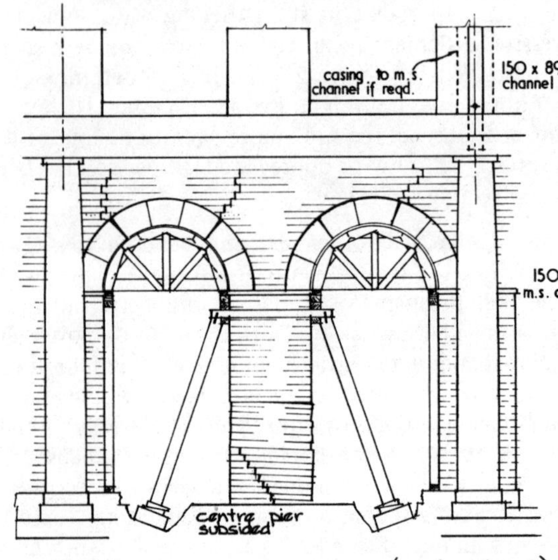

casing to m.s. channel if reqd.

150 x 89 m.s channel

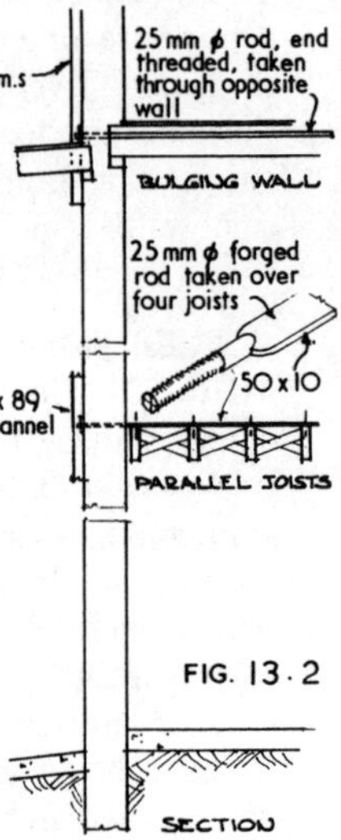

25 mm ⌀ rod, end threaded, taken through opposite wall

BULGING WALL

25 mm ⌀ forged rod taken over four joists

50 x 10

PARALLEL JOISTS

150 x 89 m.s. channel

FIG. 13·2

SECTION

centre pier subsided

FIG. 13·1   SHORING OF ARCHES   (see also volume 2.)

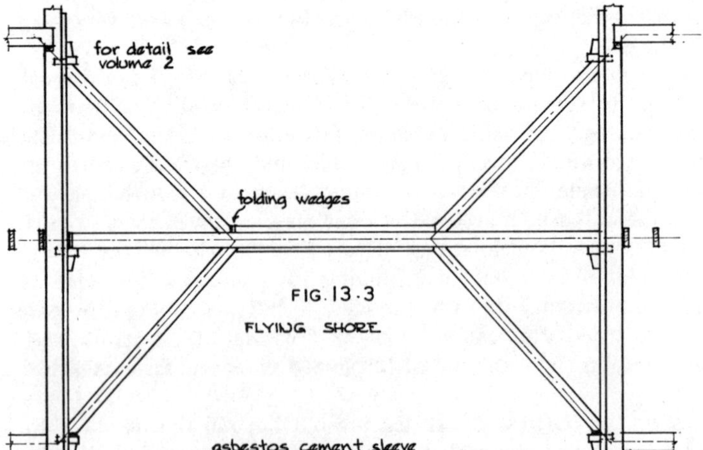

for detail see volume 2

folding wedges

FIG. 13·3

FLYING SHORE

asbestos cement sleeve

FIG.13·4

rubber rings

REPAIR OF ASBESTOS CEMENT PIPES (also new construction)

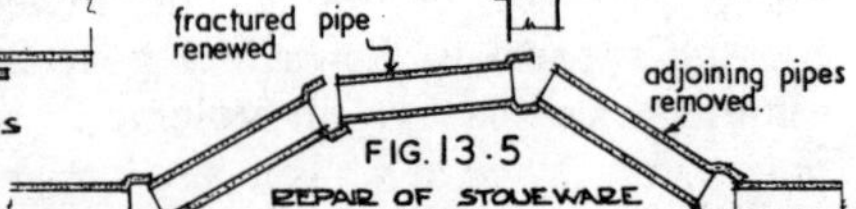

fractured pipe renewed

adjoining pipes removed.

FIG. 13·5

REPAIR OF STONEWARE PIPES

REPAIR OF WALLS AND DRAINPIPES

13.4 Flooring

In timber-growing areas it is reasonably certain that floors of older properties will be of wood. Joists are usually at the imperial 16″ centres, but frequently out of level and with a permanent sag. Floorboards can be up to 12″ (304 mm) wide and are often warped with open joints. But when in sound condition they match the surroundings and the owner may prefer to leave them undisturbed, particularly when they are highly polished. Though warping is a common cause of twist, it can also be caused by the practice of pit-sawing by hand, now becoming obsolete with the advent of machinery.

Levelling up such floors can be difficult. If the ends of the joists are supported by loadbearing partitions they can sometimes be adjusted by packing up, especially if there is no partition immediately above. At external walls where timber is in good condition, individual packing is possible; alternatively, galvanized metal shoes can be inserted and joist ends cut off. This is simpler and more effective.

Old floor-boards can be smoothed off with a sanding machine, if one is available. But the first priority is to check for decay. Door openings often cause trouble in old buildings and, provided these are not affected, floors can sometimes be trued up by means of cross-firrings nailed to the existing floor and new boards added. It may also be possible to remove the boards and level up the joists.

Wood-block and parquet flooring are frequently a source of trouble in older buildings. Many are in poor condition due to warping, shrinkage, and lifting. If the blocks are secure, a sanding machine or hand plane may be all that is needed. Joints can then be filled with a wood filler if necessary and the occasional loose block refixed, using a modern resin glue. Wood floors laid on bearers in concrete are usually satisfactory, though damp and driving rain, watering of indoor plants, or other causes can cause rapid decay of some secondary species.

Thermoplastic tiles are now in use everywhere, not only in new buildings but on smooth cement finish which was once painted or polished. Provided the surface is clean and the adhesive applied according to manufacturers' instructions, such floors are excellent. They should not be exposed to sunlight, however, particularly when shining through glass windows, as this causes them to curl and they are difficult to replace. Matching of tiles after a year or two is not always easy, especially if supplies have run out.

Finishes to concrete. With soft-soled shoes and light traffic, cement and sand finish, trowelled smooth, is usually quite satisfactory. Loose or worn patches can be repaired by cutting out the affected part, trimming the edges, cleaning the sub-floor, and using grout or additive to bind the new patch firmly in place. The same remedy would apply to granolithic flooring. Most floors, excluding wood, need only sweeping and washing with soap or mild detergent, or polishing for better appearance. Hardwood is normally finished with wax polish. Care should be taken with polished floors to keep them dry; a leaking ice box or flower container can leave a permanent stain.

13.5 Roofs

One of the main difficulties with roof inspection, especially the inspection of pitched roofs, is one of access. Long ladders are uncommon, though the introduction of tubular tower scaffolding has made the problem easier. Trap doors to attics provide good access internally, and a good torch is necessary. Flat roofs of timber, with coverings other than bitumen, are not popular in areas of high rainfall because of the skill required to form drips, rolls, welted joints, and outlets. Expansion problems arise, and lifting by suction in high winds is not unusual. Heavy storms also allow water to rise above the drip or welt before it can drain away. Bitumen, however, needs sun protection which is usually supplied by light gravel, as described earlier.

Pitched roofs in older buildings may be of shingles, clay tiles, bitumen felt, corrugated sheeting, etc. Shingles deteriorate quickly unless creosoted before fixing and clay tiles tend to snap off at the tail. These may be replaced by new tiles slipped up into the place of the old and held in place with copper or aluminium tacks or tingles.

Corrugated iron is very unsightly when rusted, and is now being replaced by aluminium, though high expansion of this sheeting can cause movement around nail holes, despite the use of neoprene washers. Only aluminium nails should be used. Corrugated asbestos sheets also make good cover and have been in use for many years. They have the advantage of taking cement mortar fillets as flashings and also as ridges, hips, and verges, when accessories or metal flashings are not available. Diamond-shaped asbestos tiles are also used though these tend to become soft after some years. In humid areas asbestos needs to be cleaned of moss and lichens occasionally.

Pitched roofs screened by parapet walls and copings are fortunately rare. The gutter at the junction of the wall and roof is rarely water-tight, often blocked by leaves and debris, and difficult to repair. Outlets get blocked easily and cause flooding. A major source of trouble with pitched roofs is with flashing. Asbestos cement and corrugated iron usually have cement mortar fillets. On other materials, bitumen sealed with tar compound or proprietary sealant is commonly used, though soft aluminium or copper sheet is now replacing it. Projections of pipes through roofs are best flashed with purpose-made cowls, either made as part of the pipe or secured with a metal strap and turnbuckle with sealant added as necessary. Openings in roofs are best avoided in new work and covered if possible in existing buildings.

Bituminous roofs often fail through not using the necessary three layers and flashings recommended in Chapter 8. Tears and small holes can be repaired by patching and clout-nailing after sealing in position. Bituminous lap cement is also available for cold application. Sometimes flat roofs which have deteriorated due to wear can be given a fluid dressing and then blinded with sand and gravel.

13.6 Internal and external finishings

Internal renderings in older buildings may contain lime plaster or cement, depending on the locality. Lime prepared under local conditions is often improperly burned or slaked, causes pitting, popping, or blowing and crazing, and cracking and bond failure frequently result. Modern factory-made limes avoid these defects. Apart from indigenous methods, internal and external renderings in many areas consist mainly of cement and sand applied to brick, stone, concrete, or expanded metal. It gives a dense impervious finish when smoothed off with a mason's trowel. This is normally treated with emulsion paint which is satisfactory for most purposes.

Mature buildings in large towns now often undergo a change of use requiring a more sophisticated finish to walls. Dry finishes can be applied as already described or tiles added as in Chapter 10. Should the wall be in sound condition, it can be rubbed down by hand or mechanically to remove blemishes and old paint, and a finishing coat of plaster applied as in Volume 2. Alternatively the wall can be coated with a variety of paints and plasters applied by brush or spray to give a coarse textured finish.

Suspended ceilings are often used to modernize buildings. This is a straightforward task with wood-joisted floors, but to obtain a fixing in old concrete is sometimes difficult. Electric hammers and cartridge guns are not always successful, despite the claims of the suppliers. Where hollow clay or concrete floors have been used, they can usually offer sufficient support, but not always. One method of increasing the bearing is to knock a small hole in the soffit of the hollow ceiling, insert a short length of reinforcing rod horizontally into the floor block, and hook the suspender on to this. Ceiling fans and light fittings can be fixed in this way, though the rod may need to be longer to spread the load.

Ceilings. Early plaster ceilings are rarely worthy of preservation except where of historical interest and where specialist craftsmen can apply traditional skills. These will not be discussed here. In some earlier traditional buildings of three or four storeys, floors consist only of polished joists supporting wide boarded flooring. Here ceilings can be added, dry lined with plasterboard or fibreboard and also insulated if necessary. They may also have suspended ceilings if preferred. Cornices or mouldings can be added if required.

Internal partitions. Altering or repairing internal partitions can be difficult, especially if they are loadbearing. Some indication of the load carried can be gained from the direction of the floor joists, position of the partitions on the floor above, and the location of beams and trimmers and point loads. Sagging, buckling, or sloping partitions could be due to inadequate ground support. Overloading could also be due to the removal of bracing to permit openings to be inserted or to the deterioration of members. Alteration should not be left solely in the hands of the local craftsman. He may be ignorant of the elementary principles of triangulation and bracing. Sizes of beams and joists should be calculated to carry the loads required of them and it is assumed that the reader now has sufficient knowledge of structures to do this. Before work is commenced, propping and strutting may be necessary and this should be taken down to firm ground.

External finishes. Although the interior involves the most study as a rule, the exterior must also be examined. The main causes of deterioration are hot sun, damp or corrosive conditions, insects, and erosion. Sun causes bleaching of paint and woodwork, and plastics and wood varnishes, unless well-shaded, are not to be recommended. Damp is not a problem in all tropical countries, though in some areas it can be severe. Scour can be caused by high winds and erosion by the atmosphere, and the use of small windows and thick walls in some climates reduces the effect of these.

Metal windows and doors are in wide use for buildings and serve their purpose excellently with normal maintenance. Putty may need replacement, rust may need to be removed, stays tightened and hinges oiled but generally they are reliable and trouble free. They can be modified or replaced by louvred windows, and adaptor kits are available if desired. Timber windows and doors can be similarly treated.

Timber windows and doors, when made by skilled men from good-quality materials, will last for years. Trouble does develop, however, through warping, sagging, faulty hinges, furniture, loose glass, and cracked putty. Top rails of windows and doors, when opening outward, do fail through exposure and lack of proper maintenance.

13.7 Services

Cold water in older buildings sometimes gives trouble due to corrosion of the original steel piping. In such cases it is better to replace it with copper piping with an appropriate gauge for external work or for underground use as necessary. Internal cold water distribution pipes are usually renewed with polythene if subject only to normal pressure. Periodic checks of joints, junctions, stop-valves, and fittings could prevent loss during drought, as could inspection of old cold-water cisterns of galvanized steel, including ball valves and overflows. If cisterns are replaced by plastic tanks, these should be on a solid base of stout plywood or similar material and not simply on bearers. Hot water distribution pipes should always be of copper.

Drainage. As main drainge is by no means universal, some institutions such as hospitals, schools, and colleges with halls of residence have their own sewage disposal plants. These are usually soundly built using stoneware pipes and standard accessories. Where traditional septic tanks are used, soil drains are often of asbestos cement pipes usually in the old 6 ft lengths and jointed solidly in cement and sand mortar. Owing to solid backfill and lack of flexibility, such pipes crack and leaks occur. Fractured pipes should always be replaced by ones capable of articulation either through a flexible joint alone, or pipe material, or both. Testing should be carried out as explained in Volume 2.

In some earlier buildings the salt-glazed stoneware pipes used were either locally made or imported. Fracture occurred in these due to overloading of backfill, burst sockets, shear fracture, or leverage due to uneven settlement. One method of replacing stoneware pipes is shown in Fig. 13.5. The fractured pipe is replaced by a slightly shorter one and adjacent pipes lifted, joints mortared, and the whole replaced and pointed. The inside is then cleaned out with rods to remove surplus mortar. Asbestos cement pipes can be replaced by using a special rubber ring as shown in Fig. 13.4. New flexible cement pipes 4 metres long with such joints are now used for normal drainage.

Electricity. As mentioned earlier, the reader needs some knowledge of electricity. He cannot always trust the electrician or engineer responsible for the inspection who may accompany him. It would be advisable for him to know what is required by way of tests and whether they have been correctly carried out. He need not be an expert but should have sufficient acquaintance with the subject to understand and discuss the report. Before any tests are made he can examine visually the condition of the wiring, switches, and flexible wire pendants. The latter are sometimes suspect in exposed places.

Without getting too involved in detail, the basic tests are:

a) to ensure that all sockets, outlets, switches, etc. are correctly connected. To test this the electrician usually uses a simple device of a torch battery and bell wired in series with two leads attached. *Having switched off the current* at the mains he tests each outlet by closing the switch and shorting the circuit to make the bell ring.

b) to ensure that the circuits from the meter are correctly wired up to the outlet he uses the same device, but one of the leads must be long enough to connect to the live wire at the meter. After connecting the other, short lead to the live terminal, the bell should ring.

c) to ensure that continuity of the earthing system is effective and

d) that the insulation resistance is high enough to prevent leakage, he normally uses a small hand or battery driven generator called a Megger which can produce a charge of 1000 V. Tests of insulation resistance are usually made by applying a voltage charge of twice that in normal use. The insulation resistance to earth must not be less than one megohm.

As a general rule an electricity system between 20 and 30 years old should always be suspect; its life could be shorter than this in humid atmospheres. In any case the client would probably want a modern ring circuit with 13 A fused plugs and with more sockets than he now has. Cartridge socket fuses which fit into plugs are much more convenient than the porcelain link type in the C.S.U. which have to be repaired with fuse wire.

13.8 Handing over inspection of a small bungalow

An illustration is given of a straightforward inspection before occupation between the client and builder based on personal experience using the plan shown in Chapter 3, Volume 2. The builder with a competent staff, was responsible for the erection of a small estate of houses and bungalows of which this was one. No subcontracting took place. The walls were of concrete block, the low-pitched roof of long length corrugated aluminium sheeting, and there was thermoplastic flooring throughout, except for the porch which was terrazzo. All joinery was made by the contractor in an up-to-date shop. The following points were noted:

Entrance porch. The terrazzo porch floor was stained with laterite due to the mix having come in contact with the soil before placing.
A large opening left in the porch roof to permit light, ventilation,

and rain for the flower bed below allowed driving rain to penetrate the louvred window, and also caused the frame to decay. The louvres in such exposed positions were inadequate in the wet season.

The entrance door to the living room was in three folding leaves, two of which were hung together; and all three opened inward. The water bar was made to act as a doorstop and so placed on the outside of the door. Thus heavy rain running down the door entered behind the doorstop creating large pools of water on the floor.

Living room. The light and fan switches on the wall were moved after fixing in order to clear the folding doors when they were folded back. Making good was badly done and a permanent scar in the plastering remained.

Though all windows were mosquito-proofed, security mesh was not fixed and the house was later burgled.

Thermoplastic floor tiles curled where the sun entered the room. Also the tiles were laid in too thick an adhesive which oozed up between the joints. Similar faults were noted in other rooms.

The dining room was too small for its purpose.

Kitchen. Insufficient electric outlets were provided and not all were in convenient positions. Alterations and additions led to scarred walls. The refrigerator space against the wall allowed no room for the fridge door to open.

The gas cooker was in line with cross ventilation causing the flame to blow out. No provision was made for gas cylinders.

Steward's quarters. Flexible light pendants were buffetted by the wind.

Pierced garden wall permitted air circulation but allowed illegal entry.

No power points provided. Centre light in steward's room became overloaded with electric iron, radio, etc.

Walls had to be rawl-plugged for hooks, shelves, etc.

Bedrooms. Built-in fitments warped and drawers stuck. Floor paint in wardrobes, etc. would not set.

Bathroom. Electric storage heater thermostat faulty. Inadequate water supply in dry season. No extra storage tanks provided.

Garage. Underground main electric cable entered through stoneware bend to C.S.U., permitting easy access for rodents and vermin.

Framed, ledged, double-doors incorrectly constructed with frame in two thicknesses instead of being solid.

Roof. Long length aluminium sheeting creaked badly with slightest change of temperature; also noisy in rainstorms.

Lizards and vermin gained entry despite precautions.

Fibreboard ceiling sagged in absence of noggins.

Generally. The bungalow was generally well-built and of good design. The defects illustrated denote the type of fault to be expected in work of this kind. Faults were corrected where possible and later designs amended.

This account is not given as an example of presentation but to show what can be expected with housing contracts and similar work.

13.9 Report writing

Having completed his notes of inspection the surveyor will then be required to produce his report. It may be an interim statement of progress during construction, a handing-over document on completion of the Defects Liability Period, or a detailed account of the state of an older property for buying, selling, change of use, or fitness for occupation. Whatever the purpose, it will be necessary to write it in clear unequivocal English. This may be difficult for some whose native tongue is other than English, but such a document is often judged not only by its factual content, but by the way it is presented.

Style. A concise bald statement of facts can be disconcerting and if presented without amplification, can be misleading. Sentences should not be dogmatic but lead in to the point the writer wishes to make, with exceptions or modifications stated with as little ambiguity as possible. Adjectives and similar means of emphasis should be used sparingly and any tendency to overstate or impress, such as by use of legal jargon, must be avoided.

Order of the Report. This varies according to the nature of the document but the usual headings for a straightforward statement are:

> Introduction
> Scope of the report
> Schedules of accommodation
> Investigation
> Summary

The introduction usually describes the purpose of the report. It often contains the client's brief in full with further instructions or emphasis on important points. The surveyor will generally enlarge on this by stating how the survey was carried out, the problems that were encountered and any relevant information.

The scope of the report would define the limits worked to: what has been included and what left out. It might also include details of physical boundaries, ground covered, details of property surveyed and other relevant information.

Schedules of accommodation are not always included but are convenient when dealing with repetitive or well-defined units such as rooms, workshops, wards, stores, etc. Such breakdowns of items into lists, tables, charts, or diagrams giving information in an orderly manner are always useful.

The investigation covers the body of the report, and the accuracy and success of the surveyor's efforts depend largely on how his information was obtained and presented and on his knowing exactly what was required. Preplanning is necessary to make this a success.

The summary of findings is given either at the beginning or the end of the report. It assists the recipient in gaining an overall picture quickly but not in detail. It may also contain conclusions and recommendations.

Finally. This account of report writing is presented only as a guide and is by no means exhaustive. A professional surveyor would need to study the subject in greater detail.

14 | Regulations and protection

14.1 Building Regulations

Building regulations are made for the safety, security, and welfare of those using buildings, though they often prove irksome to the architect and builder responsible for the erection and the client responsible for payment. They can sometimes be difficult to interpret, even with the co-operation of the local authority through its building inspectors whose duty it is to ensure that the regulations are complied with.

Apart from characteristic architecture and indigenous construction, the bulk of post-war building is now designed on the European pattern adapted to suit the needs of the tropics. Consequently many warm countries have regulations based on the Western model but tailored to suit specific needs. Such regulations are statutory instruments covering all aspects of building as far as they apply. But whilst the main provisions are usually clear, legal definition and phraseology sometimes tend to blur their meaning: this is unfortunately necessary where simply clarity could permit loopholes in the legislation.

Model building regulations are usually presented under main headings or *Parts* each dealing with a separate function. These Parts are further clarified and enlarged by the use of appendices or *Schedules*. A typical arrangement of Building Regulations is as follows:

PART A. *Interpretation and General* deals with definitions and meanings, applications to build, exemptions permitted, and tests to be carried out.

PART B. *Materials* deals with fitness of materials and special treatment where required.

PART C. *Preparation of site* and resistance to moisture deals with clearing of site; subsoil water; protection of ground floors; suspended floors; protection against damp; and weather-resistance.

PART D. *Structural stability* deals with dead-, imposed-, and wind-loads for various classes of building; construction of foundations and superstructure; structural work in steel, aluminium, concrete, timber, brickwork, stonework, blockwork, and structural failure, etc.

PART E. *Structural fire precautions* deals with provisions for buildings grouped as residential, institutional, office, shop, factory, assembly, and storage, etc. (Fire-precautions are discussed further under 14.2).

PART F. *Thermal insulation* deals with thermal transmittance of materials and components, and ventilation of the space on either side of the structure.

PART G. *Sound insulation* deals with the resistance to the transmission of sound through the structure.

PART H. *Stairways and balustrades* deals with the dimensions and shapes of steps and nosings in private and common stairways and the guarding of stairways and landings.

PART J. *Refuse disposal* is concerned with refuse storage, containers and chambers, chutes, shafts and hoppers.

PART K. *Open space, ventilation and height of rooms* is concerned with sizes of windows and spaces open to the sky; rising ground; other structures; obstacles; common land between houses; preservation of open spaces; ventilation of rooms, stairs and common stairways; height of habitable rooms.

PART L. *Chimneys, fluepipes, hearths and fireplace recesses* refer to general structural requirements, non-combustible materials, fluepipes, and appliances. (This clause is not of great importance in warm countries).

PART M. *Heat-producing appliances and incinerators* refer to types of appliances, smoke emission, flue clearing and construction.

PART N. *Drainage, private sewers, cesspools* deals with soil waste, vent traps, overflows, rainwater gutters; materials for, and construction of drains and sewers; tests; means of access; inspection chambers, trenches, junctions, cesspools, septic tanks, and similar structures.

PART P. *Sanitary conveniences* cover urinals, water closets, sanitary accommodation, earth closets, and similar structures.

PART Q. *Ash-pits, wells, tanks, and cisterns* deals with distances from buildings; construction; ventilation; purity of well water; tanks and cisterns for pure water storage. (Deleted from 1976 Regulations.)

Schedules

SCHEDULE 1. (Reg. A) *Amendments to publications* to which specific reference is made in the regulations.

SCHEDULE 2. (Reg. A) *Partially exempted buildings* is divided into various clauses ranging from small residences to large workshops. The Schedule deals separately with buildings, and works and fittings, and states the provisions by which compliance is reached.

SCHEDULE 3. (Reg. A) *Giving of notices and deposit of plans* covers the erection of buildings, alterations, works and fittings and change of use.

SCHEDULE 4. (Reg. A) *Forms of application for dispensation* or relaxation shows facsimiles of forms to be completed.

SCHEDULE 5. (Reg. A) *Short-lived or otherwise unsuitable materials.*

SCHEDULE 6. (Reg. D.) *Rules for determining dimensions of certain timber members* cover only softwood which is used mainly in Europe. It gives tables of dead loads in kilogrammes per square metre for floor joists, ceiling joists, beams, flat roof joists, purlins, roof rafters, tongued and grooved boarding, etc. relative to the size of the joist and spacing, enabling the span to be determined. A reader with an elementary knowledge of structures and knowing the properties of his own local timbers could usefully adapt these tables.

SCHEDULE 7. (Reg. D) *Rules for satisfying requirements as to structural stability of certain walls* deals with strength relative to length, height, and thickness various walls with rules governing parapets, openings, recesses, and chases.

SCHEDULE 8. (Reg. E) *Notional periods of fire-resistance* covers loadbearing and non-loadbearing walls, beams, and columns of different materials, construction, and finishes.

SCHEDULE 9 (Reg. E) *Notional designations of roof coverings* deals with covering materials and the supporting structure. Designations of AA, BB, etc. are laid down in B.S. 476 pt. 3, (1958). Flat and pitched roofs covered with bitumen felt are designated in B.S. 747 pt. 2, (1970).

SCHEDULE 10. (Reg. E) *Calculation of permitted limits of unprotected areas* gives tables showing distances from relevant boundary for unprotected percentages of walls relevant to the width and height of the enclosing rectangle enabling a period of fire resistance to be determined. (An unprotected area refers mainly to door and window openings but other conditions may apply. Fire precautions will be discussed under 14.2 below.)

SCHEDULE 11. (Reg. F) *Thermal insulation* gives thickness of different types of insulation necessary for various roofs, walls, and floors.

SCHEDULE 12. (Reg. G) *Sound insulation* gives specifications for construction of walls and floors to provide for the transmission of both airborne and impact sound.

14.2 Fire precautions

The subject was introduced briefly in Volume 2, for staircases. As a means of escape, however, stairs form only part of the solution to the problem. The architect must ensure at the design stage that ways of escape are planned so that a person has means of retreating from a fire in a building to a *protected area* or *final exit* and that he does not finish up in a blind alley. In many tropical buildings an alternative means of retreat is by the balcony, though this is not always the case. Fig. 14.1 shows plans of floors with protected staircases and lobbies. Unfortunately the fire-resisting, self-closing doors, which should normally be closed, are frequently wedged open or fastened back to improve ventilation and to permit freer passage. Electro-magnetic, fail-safe door holders are now obtainable, which hold the doors open and release them automatically in case of fire.

Fire regulations in most large cities are now very strict. Frequently it is required that the owner of an old building which is over 12·77 m and used as an hotel, restaurant, or boarding house can, subject to certain provisions as to numbers of occupants and floors, have notice served on him to provide adequate means of escape. This is already

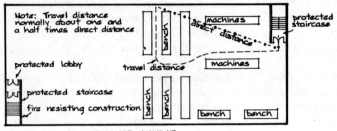

Note: Travel distance normally about one and a half times direct distance

protected staircase

protected lobby

protected staircase

fire resisting construction

bench

bench

bench

bench

bench

machines

machines

direct distance

travel distance

bench

bench

WORKSHOP LAYOUT

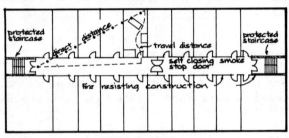

protected staircase

protected staircase

direct distance

travel distance

self closing smoke stop door

fire resisting construction

OFFICE LAYOUT

escape in one direction only

protected staircase

fire resisting construction

DEAD END SITUATION

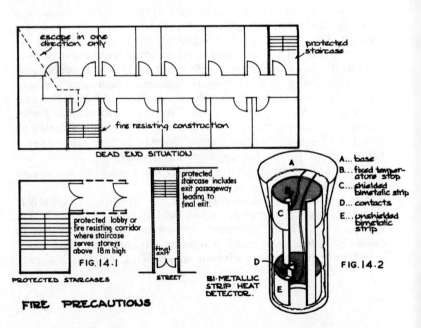

protected lobby or fire resisting corridor where staircase serves storeys above 18 m high

FIG. 14·1

PROTECTED STAIRCASES

protected staircase includes exit passageway leading to final exit.

final exit

STREET

FIRE PRECAUTIONS

A... base
B... fixed temperature stop
C... shielded bimetallic strip
D... contacts
E... unshielded bimetallic strip

FIG. 14·2

BI-METALLIC STRIP HEAT DETECTOR.

affecting thousands of building owners. In Europe it is unlawful to use any premises for the public performance of stage plays, or films, and dancing or entertainment, unless licensed by the appropriate authority. In many warm countries where such places are open-air structures regulations can be modified.

In places where people work the siting of exits and protected staircases will depend on the *direct* and *travel* distances from the furthest point, and also the maximum number of persons to be accommodated. Dead-ends must be avoided where possible, Fig. 14.1. It is essential that the staircase should be able to take all those who will use it in an emergency. Regulations are laid down as to the width of staircases, numbers of floors served, and the floor area involved, but these are too detailed to give here.

Other considerations must also be taken into account. For instance, lifts, which connect with high-risk areas such as car parking, or storage, or packing areas need to be separated from such areas by a protected lobby, Fig. 14.1. In such areas, wall and ceiling linings, and acoustic or thermal-insulation lining must be of non-combustible material.

FIRE CODES. The construction and general requirements of fire codes are primarily concerned with the protection of escape-routes from fire and smoke, and with the people who use them. The main requirements are:

Barriers should not be provided on the escape-route.

Glazing in doors of protected staircases should not exceed $0 \cdot 37$ m^2.

Glazing of fire-resisting corridors servicing dead-ends should not be under $1 \cdot 372$ m from the floor.

Cupboards within protected staircases should be of fire-resisting construction.

Final exits should have direct access to the street.

Dispersal areas should be adequately lit and free from traffic.

Doors should open outward. Final exit doors must have landings on the outside if steps are required to reach street level.

Doors opening in both directions must have glass panels.

Exit doors to be used by the public should be hung in two leaves of equal width.

Doors opening on to escape-routes and not used by the general public must be self-closing and when open must not obstruct the stream of people going towards the exit.

Ducts in protected staircases must be of fire-resisting construction.

Lobbies and corridors in escape-routes must be not less than $2 \cdot 057$ m high.

Fastenings and locks in final exits must be easily opened without using any key. Panic bolts are to be used in places of assembly. Self-closing doors must not be fitted with cabin hooks.

Unoccupied buildings may have some doors fitted with key-operated locks but these must be unlocked when the premises are occupied. Vertical ladders as a means of escape are only permitted in exceptional circumstances.

Notices. Strict regulations are in force in most countries on the size and location of EXIT and similar notices. They must be provided throughout the building in conspicuous positions adjacent to the exit doors in letters that are clearly legible. Directional signs must indicate the route to the final exit. The standard size of the exit notices is usually as follows:

Not less than 125 mm high where it must be seen at a distance greater than 25 metres.

Not less than 75 mm high where it must be seen at a distance between 15 and 25 metres.

Not less than 50 mm high where it must be seen at a distance less than 15 metres.

Other notices may include such words as 'inflammable materials', 'fire precautions', 'private', 'push bar to open', etc. and vary in height according to circumstances. Details of these and other fire matters may be had from the fire authorities.

Other regulations are concerned with:

Obstruction. Clear gangways to exits must always be maintained.

Openable windows. A building with a protected staircase may be required to have a portion of a window facing the street or facing an open space made openable within 300 mm of sill level for emergency purposes.

Loading doorways. These may be used in emergency instead of openable windows, with certain provisos.

Partitions, linings, etc. Where these are within a protected staircase they must be of non-combustible material.

Ramps. Where these afford a means of escape they must not be inclined at a gradient of more than 1 in 10.

Staircases. The width of a staircase as laid down in the building regulations must not be decreased by the introduction of projections other than handrails. Treads and risers may vary in size according to the accommodation afforded by the building and whether it is accessible to the public. Residential and private staircases are discussed in Chapter 9 and also Volume 2. Special provisions are made for open treads and risers, straight flights, landings, metal steps, winders, etc., also headroom, balustrading, etc. Details can be found in the Code of Practice of the

Greater London Council on *Means of escape in case of fire*, or in appropriate regulations. As requirements vary for different countries precise rules cannot be quoted here.

14.3 Fire detection and alarm systems

The purpose of fire detection systems is to give early warning in case of fire. These may be of the manual type, i.e. gongs, handbells, etc., or automatic, usually electric alarms. Automatic detectors are sensitive to smoke and heat and can sound off audible devices, shut off or close down plant, or activate devices of various kinds.

Bimetallic detectors consist of two strips of dissimilar metals which curve when subject to heat owing to unequal expansion. The movement causes electric contact to be made which operates the detector, Fig. 14.2.

Expanding fluid is also used. A rise in temperature causes the fluid to expand and so activate the electric switch to operate the alarm.

Fusible links which can either melt and so make or break the circuit, or which can change from a solid to a liquid state and flow to complete the circuit, are also widely used.

Laser beams are also used as detectors. Heat waves created by a rise in temperature can cause the laser to be disturbed so that the voltage output from a receiving photocell fluctuates to set off the alarm.

Smoke detectors which employ an ionization or optical system are also in use. With ionization, the radiation on air molecules causes positively charged ions to be attracted to charged plates set in a charged chamber. When airborne, combustion particles from the fire enter the chamber; the movement of ions is upset; and the change in current sets off the alarm.

Optical smoke detectors consist of a chamber with a light source and a photocell. In one type, smoke entering the chamber causes the light to scatter and by so doing affects the cell. In another, obstruction of light by means of a screen can achieve the same end. Heat detectors are generally less sensitive than smoke detectors, but are also less prone to false alarms.

Flame detectors are also used as alarms. There are two types – infrared and ultra-violet. They need more maintainence than other types and may not be as suitable in the tropics.

Portable fire equipment is in use everywhere and includes fire extinguishers, hose-reels, and sprinkler systems. Hose-reels should not be sited in staircases, if possible; the reels must be long enough to provide adequate floor coverage. Water pressure must also be sufficient and pumps provided to ensure this.

Extinguishers may be divided into five groups — water, foam, carbon dioxide, dry power, and vapourizing liquid. Water is provided through hose-reels joined directly to the mains so a constant supply is necessary.

Foam extinguishes fires in liquids by blanketing the surface and allowing the liquid to cool. Carbon dioxide extinguishes the flame more rapidly than foam and is good for dealing with escaping liquids. Dry powder and foam can leave deposits which stain.

Portable extinguishers must be supplied to locations in sufficient numbers to satisfy local fire authorities. Adequate means of vehicular access for fire-fighting purposes must be provided in all buildings.

14.4 Fire hazards on building sites

Open fires in or near timber sheds are commonly used for cooking and can be dangerous where cooking oil is used and left unattended. Temporary sheds of this nature should be at least 6 metres from other buildings.

Liquified petroleum gas for heating water and cooking is being used increasingly on sites either in camping stoves or pressure cookers with cylinders. Where cylinders are used they should be placed outside the building and supply taken through the wall by means of rigid metal piping, not rubber hose. Cylinders of acetylene, oxygen, butane, or propane should also be stored in a compound away from the site. Oxygen should be kept away from other cylinders.

Rubbish is always a hazard, especially if left in sunlight where flames are not easily seen. Bonfires should be lit well away from buildings, the direction of the wind should be noted, and they should not be left unattended, especially in the dry season. Combustible refuse, oil rags, etc. should be stored in drums and then burned; wood-shavings, off-cuts and similar waste should be cleared away frequently.

Empty petrol and paraffin cans should not be allowed to get mixed up. They should be stored in wire-mesh fenced compounds with earth floors to prevent risk from sparks or pools of oil.

When flame-producing equipment is used it is essential to see that all litter and rubbish is removed from the immediate area. Portable fire extinguishers and sand should be located in temporary workshops or work areas. Flames from blow torches and similar tools should point towards open space when they are put down momentarily. Flames and sparks should not shoot into pipes, ducts, or openings. The surrounding area should be checked for fire when the work is completed.

14.5 Security and protection

Security in buildings is governed by three considerations: danger from the neglect of building regulations; danger from theft by intruders; and accidents not covered by regulations such as health hazards, pets, carelessness, and acts of God. Theft from older buildings creates the biggest problem. Grilles and shutters have already been mentioned in Volume 2. Metal casement windows and doors, which served their purpose excellently when installed, now require mesh or bars welded on. Mortice locks, even when changed for a new tenant, are now suspect, as so many keys fit them. A stout padlock is a good deterrent but inconvenient in daytime. Long bamboo poles with hooks at the ends are used by thieves to lift small objects which can be extracted through barred, but open, windows. Leaving keys in glass doors seems to be inviting entry, though thieves prefer to work by stealth without the noise of breaking glass.

Neighbours and building occupiers can sometimes give helpful advice to newcomers. Watchmen cannot always be relied upon. Lighting the exterior at night is partly a deterrent but this would not prevent an intruder with a duplicate key. Wooden louvred window-shutters are good if well fitted and lockable, but do tend to reduce ventilation. Modern glass louvres, mosquito proofed on the outside and then covered with mesh, provide security together with fresh air. These can be adapted to the older casement windows. Around valuable property, chainlink perimeter fencing is often used, with gate padlocks for added security.

With new building, architects should consider security as a priority. Vulnerable building should, if possible, be visible from a distance and lit at night, with a reliable wide-awake watchman. The average sneak thief is unsophisticated, and has little break-in equipment, but buildings housing high-risk goods need adequate protection against the determined criminal. Internal stores or strong-rooms free from external openings are desirable. All windows and doors should be locked and securely meshed or barred. The use by thieves of telescopic aerials stolen from cars should not be overlooked; they are more convenient than bamboo poles in large towns. Insignificant back doors and windows should not be overlooked but made secure.

Panes of glass to be small enough to prevent entry must be less than $0.05\,\mathrm{m^2}$, but if this is to be exceeded then the larger they are the better. Letter boxes in entrance doors large enough to take the arm of a small child can provide access to inside locks and bolts. By leaving the front door open, a thief can make many trips without disturbing the occupants. Safes on ground floors, if not anchored or built in, are vulnerable and should be lit at night.

BUILDING SITES. Thefts from sites can be troublesome, though not usually of serious dimensions unless they are well organized. Expensive equipment such as theodilites are usually ignored. Imported cement, ironmongery, piping, timber in short lengths, head pans, machetes, tools, and equipment can all find buyers in the open market-place and petty thieving is usually rife. Portland cement is usually a good haul as it can be sold in the local market in cigarette tin measures. Good perimeter fencing and a sound watchman is the best answer.

14.6 Theft security systems

The incidence of thefts and break-ins has now reached such proportions that even small premises, such as shops, stores, and houses, are using some form of alarm system. No architect today can afford to ignore the security aspect of design, either structural or mechanical, and a range of devices is available for use in occupied as well as partially occupied premises. A detector system usually consists of a wired circuit, control centre, and signalling equipment. Modern devices can not only detect the presence of an intruder but can also indicate a fault in the system and so reduce false alarms. Care must be taken in choosing the most appropriate method of detection. The equipment must do what is required of it in its special location and not be liable to accidental or deliberate damage.

The simplest electrical device is probably the open or door-bell type, where opening a door makes or breaks a circuit to set off the alarm. Open windows for ventilation are not usually wired but are barred or meshed. Such circuits are useful in private houses or isolated premises; in large towns, however, the noise of a bell is familiar, particularly at opening and closing times of premises; they are also vulnerable to having their wires cut which renders them inoperative. Open circuits are not now installed, except in conjunction with closed circuits where the control equipment is arranged to activate the alarm if the circuit is cut.

Devices. The most common sort of device is the *linear device* which operates through micro-switches which make or break contact when doors or windows are opened. This is done by means of a spring-loaded plunger similar to that used on car doors for interior lights. They give no protection against cutting or breaking the panel or glass, however. Other types of switches are used to overcome these drawbacks.

Continuous wiring systems give protection through one or more planes as well as at given points. B.S. 4737 gives recommendations for the use of such systems. Metal foil is also used and can be securely bonded to glass or panels liable to breakage in the course of entry. Security glazing is used in conjunction with electrical systems and it also gives protection against breakage. This may be wired double-glazing, vacuum glass, or alarm glass, all of which activate alarms if broken.

Optical systems consisting of transmitters projecting a beam of light can, if the beam is broken, operate the alarm. Infra-red light is preferable to visible light for this purpose. Ultrasonic and micro-wave systems may also be used, but some tropical conditions such as mild earth tremors, wind movement, vibrations, electric fans, swinging light pendants, atmospheric changes, voltage drop, power failure, etc. can cause some systems to fail through false alarms.

The controlling equipment is used to activate the signalling system which indicates the presence of an intruder. Signalling can be either audible or silent: the former is used as a deterrent and the latter to raise a signal at a distant point. Usually the audible system is favoured as being the most direct, and, generally, the simpler the system the better. It must be capable of defeating the efforts of the intruder who attempts to disconnect or bypass alarms during the daytime, even if the potential burglar can rely on the help of an accomplice inside the building.

14.7 Safety on building sites

Up to a decade or so ago, site safety of personnel was not considered a problem, as accidents were rare; compared with Europe, they still are. In Britain and elsewhere, Health and Safety Regulations Acts have been drawn up which impose severe penalties on employers and other persons who fail to observe their provisions. The Acts place a general duty on employers to ensure the health, safety, and welfare of their employees while at work; to ensure that their activities do not endanger third parties; that the workplace, plant, and machinery do not endanger people using them; that articles and substances used in the course of work are not dangerous; that plant used at work is safe and safely installed; that employees take reasonable care not to endanger themselves; and the Acts also provide that no employee shall be required to pay for any measures taken for health or safety reasons.

Much of the above may not be strictly relevant in warm countries. Working places involving e.g. special scaffolding such as slings, boatswain chairs, suspender cradles, etc. are rare; tower blocks and tall buildings are not all that common; dangerous and noxious materials are not often handled; electrical and mechanical plant and equipment is not in extensive use, and noxious fumes which are dangerous in confined areas are

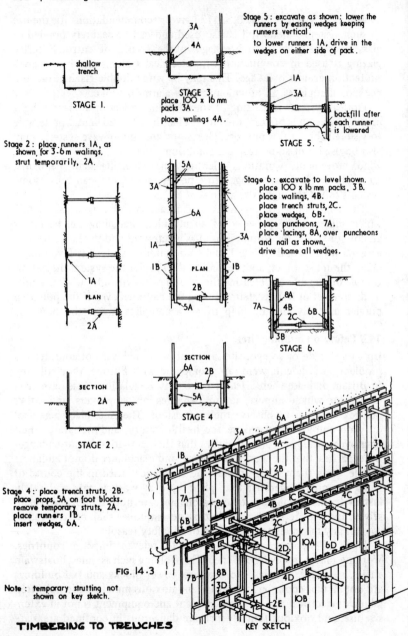

shallow trench

STAGE 1.

Stage 2: place runners 1A, as shown, for 3-6 m walings, strut temporarily, 2A.

STAGE 3.
place 100 x 16 mm packs 3A.
place walings 4A.

Stage 5: excavate as shown; lower the runners by easing wedges keeping runners vertical.

to lower runners 1A, drive in the wedges on either side of pack.

backfill after each runner is lowered

STAGE 5.

Stage 6: excavate to level shown.
place 100 x 16 mm packs, 3B.
place walings, 4B.
place trench struts, 2C.
place wedges, 6B.
place puncheons, 7A.
place lacings, 8A, over puncheons and nail as shown.
drive home all wedges.

STAGE 6.

PLAN

PLAN

SECTION

STAGE 2.

SECTION

STAGE 4

Stage 4: place trench struts, 2B.
place props, 5A, on foot blocks.
remove temporary struts, 2A.
place runners 1B.
insert wedges, 6A.

FIG. 14.3

Note: temporary strutting not shown on key sketch.

KEY SKETCH

TIMBERING TO TRENCHES

not yet a hazard. Furthermore, the site worker is not usually encumbered by protective clothing and equipment; many wear only shorts which enable them to move freely in tight situations. Falling from high places is comparatively rare and danger is often spotted quickly and averted through mobility. But as buildings become larger, and more sophisticated plant is used, so the need for site safety and construction precautions become more necessary.

Protective clothing and equipment. The time is approaching in some countries when the site worker will need to be provided with suitable clothing and equipment to protect him from dangerous conditions at work. He may have to use respirators to avoid inhaling injurious dust, for instance when working with asbestos or asbestos-based materials; wear gloves or use screens to prevent electric shock; wear goggles or other eye-protectors where certain processes are being carried out; use ear-protectors in noisy situations; wear harness or have safety nets where needed or have protection from the sun.

ACCIDENT PREVENTION. Accidents, where they occur, should be investigated and not assumed to be the victim's fault. Types and trends of accidents should be observed, whether they are the result of people falling, handling objects, using hand tools, using transport, or other causes. First-aid boxes should always be available, suitably stocked, and in charge of someone reasonably capable of giving early treatment. Proper instruction must be given to anyone connected with the following:

Machinery and plant. Driving mechanical plant and equipment; advice on machines in operation; protection against moving parts; guards, safety measures.

Transport. Hazards connected with site transport, particularly dumpers, tipping lorries, earth-moving equipment, steep excavations, tractors and trailers and riding on the drawbar.

Site tidiness. Dangerous projections, handling and stacking of materials, fire precautions, clear site access, pitfalls, etc.

Electricity. Hazards of electricity; proper use of power and tools; danger from underground cables and overhead lines; safety precautions.

Excavation. Methods of timbering; sheet piling; danger from existing structures; edges of excavations; accidental collapse. Deep trenches must be supervised by a competent person and carried out as shown by Fig. 14.3.

Access and place of work. Hazards connected with the use of ladders; faults in ladders; scaffold overloading; work on pitched and flat roofs; fragile materials; openings in walls and roof.

Cranes and lifting machines. Appreciation of hazards; accident prevention; correct sling methods; signalling to drivers; erection and use of hoists; maintenance and inspection.

Lifting gear. Slings, single and multi-leg; properties of ropes; safety hooks; eye bolts; sockets. Hazards connected with chains, ropes, lifting gear; common causes of failure such as corrosion, overloading, fatigue, lack of maintenance, storage.

Health. Hazards to health on site; dust, gas fumes; employees state of health; local and indigenous infection; contagious diseases; health problems; diet, nutrition.

Finally. It should be remembered that many employees, especially labourers and tradesmen, may be strangers to large towns or even building sites. Some may be unable to read, may never have used electricity, be working near water and unable to swim, or be unaware of most of the hazards mentioned above. It is essential that the reader should learn these for himself through site experience, in order to inform others of the need for safety at work.

Index

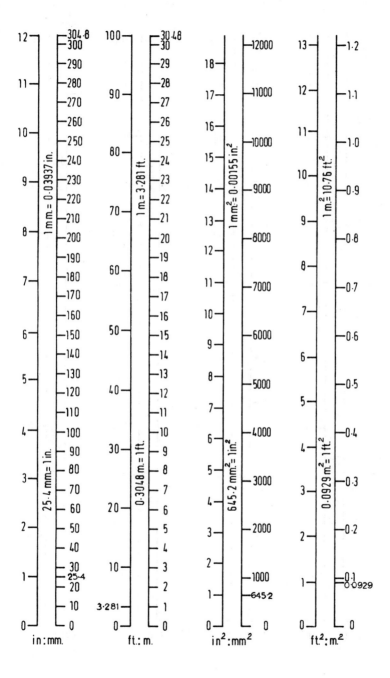

in:mm. ft.:m. in²:mm² ft.²:m.²

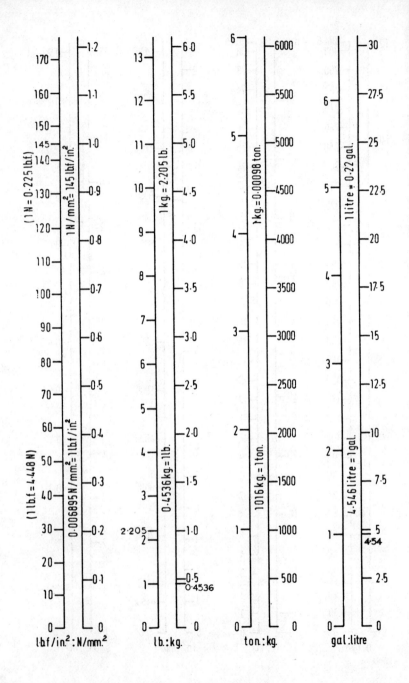

TONY
BIANCHI

-esgyrn bach-

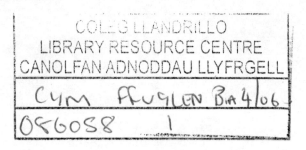
Argraffiad cyntaf: 2006

© Tony Bianchi a'r Lolfa Cyf., 2006

*Mae hawlfraint ar gynnwys y llyfr hwn ac mae'n anghyfreithlon i
atgynhyrchu unrhyw ran ohono trwy unrhyw ddull ac at unrhyw
bwrpas (ar wahân i adolygu) heb ganiatâd ysgrifenedig y
cyhoeddwyr ymlaen llaw.*

Cynllun clawr: Marc Jennings

Rhif Llyfr Rhyngwladol: 0 86243 862 4

Cyhoeddwyd, argraffwyd a rhwymwyd yng Nghymru
gan Y Lolfa Cyf., Talybont, Ceredigion SY24 5AP
e-bost ylolfa@ylolfa.com
gwefan www.ylolfa.com
ffôn (01970) 832 304
ffacs 832 782

Nodyn gan yr Awdur

Diolch i'r Academi Gymreig am yr ysgoloriaeth a brynodd amser imi i orffen y nofel hon a dechrau'r un nesaf; i Ellie Glover am y llun o'i gwaith ar dudalen 18; i ITV a London Features International am ganiatâd i atgynhyrchu'r llun ar dudalen 189; i Ruth am greu nifer o'r lluniau eraill; ac i Dafydd, gwasg Y Lolfa, am arllwys coffi, yn gelfydd iawn, dros y RICECAR, a llawer cymwynas arall.

Dychmygol yw holl gymeriadau'r nofel hon.

1 Steffan!

– STEFFAN!

Roedd Steffan wedi cyrraedd Rhif 85 yn y pentwr o becynnau. Union hanner ffordd a digon, efallai, i gyfiawnhau rhoi'r gorau iddi am heddiw. Tu ôl iddo, yn grugynnau blêr ar y llawr, roedd yr 84 cais a ddarllenodd eisoes: 54, y rhai aflwyddiannus, ar y llaw dde, dim ond 12 o rai cymeradwy ar y chwith, a rhyngddynt, 17 o geisiadau nad oedd eu tynged eto wedi'u penderfynu.

– Steffan! Wyt ti 'na?

Roedd 85 yn edrych yn addawol o ran plot. Prin oedd y nofelwyr Cymraeg a fentrai ddilyn criw o bobl ifanc nwydwyllt i draethau a chlybiau nos Ibiza. Oedd, meddyliai Steffan, roedd angen cyffyrddiad bach o haul Môr y Canoldir ar ein dychymyg gwelw, gaeafol. Gresynai, am eiliad, nad oedd wedi profi tipyn mwy o'r gwres hwnnw ei hunan ac yntau, bellach, yn rhy hen i fanteisio ar y cyfle. Os na bu'n rhy hen erioed. Ond fe wyddai'n iawn mai haul o natur wahanol a oleuai ei fyd ef ac mai hel mwg i sachau, ys dywedai ei hen athro Cymraeg, oedd cenfigennu wrth eraill. Pawb at y peth y bo. Rhyw dorheulo'n hamddenol yn y meddyliau hyn roedd Steffan pan ddaeth yn ymwybodol o weiddi uchel o'r swyddfa nesaf.

– Steffan! Mae Ioan Thomas ar y ffôn!

Llais Audrey.

– Twll ei din e!

Ie, plot addawol. Ond roedd y plot yn rhagori ar y

mynegiant. Oedd, wir. '*Syrthiodd ei lygaid ar y bronnau noeth.*' Pa fath o Gymraeg oedd hynny? Yn nrych ei feddwl, gwelai Steffan lun byw o'r arwr cegrwth yn syllu'n flysiog ar y ddawnswraig osgeiddig: syllu mor benderfynol ac mor egnïol nes bod ei ddau lygad yn saethu allan o'u tyllau a thrybowndian yn erbyn y corff lluniaidd. *Ping!*

– Ioan Thomas o'r Cynulliad, Steffan! Cynorthwy-ydd i Elin Puw!

Ond na, nid *ping* fyddai'r sŵn, chwaith. Mewn cartŵn, efallai, ond ddim yn y byd iawn. Yn wir, roedd Steffan yn amau a oedd llygad, nad oedd yn fwy na rhyw dalp o gyhyrau a nerfau, yn meddu ar y nodweddion angenrheidiol i wneud sŵn *ping* hyd yn oed pe câi ei hyrddio i fwced sinc yn hytrach na chnawd meddal. Ond o'i saethu'n hynod nerthol, efallai, fel bwled o ddryll...? Ond beth, wedyn, fyddai tynged y llygad druan? Heb sôn am y bwced. Heb sôn am y fron! A dyna lle bu Steffan yn pendroni ac yn ymboeni ynglŷn â sŵn tebygol llygad yn bwrw dawnswraig noeth ar ynys Ibiza, tra parhâi Audrey i weiddi nerth ei phen am ofynion diweddaraf y Cynulliad.

– Elin Puw, Steffan! Mae hi eisie gwbod...

– Twll ei thin hithau hefyd!

Yn y diwedd, wedi ystyried nifer fawr o synau posibl, gan gynnwys, ymhlith eraill, *ffwap*, *wff*, *poc*, *shlop*, *nghow* a *pwd*, a methu'n lân â darganfod un a fodlonai ei ddychymyg, rhoes Steffan y gorau i'w fyfyrdod a dodi Rhif 85 ymhlith y ceisiadau amhendant eu tynged. Dyna 18 o'r rheiny, felly...

– Elin Puw, Steffan! Ar ran y Gweinidog!

... 18... sydd, o'u hychwanegu at y 54 a'r 12, yn gwneud cyfanswm o 84. Ie, 84, nid 85, oherwydd yr oedd un cais na fedrai Steffan ei wthio i'r un o'r tri chategori. Dyma'r cais y bu'n rhaid iddo ei anfon yn ôl i'r Adran Oruchwylio i'w ddilysu gan fod yr ymgeisydd yn honni ei bod hi'n 139 mlwydd oed, yn frodor o blaned Romula ac yn awyddus i sgrifennu nofel yn Rihannsu, sef un o ieithoedd y blaned honno.

– Twll ei… Pwy, wedaist ti?

– Y Gweinidog, Steffan. Eisie gwbod ma fe…

Na, doedd Steffan ddim yn siŵr ai jôc sâl oedd y cais o Romula, ynte gwallgofrwydd, ond ni thalai fod yn fyrbwyll wrth ffurfio barn ar y mater, o gofio profiad cas yr haf blaenorol. Câi gyngor gan y Swyddogion Dilysu. Ni fyddai oedran nac iaith yr ymgeisydd yn peri anhawster iddyn nhw: roedd gan y Weinyddiaeth er Hyrwyddo Buddiannau Strategol y Celfyddydau bolisïau goleuedig ar gyfer meithrin gweithgarwch diwylliannol ymhlith yr henoed ac ieithoedd lleiafrifol, ac yn wir, amheuthun o beth fyddai cael cyfuniad o'r ddau mewn un cais.

– Beth wedest ti, Audrey?

– Eisie gwbod ma fe, a yw Colum Hayes yn iawn…

Na, nid yr iaith na'r oedran oedd y maen tramgwydd. Ond, pe gellid coelio ei chyfeiriad, doedd dim dwywaith amdani: roedd yr ymgeisydd hwn yn torri'r rheolau preswyl.

– Wy ar ganol y ceisiadau, Audrey…

Nid yng Nghymru roedd planed Romula, siawns? Ar y mater hwn roedd Steffan am gael arweiniad pellach.

– A ma fe eisie gwbod ar unwaith, Steffan… Ar unwaith, medde fe… Wyt ti'n gwrando, Steffan?

– Mm?

– Ma' fe eisie gwbod faint o wirionedd sy yn honiadau Colum Hayes taw dim ond tri gweinidog a chi defaid oedrannus sy'n darllen llyfrau Cwmrâg y dyddie 'ma…

Yn ei awydd i groesi'r hanner ffordd a gadael llai o

orchwyl ar gyfer trannoeth, dechreuodd Steffan fyseddu pecyn 86, a oedd yn anghyffredin o denau.

– Gad e i fi, Audrey. Gad e i fi.

Ie, byddai'n braf cael 86 o'r ffordd. Os mai dim ond er mwyn rhoi rhyw bellter rhyngddo a'r llygaid sbonclyd. Ond callach, dan yr amgylchiadau, fyddai llunio rhyw bwt bach o ymateb i'r Gweinidog, neu Elin Puw, neu Ioan Thomas, neu Colum Hayes, neu pwy bynnag oedd yn holi. Ymateb ystadegol cytbwys, cadarn.

– A ma Llio eisie gwbod pryd rwyt ti'n disgwyl bod gartre heno.

Troes Steffan at ei gyfrifiadur.

– Ar ôl y theatr, Audrey… rwy'n gorfod mynd i'r…

Ond roedd ei feddwl eisoes yn crwydro llwybrau dyrys ei gronfa ystadegau. A pharhaodd i grwydro'r llwybrau hynny am y ddwy awr nesaf, er mai rhyw saib byr ar ganol brawddeg fu'r cyfan i Steffan.

–… rwy'n gorfod mynd i'r theatr heno, Audrey…

Gwasgodd fotwm ar ei gyfrifiannell a daeth gwên o foddhad i'w wyneb.

–… a dyma'r wybodaeth i Mr Ioan Puw Colum be ti'n galw…

Pesychodd, ac yna cyhoeddi mewn llais uchel, buddugoliaethus:

– Dwedwch wrtho, Audrey… dwedwch wrtho fod Cymry Cymraeg, yn ôl yr ystadegau diweddaraf, yn profi 6,487,000 awr o bleser rhywiol mewn blwyddyn. Y cyfanswm yw hwnna, wrth gwrs, nid y ffigwr i bob un. Ha! Gwaetha'r modd, yntife, Audrey…? Mm…?

Pesychodd eto.

– Ond… OND! Gwrando ar hyn, Audrey. Mae'r ystadegau'n dangos bod Cymry Cymraeg… Ddim yr un rhai, o reidrwydd, ych chi'n deall. I'r gwrthwyneb, 'falle, o ystyried. Ond ta waeth. Maen nhw'n dangos bod Cymry Cymraeg yn treulio dros saith… Ie, SAITH!… dros SAITH MILIWN o oriau'n darllen llyfrau Cymraeg bob blwyddyn. Sy'n profi… sy'n profi… Wyt ti'n copïo hyn lawr, Audrey…? Audrey…?

Ond roedd Audrey wedi mynd tua thre awr a hanner ynghynt. Yn wir, yr oedd pawb wedi hen ymadael â'r swyddfa: pawb ond Steffan. Atseiniai ei lais trwy'r coridorau gwag.

– AUDREY!

2 Y Theatr

PAN GYRHAEDDODD Steffan y theatr yr oedd llais coeth yr
uchelseinydd eisoes yn rhoi ei rybudd olaf i'r boneddigion a'r
boneddigesau i gymryd eu seddau am fod y perfformiad ar
fin dechrau. Ymlwybrai'r dorf at y ddau ddrws a arweiniai i'r
awditoriwm. Dilynai Steffan, ychydig yn ffwndrus, a'i wynt
yn ei ddwrn. Taflai gyfarchiad brysiog at hwn a'r llall, at gyd-
weithwyr, at gyfeillion, at yr wynebau hanner cyfarwydd,
hanner dieithr hynny a welir ar achlysuron fel hyn. Byddai
Steffan bob amser yn defnyddio'i wên a'i sirioldeb parod i
dynnu sylw oddi ar yr hyn a deimlai, iddo fe, fel cwmwl o
anhrefn: cwmwl a gylchai ei ben yn barhaus megis haid o
wybed.

Aeth i'w sedd a thynnu'r Ffurflenni Monitro (**At Ddefnydd
Swyddogion y Weinyddiaeth er Hyrwyddo Buddiannau Strategol
y Celfyddydau yn Unig**) o'i boced. Byddai goleuadau'r theatr
yn cael eu diffodd ymhen ychydig eiliadau ond ni phoenai
Steffan am hynny oherwydd roedd ganddo fflachlamp bach
yn ei boced arall. Gofalai bob amser y câi sedd ar ben y rhes
gefn, fel na fyddai ei smotyn bach o oleuni'n tarfu'n ormodol
ar ei gymdogion. A hefyd, ac yn bwysicach, efallai, fel na
fyddai dieithriaid yn cael dwyn cip slei ar gyfrinachau ei waith
monitro heb yr awdurdod priodol.

Trodd at y cwestiynau ar y dudalen gyntaf.

1. Nifer yn y gynulleidfa

Anodd gweld. Tri chwarter llawn? 300…? Na. 307.

2. Croestoriad oedran

Hwn oedd y cwestiwn anodda bob tro, gan na fyddai'r ifanc,
y canol oed na hyd yn oed yr henoed fel arfer yn ymgarfanu'n
daclus. Mynnent eistedd blith draphlith, yn bennau gwyn
a phennau du a phennau brith a phennau moel, i gyd yn
gymysg yn un gybolfa o ddynoliach. Teuluoedd oedd y
gwaethaf. Roedd ffrindiau, at ei gilydd, yn fwy ystyriol o
gyfoed. Bras amcan? 23% dan 25… 46% rhwng 26 a 55…
29% dros 55. A'r 2 % a oedd yn weddill yn brawf bach o
drylwyredd yr ystadegwyr yn y swyddfa. (Roedd gan Steffan
amheuon mawr yn y cyfeiriad hwnnw.) Byddai'n rhaid aros
tan yr egwyl i gael rhyw amcan o'r amrywiaeth ethnig a'r
nifer o bobl anabl.

Diffoddwyd y goleuadau a chododd y llen. Ar ganol y
llwyfan cyrcydai ffigwr bach tenau – merch, mae'n debyg
– mewn pwll o oleuni. Plygai ei phen nes bod ei thalcen yn
pwyso ar ei phenliniau, a'i hwyneb yn gyfan gwbl o'r golwg.
Roedd ei breichiau wedi'u lapio am ei choesau. Yn ystod y
pum munud nesaf, dechreuai'r corff bach eiddil siglo yn ôl ac
ymlaen ar ei sodlau. Yn araf, araf, dadblethodd ei breichiau.
Yna, agorodd ei dwylo, y cledrau i fyny, tuag at y golau.
Yn ddirybudd, ar amrantiad, symudodd y cylch golau i ran
arall o'r llwyfan, gan adael y ffigwr mewn tywyllwch llwyr.
Gyda hynny, clywyd sŵn cyfarth a lleisiau plant yn chwarae.
Camodd menyw ganol oed, mewn ffrog flodeuog hen ffasiwn,
i'r pwll goleuni. Tawodd sŵn y plant a daeth alaw *jazz* yn ei
le, rhyw hen gân serch, a dechreuodd y fenyw ganu'n llawn
angerdd a hiraeth er bod ei breichiau'n dynn wrth ei hochrau.
Stopiodd y gerddoriaeth ar ganol llinell, ac felly hefyd y canu.
Ar ôl saib, clywyd babi'n llefain.

Trodd Steffan ei fflachlamp personol ymlaen.

Arall, mae'n debyg… Ond pa fath o arall…?

Dechreuodd y fenyw ganu eto, rhyw hwiangerdd neu'i gilydd yn ôl eu hystumiau maldodus a'i llais cwtshi-cŵ, ond daeth y gân honno i ben yn ddisymwth hefyd. Ar hynny, diffoddwyd y golau eto a llyncwyd y llwyfan gan y gwyll am funud gyfan. Pan ddychwelodd y pwll goleuni, roedd y fenyw wedi diflannu a'r ferch denau, eiddil wedi cymryd ei lle. Dechreuodd siglo drachefn, yn ôl ac ymlaen, yn ôl ac ymlaen. Siglo, ac yna, unwaith yn rhagor, ymagor fel blodyn. Ond y tro hwn, wedi troi'i dwylo, fe agorodd ei llygaid hefyd. Yna'r tywyllwch sydyn.

Sut roedd gwybod? Edrychodd Steffan o'i gwmpas. Ai hon, tybed, *oedd* y gynulleidfa arfaethedig? Onide, ble'n union *roedd* y gynulleidfa honno, fel y gallai eu holi am addasrwydd y cynhyrchiad? Ond sut y gwydden nhw, heb weld… Ac ai nhw fyddai'r rhai gorau i ddweud beth oedd yn addas a beth nad oedd…?

Ni fu egwyl, ac felly ni chafodd Steffan gyfle i gyfri rhagor o bennau na chlustfeinio wrth y bar er mwyn mesur ymateb y gynulleidfa. Aeth y ddrama ddieiriau yn ei blaen am ryw awr arall, gan gyflwyno wyth o amrywiadau ar yr un patrwm ond bod y golau, yn y diwedd, yn aros yn ei unfan, yn diffodd am ychydig eiliadau ac yna'n disgleirio drachefn, diffodd a disgleirio. Y tro hwn, ni symudai'r corff bychan. Aeth y pwll goleuni'n llai ac yn llai, fel y byddai'r teledu yn ei wneud ers llawer dydd.

Goleuwyd y theatr.

Roedd Steffan wedi gweld y diwedd yn dod ers y trydydd cwrcwd, a gweddill y gynulleidfa hefyd, fwy na thebyg. On'd oedd e wedi gwylio'r ddrama hon, neu rywbeth tebyg iddi, ganwaith o'r blaen, yn troi yn ei chylch bach o ing, yn codi gobeithion dim ond er mwyn eu chwalu? Gwaith cynnil, diaddurn, dwfn hyd yn oed: allai Steffan ddim dadlau. Ond dyheai, weithiau, am ddiweddglo llai tywyll. Am ychydig o ymgom ffraeth. Am ryw stori fach afaelgar. Am stori oedd yn ddigon dewr i ddathlu bywyd y Dyn Cyffredin, fel y byddai pobl yn arfer dweud, neu'r Fenyw Gyffredin, wrth reswm, a'i ddathlu heb gywilydd. Heb ddweud, o hyd ac o hyd, *cystal i chi roi'r ffidil yn y to, bois bach — mae hi ar ben arnoch chi ta beth wnewch chi!* Bywyd tebyg i'w fywyd e ac i fywydau'r rhan fwyaf o'r gynulleidfa hefyd, siawns. Bywyd digon cachlyd ar brydiau, fe wyddai hynny. Ac roedd rhai yn fwy cyffredin na'i gilydd, wrth gwrs. Ond bywyd na welodd ei ddechrau a'i ddiwedd yn ei fotwm bola ei hunan. Petai ganddo'r ddawn, meddyliai Steffan, byddai'n troi ei law ei hun at lunio'r fath stori. Ond, yn anffodus, nid oedd ganddo'r ddawn. Ac ar hyn o bryd, doedd ganddo mo'r amser chwaith.

5. Beth oedd ymateb y gynulleidfa i'r perfformiad?

Bu'r gynulleidfa'n cymeradwyo am funud a deg eiliad, a hynny gyda sêl gymedrol, ym marn Steffan. Dim mwy. Dim llai. A gawsant eu bodloni? Ni wyddai.

6. Ym mha fodd y cyfrannodd y cynhyrchiad at wireddu polisïau'r Weinyddiaeth er Hyrwyddo Buddiannau Strategol y Celfyddydau parthed gwella hygyrchedd ac ymestyn cyfleoedd i gymryd rhan?

Penderfynodd Steffan y byddai'n haws cwblhau'r ffurflen werthuso dros gwpanaid yn y bore pan fyddai anobaith heintus y sioe wedi cilio rywfaint. Gwrthododd beint gan un o'i ffrindiau a mynd adref ar ei union.

3 Carnforth

ROEDD STEFFAN yn byw gyda'i gymar, Llio, mewn tŷ teras bach heb fod ymhell o ganol y ddinas. Rhif 71 i chi a fi, ond **Carnforth** i rywun o'r oes o'r blaen a fu'n byw yn y tŷ hwn, mae'n debyg, rhyw alltud a geisiai gadw'n fyw adlais bach o'i gynefin pell. Neu, a barnu yn ôl arddull yr ysgrifen, rhywun a gredai ei fod wedi dod ymlaen yn y byd wrth symud i ardal fel hon. Ta waeth, ni fyddai Steffan yn arddel yr enw, rhag swnio'n ymhonnus. Ar yr un pryd, ofnai ei ddileu, a rhoi rhywbeth amgenach yn ei le, rhag tynnu mwy o sylw ato'i hun. Byddai ei ffrindiau, a oedd yn hoff o dynnu'r goes, yn ei ddefnyddio'n ddi-ffael. Trueni, meddyliai Steffan, iddo gael ei lythrennu yn y fath arddull flodeuog. Petai e'n **Carnforth** neu'n *Carnforth* hyd yn oed, efallai na fyddai pobl yn gwneud môr a mynydd o'r peth.

Roedd Llio, ar y llaw arall, yn eitha hoff o'r enw ac o'r gwaith llythrennu. Câi ryw bleser dirgel, hefyd, o'r ffaith ei fod yn mynd dan groen ei chymar. Yn gerflunydd wrth ei galwedigaeth, credai Llio mai ei chenhadaeth yn y byd oedd ei addurno, ei harddu a'i gyfoethogi, ac ymhyfrydai ym mhob ymgais i wneud hynny, ni waeth pa mor syml a di-nod y canlyniad. Câi wefr arbennig o guddio hen flychau sgidiau â phapur lliwgar a'u llenwi â'i thrysorau bach: cregyn a cherrig o'r môr, lluniau o Steffan a Llio ar eu gwyliau, lluniau o Llio a'i chyn-gariadon, lluniau o Llio'n fabi, yn groten ysgol, yn fyfyrwraig, brigau bach, dail, blodau, tocynnau trên, bws ac awyren, corciau poteli gwin. Ac yn y blaen. I Llio roedd bywyd yn fath o frethyn diderfyn, brethyn a fodolai dim ond

Ffig. 1   Enghraifft o Waith Llio

iddi hi gael gweithio ei brodwaith arno. Po fwyaf o batrymau, o blethiadau, o gyfuniadau tlws, mwya'r pleser a ddôi i'r brodiwr; neu, yn yr achos hwn, y frodwraig.

Nid brodiwr mo Steffan. Ac nid brethyn oedd bywyd iddo chwaith. Rhyw glytiau oedd realiti iddo, clytiau a ddôi ato ar hap, ar awelon o wynt, ym mhob siâp a lliw a'r rheiny gan amlaf yn glytiau digon di-raen. Pe meddai'r dychymyg, dichon y gallai fod wedi gwneud clytwaith gweddol bert ohonynt. Ond ni fedrai weld y posibiliadau mewn defnyddiau mor llwm, mor bitw. Annibendod oedd y bratiau hyn a phriod dynged annibendod oedd cael ei dacluso – neu ei luchio. Ac ystyr tacluso oedd rheoli, nid maldodi. Golygai ddosbarthu'r clytiau'n bentyrrau cymen, cant ymhob un, yn ôl siâp, lliw, natur y defnydd ac ati. Nid oedd gwnïo'n rhan o'r gorchwyl.

Serch hynny, yr oedd yr unig ran o'r tŷ y gallai Steffan gilio iddo i fod ar ei ben ei hun – y stydi fach – yn parhau'n llanast llwyr, tra byddai'r gofodau a ddaeth o dan oruchwyliaeth Llio, er gwaethaf ei hoffter o bethau, yn gymen ac yn lân. Roeddent hefyd, rhaid cyfaddef, yn llawn i'r ymylon. Roedd pob stafell dan orchudd o luniau nes bod y muriau bron yn anweladwy. Roedd cerfluniau, crociau, dysglau, potiau, canhwyllau, blodau sidan ac amryfal be'chi'n galws eraill yn harddu pob bwrdd, cwpwrdd a silff, a phan âi'r trugareddau'n rhy niferus, fe brynai Llio ragor o gelfi ar gyfer eu harddangos.

– Mae angen bord newydd, Steffan, i ddodi'r pot thus arni.

– Oes angen pot thus?

Wedi dweud hynny, roedd pob eitem wedi'i gosod gyda'r fath ofal a chwaeth, fel nad oedd y lluosi diflino'n amharu dim ar harddwch a chydbwysedd y cyfan. Hyd yn ddiweddar, beth bynnag.

Edmygai Steffan athrylith ei gymar o bell, heb fedru

cyfranogi ohono. Cawsai'i fagu i gyfri'r ceiniogau ac i weld
gwerth pethau naill ai yn eu defnyddioldeb neu yn yr arian
a geid o'u gwerthu. Ni wyddai ef, na'i fam o'i flaen, na neb
o'r teulu, ymhle y claddwyd yr un o'u perthnasau, gan na
thrafferthai neb i gofnodi bedd na gosod carreg. Nid cybydd-
dod oedd hynny, ond yn hytrach y gred mai llwch oedd llwch
ac na haeddai llwch un eiliad o'u sylw. Prin bu'r sôn am linach
a phrinnach fyth yr eiddo a drosglwyddwyd o genhedlaeth i
genhedlaeth. Cafodd hen jygiau Fictoraidd Mam-gu eu troi'n
fresych a moron a glo yn ddigon diseremoni, diolch yn fawr.
Ni fu erioed albwm lluniau teuluol; ni chadwyd llythyrau;
a phe byddid weithiau'n dychwelyd i fangre rhyw hen
ddedwyddwch, ni ddôi'r un atgofus wên i'r wyneb na'r un
pwl o hiraeth. Yr hyn a fu, a fu.

· Ac efallai mai dyna pam na wyddai Steffan beth i'w wneud
â broc môr ei fywyd. Ac yntau'n ddeugain mlwydd oed ac,
hyd yn hyn, yn ddi-blant, roedd yn fwy ymwybodol nag
erioed o'r eiliadau'n tic-tician eu hoelion i'w arch. Roedd
ei hen siacedi a'i siwmperi, ei sgidiau tyllog, y darnau o
gerddoriaeth na fyddai byth yn eu dysgu, y miloedd o lyfrau
– roedd y cwbl yn cronni o'i gwmpas fel pridd ei fedd ei hun.
Pam, felly, na fyddai'n eu taflu i'r bin sbwriel, neu'n eu rhoi
i'r siop elusen, neu'n eu gwerthu er mwyn prynu tato, yn
unol ag arferion ei deulu?

Gwyddai Steffan, wrth gwrs, na fyddai neb yn ei iawn
bwyll am brynu gweddillion truenus ei febyd, pethau fel *The
Observer's Book of Pond Life, The Observer's Book of Lichens,*
a dros ugain o deitlau eraill yn yr un gyfres, heb sôn am y
nodiadau helaeth a sgrifennodd ar y pethau hyn. Dyna ei restr
o nodweddion y widdon dŵr (*Hydracarina*), er enghraifft: y
rhai egnïol, rhwyfus, a hoffai nofio'n wyllt i bob cyfeiriad,
a'r rhai pwdwr yr oedd yn well ganddynt lolian ym mwd eu
cynefin. Dyna ei lun o bryfyn y dŵr (*Hetroptera*) yn sugno,

trwy ei big main, hylifau bywyd rhyw *Daphnia hyalina* anffodus. Dyna ffotograff a dynnodd – ar wyliau gyda'i fam a'i dad, fwy na thebyg – o fursen las gyffredin yn hela gwybed ar lan rhyw afon neu'i gilydd. A dyna restrau maith o'r holl greaduriaid mân, mân, y protosoa, yr amebâu, y coleps, y difflwgia. A'r algâu, hefyd: y diatomau, y desmidau, y pandorina a'r eudorina, ac yn y blaen.

Beth oedd diben y papurach yma? Nid creiriau oeddent. (Nid oedd genynnau Steffan yn cydnabod creiriau.) Nid oeddent, erbyn hyn, o unrhyw werth ymarferol. Yr ateb, efallai, oedd bod Steffan yn gobeithio – yn ddiarwybod iddo'i hun, o bosibl, gan ei fod yn obaith hurt – y câi hyd, ryw ddiwrnod, i ryw fan gwyn lle câi wared â'i fflwcs, fel neidr yn bwrw hen groen. Yn y cyfamser, ni fedrai fod hebddynt. A dyna un rheswm pam yr oedd yr e-bost a gyrhaeddodd eu cyfrifiadur am ddeg o'r gloch y bore Sadwrn hwn ym mis Mai o ddiddordeb i'r ddau.

– Steffan, mae ateb wedi dod oddi wrth Ebbway's.

– Mm? Eb pwy?

– Mae Ebbway's, yr estate agent, wedi hala manylion tai i ni.

Edrychodd Steffan yn fyfyrgar tua'r nenfwd.

– Fallai ei fod e'n llygriad o Ebwy...?

– Sa i'n gwbod, Steffan. 'Drycha ar y tai ma!

– 'Swn i ddim yn credu 'ny, rywsut. Fuodd Ebwy erioed yn enw personol ar neb, am wn i...

– Mae pum tŷ fan hyn. Drycha. Mae pob un â thair stafell wely... Mae gardd fawr 'da hwn... Mae nant fach yn rhedeg wrth ochor hwn... A drycha, Steffan... dyna'r olygfa o'r drws ffrynt!

– Ydyn nhw'n swnllyd?

– Beth?

– Ydyn nhw'n llefydd swnllyd?

– Wel na'dyn, glei… Gofynnon ni am lefydd tawel…

– 'Smo nant yn dawel…

– Gei di ddim un man sy'n hollol, hollol dawel, cariad…

–… a'r awyrennau…

– Pa awyrennau…?

– Mae'r llefydd tawel honedig 'ma a'u golygfeydd ffantastig a'u gerddi pert a'u nentydd bach yn ymdroelli drwyddyn nhw… maen nhw i gyd ar flight paths y Jymbos 'na, yn mynd i America a Canada a Brasil a phobman… heb sôn am y jets cythreulig 'na sy'n hedfan yn isel bob tro mae'r haul yn dod mas, ac yn siglo'r ddaear gyfan…

– Je, ie, rwy'n gwbod, Steffan bach… Cawn ni holi ambwyti 'ny mas o law. Awn ni ddim i unman sy ddim yn 'siwto ni'n dau.

Ond ni fu holi pellach y penwythnos hwnnw gan na chafwyd ateb yn swyddfa Ebbway's.

4 Y Llwybr Papur a'r Hipo

ROEDD YR HOLL SIARAD am brynu tŷ wedi cynhyrfu Steffan. Ie, fe chwenychai fuchedd amgenach, ond nid oedd mor obeithiol â'i gymar y gallent ei ddarganfod rhwng pedair wal. Yr oedd dychwelyd i'w waith fore dydd Llun felly'n cynnig rhyw gysur iddo, rhyw encilfa oddi wrth yr angen i feddwl drosto'i hun.

Er mwyn lleddfu ar undonedd ei daith feunyddiol i'r swyddfa, byddai Steffan yn cymryd llwybr gwahanol bob dydd. Wel, nid bob dydd, wrth reswm. Nid oedd ond hyn a hyn o amrywiadau'n bosibl, wedi'r cyfan, hyd yn oed yn y rhan hon o'r ddinas, lle roedd y terasau a'r heolydd a'r lonydd bach rhyngddynt yn ymdebygu i wâl cwningod. Na, cymerai Steffan drywydd gwahanol bob dydd am gyfnod o wyth niwrnod. Gan mai am bum niwrnod yn unig y byddai'n gweithio mewn wythnos gyffredin, golygai hynny na fyddai byth yn mynd yr un ffordd ar ddau ddydd Llun canlynol, na dau ddydd Mawrth, ac yn y blaen. Byddai ei ail gylch wyth niwrnod yn dechrau ar ddydd Iau, y trydydd ar ddydd Mawrth, y pedwerydd ar ddydd Gwener a'r pumed ar ddydd Mercher, cyn ailafael unwaith eto yn y dydd Llun y bydd y rhan fwyaf ohonom yn ei ystyried yn fan cychwyn naturiol i'r wythnos.

Ond er mwyn lliniaru ychydig ar undonedd y patrwm hwnnw hefyd, ni fyddai Steffan yn dilyn yr un llwybr adref ag a ddilynai wrth fynd i'r gwaith yn y bore. Nid ar hap y dewisai'r llwybr hwnnw chwaith. Ar ddydd Llun y cylch cyntaf o wyth niwrnod, dychwelai adref gan ddilyn y llwybr

y byddai'n ei gymryd y dydd Iau canlynol (sef, tri diwrnod
o wahaniaeth) i fynd i'r gwaith. Ddydd Mawrth, cymerai
lwybr dydd Llun, ac yn y blaen. Ond ar ddiwrnod cyntaf yr
ail gylch, âi adref ar lwybr boreol dydd Mercher (sef, pedwar
diwrnod o wahaniaeth). Erbyn Cylch 6 yr oedd yn rhaid iddo
dderbyn yr anochel a mynd a dod gan ddilyn yr un strydoedd,
coedlannau a therasau. Trwy oddef yr anghyfleuster bach
hwnnw, gallai ddiogelu'r system. Yn ddiweddar, roedd Steffan
wedi llunio tabl o'r amrywiadau hyn, er mwyn peidio â mynd
ar gyfeiliorn.

	CYLCH 1	CYLCH 2	CYLCH 3	CYLCH 4	CYLCH 5	CYLCH 6
1 llwybrau	Llun 1/4	Iau 1/5	Mawrth 1/6	Merch. 1/7	Llun 1/8	Iau 8/8
2 llwybrau	Mawrth 2/5	Gwener 2/6	Merch. 2/7	Iau 2/8	Mawrth 2/1	Gwener 1/1
3 llwybrau	Merch. 3/6	Llun 3/7	Iau 3/8	Gwener 3/1	Merch. 3/2	Llun 2/2
4 llwybrau	Iau 4/7	Mawrth 4/8	Gwener 4/1	Llun 4/2	Iau 4/3	Mawrth 3/3
5 llwybrau	Gwener 5/8	Merch. 5/1	Llun 5/2	Mawrth 5/3	Gwener 5/4	Merch. 4/4
6 llwybrau	Llun 6/1	Iau 6/2	Mawrth 6/3	Merch. 6/4	Llun 6/5	Iau 5/5
7 llwybrau	Mawrth 7/2	Gwener 7/3	Merch. 7/4	Iau 7/5	Mawrth 7/6	Gwener 6/6
8 llwybrau	Merch. 8/3	Llun 8/4	Mawrth 8/5	Gwener 8/6	Merch. 8/7	Llun 7/7

Ffig. 2   Cylch y Llwybrau

Ond beth, ddywedwch chi, am wyliau, anhwylder a'r troeon
hynny pan fyddai Steffan yn gorfod mynd i rywle heblaw'r
swyddfa, efallai ar y trên neu'r bws? Sut y deuai i ben wedyn
â'r difrod a wneid i'w drefniadau a'u patrymau perffaith?
Y gwir amdani yw nad oedd Steffan erioed wedi datrys y

broblem honno'n foddhaol a dyna pam y teimlai weithiau fod anhrefn yn cael y gorau arno beth bynnag a wnâi.

Ond yn fwy na hynny, perai'r siart ei hun loes meddwl i Steffan, cymaint o loes meddwl, yn wir, fel ei fod bellach yn dechrau difaru ei dyfeisio. Tybiai y byddai, mewn system mor rheolaidd â hon, ryw gymesuredd, rhyw resymeg cadarn, rhyw ddeddf fathemategol yn llechu ynddi'n rhywle. Ni wyddai ba ddeddf, gan nad oedd yn fathemategydd. Ond roedd y siart, ysywaeth, hyd y gallai weld, yn syrthio'n fyr iawn o'r hyn y dylai deddf fod. Roedd hi'n wir bod pob cyfuniad o lwybrau yng Nghylch Un yn cael ei ailadrodd yng Nghylch Tri, gan greu patrwm digon pert. Ond wedyn, roedd pob cyfuniad yng Nghylch 2 yn ymddangos ddwywaith o fewn yr union un cylch hwnnw. Ac am Gylch 4, nid oedd dim un o gyfuniadau'r cylch hwn i'w weld yn y cylchoedd eraill yr oedd wedi'u cynnwys yn y siart hyd yn hyn. A ddeuai'r patrwm yn glir petai'n ymestyn y siart? Ond am faint? Dyna'r benbleth a oedd yn bygwth mynd yn drech nag ef.

Roedd gwaith, ar y llaw arall, yn ymgorfforiad o'r drefn y dyheai Steffan amdani. **Gwyliwch y Llwybr Papur!!** oedd arwyddair y Weinyddiaeth, a thrwy flynyddoedd ei wasanaeth yno fe ddaeth yn ganllaw i Steffan yntau. Canys wrth ufuddhau i'r gorchymyn hwnnw, fel y dywedasai pennaeth y sefydliad ar ddiwrnod cyntaf Steffan yn ei swydd, gellid ymgyrraedd at bob daioni arall, ac yn y pen draw efallai, at y cyflwr godidocaf oll, sef yr hyn a elwid **Cydymdreiddiad Egwyddor ac Ymarfer**. Ychydig o weision y Weinyddiaeth a brofodd y cyflwr hwnnw yn ei lawnder. Yr oedd y rhan fwyaf, a Steffan yn eu plith, yn dal i ymrafael â phum rheol sylfaenol y Weinyddiaeth a ddatgenid, mewn llythrennau breision, dwyieithog ar dudalen flaen ei ddisgrifiad swydd:

Be accessible!	Blyddwc hygyrch!
Be transparent!	Blyddwc trylhoyw!
Be accountable!	Blyddwc at ebol!
Be local!	Blyddwc lol!
Be strategic!	Blyddwc strateglol!

Cafodd fraw pan welodd Steffan y geiriau hyn am y tro cyntaf.
A pha syndod? Doedd yr un ohonynt i'w weld yn amlwg
iawn ym mreuddwydion y Bardd Cwsg, na myrtwydd Ann
Griffiths druan, na chymylau haf Waldo, na hewlydd agored
Jack Kerouac, na llygaid Baby Blue Bob Dylan, na neb o'r
cwmwl tystion a fu'n gwmni iddo trwy flynyddoedd diniwed
ei flodeuo. Ac nid ar chwarae bach mae rhywun a fagwyd ar
laeth enwyn a chawl maip yn gwyro ei fol at ei stecen gyntaf.

Cafodd fraw, hefyd, pan roed iddo ei dasg gyntaf erioed
fel Swyddog Monitro dan Hyfforddiant. Cofiai bob gair o'r
cyfarwyddiadau manwl a thaer a gafodd gan y pennaeth ar
y pryd, Mr Homer Sincyn. Dyn diwylliedig odiaeth oedd
hwnnw, dyn a fu'n dipyn o fardd yn ei ddydd ac a lwyddodd
i gymhwyso ei ddiwylliant dwfn at ofynion yr oes. Ni welir
ei debyg byth eto. Crwydrai meddwl Steffan yn ôl yn aml at
y trobwynt tyngedfennol hwnnw, pan gefnodd ar arferion
plentyndod a thyfu'n ddyn. Cofiai'n arbennig y cyngor doeth
a gafodd ynglŷn â thrin a thrafod y beirdd anystywallt hynny a
oedd yr adeg honno'n cyniwair trwy'r wlad fel pla o lygod. Yr
oedd Steffan – ac fe ddôi pwl o gywilydd drosto wrth feddwl
am y peth – yn ddall i beryglon y giwed hon ac yn wir, yn
ei ddiniweidrwydd, byddai'n rhaid cydnabod ei fod ef ei hun
yn glynu o hyd wrth rai o hen ofergoelion a defodau'r rafins,

megis 'Darllen y Testun', 'Meistroli Crefft' a 'Gochel Rhag y Beiau Gwaharddedig'.

– Ond beth, Mr Sincyn bach... Beth, wedyn, am y beirdd cyhoeddus mae cymaint o sôn amdanynt?

Ffyrnigodd y pennaeth.

– O ba fudd, hen gywydd gau, a hidlo mil o odlau? *Ffurflenni*, nid cerddi caeth, yw gwarant ein rhagoriaeth!

Ac eto, ar yr un pryd, y tu ôl i'r ffyrnigrwydd, synhwyrai Steffan fod yno ryw anniddigrwydd, rhyw rwystredigaeth, rhyw ymrafael mewnol yn ymystwyrian. Dwysaodd yr anniddigrwydd hwnnw wrth i Mr Sincyn ddisgrifio'r dasg yr oedd wedi ei neilltuo ar gyfer ei swyddog newydd.

– Rhaid, meddai'r Prif Weithredwr, cael ffurflen i bob llenor ac i'w gorpws, gadw sgôr. Rhaid mesur pob awdur byw, bob un, yn ŵr a benyw'n ddiwahân – yn hardd a hyll, yn barchus neu'n slob erchyll, yn dew, yn denau, neu'n dal. Begio wyf...

Ac yn y fan yma, credai Steffan iddo weld deigryn bach yn gwlychu amrant ei fos.

– *Begio* wyf, â phob gofal, nodwch hyd bob troed a choes y cyfan o'n beirdd cyfoes. A hefyd, Steffan, cofiwch werthuso, yn ei dro, drwch coes ganol gŵr *Lol* ei hun... a'i hyd... '

– Be? Hyd ei bidyn...?!

Do, cafodd fraw. Beth ddywedai'r hen ddihareb? 'Nid hysbys y dengys dyn o ba hyd y bo'i bidyn'. Ond doedd dim angen poeni. Trwy ddyfalbarhad, trwy ymroddiad diwyro i achos tryloywder ac atebolrwydd a thrwy ddefnydd cyfrwys o gymorthdaliadau, fe lwyddodd Steffan a'i gyd-weithwyr, ymhen rhyw gwta ddegawd, i greu cronfa gynhwysfawr o fesuriadau personol pob bardd, pob nofelydd, pob ysgrifwr, pob dramodydd, pob Meuryn a phob meicro-lenor, heb sôn am gyhoeddwyr a golygyddion a cherddorion a chantorion ac arlunwyr a dylunwyr a cherflunwyr, a hyd yn oed (am resymau diogelwch) y tocynwyr yng Nghanolfan y Mileniwm. Cronfa wych, cronfa ysblennydd. A chronfa a allai hawlio'i lle anrhydeddus yn y Gronfa Fawr ei hun a fu'n gonglfaen i'r Weinyddiaeth gyfan.

Ac eto, er ei gogoniant, nid y Gronfa oedd uchafbwynt na phenllanw gorchestion y sefydliad. Oherwydd, er mwyn deall a dadansoddi cronfa mor fawr ac mor fanwl, a'i defnyddio wedyn er lles strategol y celfyddydau, yr oedd angen system a ragorai ar gymhlethdod rhyfeddol hyd yn oed y Gronfa Fawr honno. Yr enw ar y system hon oedd *Y Trigain Cam Namyn Un Tuag At Lwyr Effeithiolrwydd Gweinyddol*. Rhain oedd yr arwyddion, y goleuadau, yr atalfeydd cyflymder, y llinellau gwynion a'r llygaid cathod a gadwai'r gwas ffyddlon ar y ffordd gywir tuag at y nod. Crynhowyd y camau hyn ar lifsiart a ymdebygai, ar yr olwg gyntaf, i ryw batrwm Celtaidd, cordeddog, hynod gymhleth a chain. Mesurai'r siart bedair troedfedd o hyd a dwy droedfedd a hanner o led, ac roedd copi ohoni'n crogi mewn man cyfleus ym mhob swyddfa fel y gallai'r gweision gyfeirio ati yn ôl yr angen. O bellter, ymddangosai fel hyn:

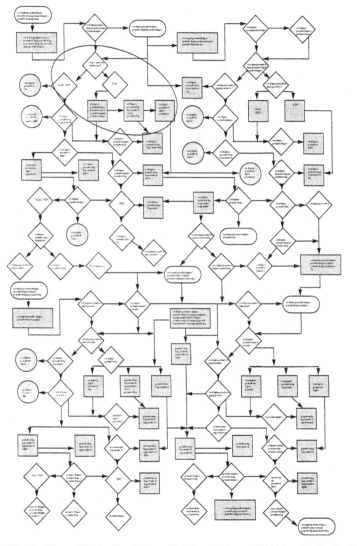

Ffig. 3   Siart y Trigain Cam Namyn Un Tuag At Lwyr Effeithiolrwydd Gweinyddol

Ond o graffu arni, gellid dirnad, y tu mewn i bob dolen yn y gadwyn, destun mewn print mân, mân, fel hyn:

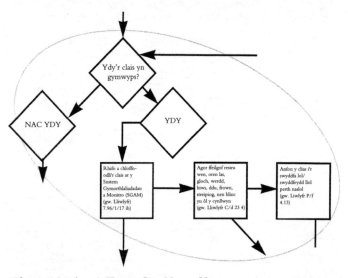

Ffig. 4   Manylyn o'r Trigain Cam Namyn Un

Ac o ddarllen y testunau hyn am rai munudau (gyda chymorth chwyddwydr) fe welid nad cyfarwyddiadau na rheolau unigol mohonynt. Gwelid, yn hytrach, eu bod nhw'n ymgysylltu â'i gilydd, bob un, ac yn ymgysylltu mewn cydberthynas berffaith o achos ac effaith, gofynion ac amodau, penderfyniadau a gohiriadau, ystyried, ail-ystyried a thrydydd-ystyried, ymgynghori a chadarnhau, gwerthuso a barnu, derbyn a dychwelyd, a'r cyfan yn sicrhau bod y nifer mwyaf posibl o bobl ac o brosesau gweinyddol yn cael eu defnyddio wrth ymdrin â phob cais am gymorth. Nid heb bwrpas, wrth reswm: trwy'r fath gydberthynas gyflawn, fe grëwyd rhwyd mor dynn, a'i rhwyllau mor fân, fel na allai nemor un gronyn o dywod siawns lithro trwyddi. A dyma'r Llwybr Papur y byddai Steffan a'i gydweithwyr yn ei ddilyn, fel rhyw Hansel a Gretel yr oes fodern, er mwyn sicrhau tryloywder ac atebolrwydd ym mhob un o'u gweithrediadau.

Braslun yn unig oedd hyn, wrth gwrs, o'r prosesau gweinyddol yr oedd yn rhaid i Steffan a'i gydweithwyr eu meistroli. Er enghraifft, petaem yn chwyddo Cam 13 o'r Trigain Cam Namyn Un, gwelem gyfeiriad at Beirianwaith Asesu na welid dim manylu arno yn y fan honno. Ond fe wyddai'r swyddog yn yr adran berthnasol (gan gynnwys Steffan ei hun) fod i'r Peirianwaith hwnnw bum llinyn mesur ar hugain a bod i bob un o'r rheiny ei isadrannau niferus. Cedwid manylion am y rhain, ac am lawer o bethau eraill, mewn Llawlyfr Asesu pwrpasol, ynghyd ag atodiadau, cywiriadau a diweddariadau a ychwanegwyd ato o dro i dro, yn ôl yr angen.

Gwelid hefyd, o dan Gam 38, sôn am Drefn Ôl-werthuso na wyddai Steffan fawr ddim amdano, dim ond ei fod yn cynnwys wyth math o ddangosyddion cyflawni a bod disgwyl iddo drosglwyddo i'r adran a reolai'r peirianwaith hwnnw, ar y dydd Gwener cyntaf yn y mis, atebion i gwestiynau ynglŷn â phob un o'r dangosyddion hynny a hefyd ynglŷn â'r berthynas rhyngddynt. Fe dybiai'r lleygwr, efallai, mai gorchwyl digon syml fyddai hynny, rhyw fater o adio a rhannu a lluosi, er mwyn profi, dyweder, bod 768 o fenywod rhwng 32 a 45 wedi mynychu gweithdai Tango ym Mhen Llŷn ym mis Tachwedd, neu fod timau talwrn y canolbarth, ar gyfartaledd, yn cynhyrchu 23.77% y pen yn llai o linellau croes o gyswllt na thimau'r gogledd. Buan y câi ei ddadrithio. Nid digon oedd nodi bod hyn a hyn o englynion wedi'u cyfansoddi (a'u tafoli yn ôl graddfeydd ansawdd 1-5) gan hwn a hwn neu hon a hon (a'u mesur hwythau yn ôl graddfeydd oedran, rhyw, tueddfryd rhywiol, anabledd, hunaniaeth ethnig, a

difreintiedigrwydd): rhaid, hefyd, mesur poblogrwydd y
gwaith hwnnw yn ôl ei dreiddiad (trwy werthiant neu drwy
ddarlleniad) i'r gynulleidfa arfaethedig. Rhaid eu gosod
yng nghyd-destun Amcanion Strategol y Weinyddiaeth ar
y pryd. Rhaid ei gysoni â pholisïau cyrff llywodraethol, yn
genedlaethol ac yn lleol. Ac yn y blaen.

'Nid ar hap y daw'r hipo,' meddai'r bardd, gan dybio, yn
ei ddiniweidrwydd, mai dyna ddiwedd ar y mater. Ond na:
roedd gan yr hipo hwnnw ran bwysicach i'w chwarae ym
mywyd ei genedl nag a dybiai'r bardd erioed. A grwydrai'r
hipo i Flaenau Gwent, er enghraifft? Ai hipo hoyw ydoedd?
A oedd lledaenu gwybodaeth am yr hipo (yn hytrach na'r
rheino, dyweder) yn gydnaws ag amcanion y Llywodraeth?

> Nid ar hap y daw'r hipo i nofio
> bob haf yn y Bermo…

A oedd y Bermo, tybed, wedi'i chofnodi'n dref ddifreintiedig?
Neu, os nad oedd, a ddylid pwyso ar yr awdur i newid y
lleoliad i Ynys-feuno er mwyn dangos ymlyniad wrth y
cymoedd? Na oedd yr ateb: byddai gormod o sillafau wedyn.
Llanwynno, te? Iawn, o ran cywirdeb. Ond sut y gallai'r hipo
nofio yn Llanwynno?

> Nid ar hap y daw'r hipo i nofio
> bob haf yn y Bermo:
> yn y llaid mae'n cofleidio
> anwylyd ei wynfyd o.

Ie, fe allai fod yn hoyw… neu'n ddeurywiol, o leiaf. Ond
'gwynfyd'? On'd oedd hynny'n delfrydu gormod? Ac yn
waeth na dim, 'ei wynfyd o'! Gog oedd yr hipo hwn! 'Wnâi
Llanwynno byth mo'r tro. Beth am Landudno? Go brin.
Penmachno? Mm.

A dyna'r math o benbleth a wynebai Steffan wrth fynd i'r

afael â'i orchwyl misol. Byddai'n gofyn iddo'i hun, weithiau, ai enghraifft, efallai, o gydymdreiddiad egwyddor ac ymarfer oedd y gwead cymhleth hwn o bwyso a mesur, o gymharu a diwygio? Ai rhoi'r hipo yn ei gyd-destun ystadegol-gymdeithasegol-wleidyddol oedd yr allwedd a agorai'r drws i'r neuadd ysblennydd honno? Ac ar adegau felly, gresynai Steffan nad oedd yn gweithio yn yr Adran Ôl-werthuso er mwyn dod i ddealltwriaeth lawnach o'r pethau hyn. Rhyw ddydd, efallai.

Ar yr adegau hynny, hefyd, ac er gwaethaf ei amheuon greddfol ynglŷn â phatrymau, fe gâi Steffan ei hun, weithiau, yn syllu ar rwyd y dangosyddion. Nid ar y geiriau y byddai'n syllu, nac ar y prosesau, nac ar y llinynnau mesur eu hunain. Syllai yn unig ar y patrwm a wnaed gan y pethau hyn.

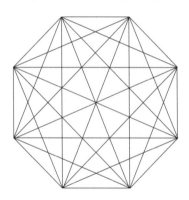

Ffig. 5   Rhwyd y Dangosyddion Cyflawni

Ac o fyfyrio uwchben y patrwm hwn yn ddigon hir, byddai meddwl Steffan yn crwydro'n bell o fyd yr hipo a'r Bermo a manion cyffelyb. Byddai'n troi, yn hytrach, at gwestiynau mwy dyrys, a mwy difyr, fel: *a oes gan y trionglau o fewn y cylch, a'r siapau eraill (siapau na wyddai mo'u henwau), ryw nodweddion arbennig?* Neu: *a yw'r ffaith bod y patrwm yn creu 28 o groes gysylltiadau yn bwysig?* Neu: *ai yn yr onglau sy'n cuddio y tu mewn i'r siapau y ceir yr allwedd i ddirgelion y cylch?* Rhyw ddydd. Rhyw ddydd. Yn y cyfamser, byddai'n rhaid rhygnu ymlaen ym myd ffeithiau a ffigurau.

5 Ymgeisydd Rhif 86

GAN AMLAF, byddai ffrwyno ffeithiau a gwastrodi ffigurau'n ddigon i fodloni Steffan, yn enwedig ar fore dydd Llun. Roedd ganddo 86 o geisiadau ar ôl i'w didoli, gan awduron gobeithiol o bob siâp a llun, ac roedd yn benderfynol o'u rhifo a'u cofnodi a'u dilysu a'u chwynnu a'u dodi yn y pentyrrau cywir cyn i'r diwrnod hercian i'w derfyn. Er na fwynhâi'r gorchwyl ei hun, câi ryw foddhad, dros dro, o gael pawb yn dwt yn eu corlannau priodol.

Pecyn 86. Amlen, nid pecyn, mewn gwirionedd. Yr un tenau. Y teneuaf o'r cyfan, yn ôl pob golwg. Ond nid, o reidrwydd, yr un salaf. A dweud y gwir, ym mhrofiad Steffan, po leiaf y dalent a feddai'r ymgeisydd mwyaf i gyd o ddeunydd a anfonai i ategu ei gais. Tristâi Steffan o weld yr un nofelau anghyhoeddedig – eu tudalennau'n melynu a breuo ac ymddatod o flwyddyn i flwyddyn – yn dychwelyd unwaith yn rhagor am eu dedfryd. Yr hunangofiannau hefyd, a'u lluniau truenus o Smot y ci, Sami'r byji a holl ffrwcs rhyw orffennol monocrom na ddeuai byth yn ôl. Ni chaent hyd yn oed rith o atgyfodiad yn nychymyg y darllenydd. Ni cheid darllenydd, nid y tu hwnt i furiau'r Weinyddiaeth, beth bynnag. Y rheiny a gyrhaeddodd gyntaf, yn archaeolegol o faith, i chwilio am rywun i gloddio'u cyfoeth cudd.

Roedd yn amlwg na pherthynai 86 i'r categori hwn. Agorodd Steffan yr amlen a thynnu allan ei chynnwys. Ffurflen gais. Ffurflen gais, a dyna i gyd. Hyd y gallai Steffan weld, doedd 86, boed yn fwriadol neu drwy amryfusedd, ddim wedi anfon yr un tamaid o ddeunydd ategol. Edrychodd

eto yn yr amlen. Dim. Troes yr amlen â'i phen i waered.
Dim. A'i siglo. Dim byd. Daliodd hi i fyny yn erbyn y
golau, rhag ofn bod rhywbeth bach wedi mynd yn sownd y
tu mewn. Dim. Gwthiodd ei law i'r gwaelod ac ymestyn ei
fysedd, i'r dde ac yna i'r chwith. Dim oll. Dim llythyr. Dim
llun o Sami'r byji. Dim.

Nawr, roedd Steffan yn ddigon hirben i wybod dau beth:
yn gyntaf, fel y gwelsom yn barod, nad da popeth maith;
ond hefyd nad syml popeth byr. Dysgodd y wers honno'r haf
blaenorol. Ac oherwydd hynny fe aeth ati i ddarllen y ffurflen
gais â mwy na'i ofal arferol.

Enw'r ymgeisydd	Kate O'Brien
Cyfeiriad	Cwm Clydach
	(A wnaiff cyfeiriad e-bost y tro?)

Wel, na wnaiff, glei!

| **Canolwyr** Ana de Mendoza | Jorge Coimbra |

A dim cyfeiriad i'r un ohonynt!

Amlinelliad o'r gwaith arfaethedig Nofel gofiannol
yn seiliedig ar g(t) = f(18-t) neu

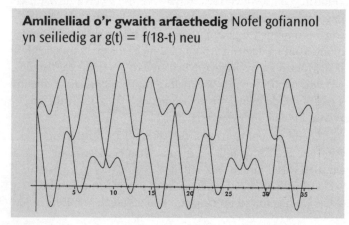

Anadlodd Steffan yn ddwfn. Doedd yr enwau'n golygu dim iddo. Ac am yr amlinelliad, rhaid taw jôc oedd hwnnw. Edrychodd yn yr amlen unwaith yn rhagor, a chael yr un gwacter ag o'r blaen. Ie, jôc. Ond pa fath o jôc? Heb wybod pam – efallai am nad oedd dim o'i synhwyrau eraill yn cynnig atebion i'w gwestiynau – cododd y papur i'w ffroenau a'i wynto. Oedd, roedd aroglau rhyw bersawr neu'i gilydd arno. Rhywbeth tebyg i fricyll, efallai? Os persawr hefyd. Ynte, tybed ai dim ond olion sebon ar ei fysedd oedd e? Plethodd Steffan ei ddwylo y tu ôl i'w gefn a phlygu at y bwrdd. *Sniff*. Rhyw awgrym o fanila, o bosib? *Sniff… Sniff*. A thwts bach o sinamon? Oedd, roedd rhyw frath bach o sinamon i'w glywed yn bendant yng nghwt y melyster. Ond faint callach oedd e o wybod hynny? Jôc sâl oedd hi beth bynnag. Doedd *e* ddim yn chwerthin, nac oedd wir. A oedd *hi*'n chwerthin tybed, draw yn ei chartref anhysbys yng Nghwm Clydach, wrth feddwl amdano fe'n ffwndro a drysu? A chwysu. Heb ddeall pam, yr oedd Steffan wedi dechrau chwysu.

A doedd dim un sampl o'i gwaith, chwaith!

Yna gwnaeth Steffan yr hyn y byddai'n ei wneud ar bob achlysur tebyg, pan fyddai'r byd yn sgwâr a'i ymennydd yn grwn, sef mynd â'i broblem drws nesaf at Audrey. Roedd Audrey yn un o'r 25 a gyflogwyd i gasglu a didoli'r wybodaeth fonitro. Hi oedd yr hynaf ohonynt, a'i chof yn mynd yn ôl i oes y teipiaduron a'r Chwe Barwn Budr a arferai reoli'r Weinyddiaeth yn yr hen hen amser, yn ôl chwedloniaeth y sefydliad, cyn bod sôn am Lwybr Papur na Strategaeth na'r Trigain Cam Namyn Un a'r bendithion lu a ddaeth yn eu sgil. Camodd Steffan i mewn i'w swyddfa hi, â'i osgo petrusgar, ymbilgar arferol.

– Audrey… ydy'r enwau 'ma yn golygu rhywbeth i ti?

Rhoes y ffurflen gais ar y ddesg o'i blaen.

Enwyd Audrey ar ôl yr actores enwog, Audrey Hepburn, a chadwai lun ohoni ar ei desg. Yr esboniad am hyn, fel yr oedd yn hoff o ddweud wrth ymwelwyr a ddeuai i'r Weinyddiaeth, ac yn enwedig rhai swil, diniwed, gwridog, oedd iddi gael ei chenhedlu, ymhell bell yn ôl yn oes Twm Siôn Cati a *fflics* du a gwyn, y tu ôl i sinema Llanymddyfri, lle'r oedd ei rhieni newydd weld y ffilm *Roman Holiday*. Ni wyddai neb ond Audrey ei hun ai cellwair roedd hi. Un gellweirus, ddireidus oedd hi, yn sicr. Ac nid oedd neb yn y Weinyddiaeth gyfan a feginai fflamau'r direidi hwnnw yn fwy na Steffan. Byddai nerfusrwydd cynhenid ei bos tybiedig bob amser yn ennyn yn Audrey ysfa anorchfygol i odro o'r sefyllfa ddwys bob diferyn posib o ddifyrrwch a thynnu coes, a hynny ar adegau pan ddylai, yn ôl y safonau proffesiynol arferol, fod wedi ymateb yn chwimwth o effeithiol. A pho fwyaf y brys a'r gofid, mwyaf i gyd fyddai'r gwatwar a'r dili-dalio.

– Wel Steffan bach! Beth sy 'da ti fan hyn?

Yn y llun ar ddesg Audrey, gwelid yr actores enwog yn cogio syfrdandod. Câi Audrey bleser arbennig o ddynwared y llun hwnnw o bryd i'w gilydd, er mai tenau iawn oedd y tebygrwydd rhyngddi hi a'i harwres. A dyna a wnaeth wrth ymateb i banic diweddaraf Steffan. Cododd ei llaw at ei cheg a gwneud soseri â'i llygaid.

– W! Steffan! Watsia di! Ma' hon mas o'i chlocs, w!

Ac ymhen munud neu ddwy roedd y merched eraill yn y stafell i gyd yn porthi fel corws Groegaidd.

– A smo fe wedi dod dros y sioc ofnatw gas e llynedd 'to, naddo…?

– O, a be ddigwyddws llynedd?

A dyma sut y cafodd y newydd-ddyfodiad, Anna-Marie,

wybod am y *ffatwah* a fwriwyd ar y Weinyddiaeth yr haf
blaenorol oherwydd gweinyddu byrbwyll Steffan. Tynnodd
Audrey'r ffeil o'r drâr, a gwên ddisgwylgar ar ei hwyneb.
Cochodd Steffan wrth feddwl am y gwatwar didrugaredd a
fyddai'n dod i'w ran yn ystod y munudau nesaf.

Ffig. 6   Audrey

6 Cofio'r Eneiniog

RHYW BWT PITW O GAIS a gafwyd bryd hynny hefyd.

Annwyl Syr

Erfyniaf arnoch chwi i ufuddhau i ewyllys DUW
i'r holl genhedloedd trwy estyn i mi gymorth
ariannol fel y gallwyf gwblhau'r 'llyfr
bychan' y soniwyd amdano yn Natguddiad Ioan
(10:10,11): 'Ac mi a gymerais y llyfr bychan
o law'r angel, ac a'i bwyteais ef; ac yr oedd
efe yn fy ngenau megis mêl yn felys: ac wedi
imi ei fwyta ef, fy mol a aeth yn chwerw. Ac
efe a ddywedodd wrthyf, Rhaid i ti drachefn
broffwydo i bobloedd, a chenhedloedd, ac
ieithoedd, a brenhinoedd lawer.'

Yn ddiffuant

Eneiniog yr Arglwydd

Ateb swta gafodd hwn gan Steffan. Eglurodd yn ei lythyr
fod y cais yn torri un ar bymtheg o'r saith rheol ar hugain
ynglŷn â chymhwyster ac felly na ellid ei ystyried. Nid oedd
yr ateb hwn wrth fodd Yr Eneiniog, fel y gellid disgwyl.
Atebodd gyda'r troad.

Annwyl Syr

Ysgrifennaf hyn o lythyr er mwyn dwyn
cyhuddiad yn eich erbyn chwi a'ch cydweithwyr
oll yng Ngweinyddiaeth y Fall. Er bod yn
eich dwylo'r modd a'r gallu i gynorthwyo
un o feibion eneiniog DUW yn Oes y Pethau
Diwethaf, sef yr hwn a ddewisodd yr Arglwydd
i ledaenu neges y 'llyfr bychan' er lles EI
bobl, chwi a ataliasoch y cymorth hwnnw, gan
gefnu ar y gwirionedd ac arddel anwiredd yn
ei le.

Myfi a ddarllenais heddiw i chwi estyn cymorth
yn hytrach i 'fardd' i grafu o waelodion
drewllyd ei lygredigaeth lond sach o
englynion a limrigau a ffieiddbethau cyffelyb
am yr hipopotamws, sef eilun yr Eiffdiaid
paganaidd; ac i'r rhigymwr hwn, trwy gyfrwng
ei ddychymyg dieflig, rithio'r cyfryw fwystfil
i'n plith ni, relyw ffyddlon Penmachno, i
ymdrybaeddu mewn glythni a thrythyllwch. Ac
oblegid hyn, chwi a ddaethoch tan ddedfryd
eithaf yr Arglwydd. Canys, 'yn gymaint ag nas
gwnaethoch i'r un o'r rhai lleiaf hyn, nis
gwnaethoch i minnau. A'r rhai hyn a ânt i
gosbedigaeth dragwyddol...' (Mat. 25:45,46)

Yr ydym ni, frodyr eneiniog Crist, yn rhoddi
gerbron gorsedd DUW ein Tad, ein gweddïau
taer, gan ymbil arno, yr awr hon, i derfynu
einioes pob un ohonoch. Amen.

etc.

Cofiai Steffan yn dda sut y'i fferwyd gan ofn pan ddarllenodd y frawddeg olaf. Yr oedd achos y *ffatwah* yn erbyn Salman Rushdie yn dal yn fyw yn y cof, yn enwedig yng Nghymru, lle'r oedd yr awdur wedi cael lloches. Pwy allai ddweud nad oedd yr hanes hwnnw wedi plannu syniad ym mhenglog rhyw hurtyn crefyddol a chwenychai enwogrwydd ac a fyddai'n ddigon parod i ddefnyddio dulliau eithafol i'w ennill? Am fis cyfan, bu dau fownser yn gwarchod prif fynedfa'r Weinyddiaeth. Cadwyd tŷ Steffan dan oruchwyliaeth barhaus, ddydd a nos. Mynnai'r heddlu, hefyd, ei fod yn dilyn ffordd wahanol bob dydd ar gyfer mynd i'r gwaith a dychwelyd adref. Esboniodd Steffan fod ganddo system gylchoedd a warantai hynny, am fis beth bynnag. Ond barnai'r heddlu'n wahanol. Sut y gwyddai Steffan, gofynnent, nad oedd Yr Eneiniog wedi cael gafael ar ei system, wrth dwrio trwy sbwriel y Weinyddiaeth liw nos, neu wrth ymweld â'i gartref yng ngwisg dyn trydan neu ymgynghorydd yswiriant? Sut y gallai fod yn siŵr nad oedd yr ymgeisydd piwis yn ei wylio beunydd er mwyn canfod, ymhen hir a hwyr efallai, ie, ond canfod yn y pen draw, gyfrinach y cylchoedd? Na, byddai'n rhaid iddo ddyfeisio rhyw system amgenach neu, yn well byth, rhyw system nad oedd yn system o gwbl, fel na fyddai gobaith hyd yn oed i'r dihiryn mwyaf cyfrwys a phenderfynol amgyffred patrwm ei symudiadau.

– Iawn… Ailwampio'r cylchoedd, te.

Ond na, meddai'r heddlu, rhag ofn i'r Eneiniog dreiddio i hanfod yr ailwampiad hefyd.

A dyna sut y bu i Steffan ddarganfod pwysigrwydd haprifau.

7 Trafod Haprifau gyda'r Athro

SYLLAI STEFFAN ar ei siart cylchoedd. Onid peth hawdd fyddai diddymu unrhyw arlliw o batrwm yn y rhifau? Yn lle trefn ddisgwyliedig 1, 2, 3, 4, 5, 6, 7, 8, beth am 7, 3, 5, 1, 4, 6, 2, 8? A'r tro nesaf… 8, 4, 2, 5, 6, 3, 1, 7. Ac yn y blaen.

Dewisodd gornel tywyll ym mar Incognito, nid nepell o'r Weinyddiaeth, i drafod ei syniad gyda Juan, cyfrifydd wrth ei alwedigaeth ac un o'r ôl-werthuswyr mwyaf profiadol. Aethant yno ar adeg nad oedd yn amser coffi nac ychwaith yn amser cinio, er mwyn lleihau'r perygl o gael eu darganfod. Rhoddai enw'r bar ryw gysur seicolegol i'r sawl a fynnai breifatrwydd, ac roedd ei gilfachau cysgodol yn ddelfrydol ar gyfer seiat sibrwd. Ond prin y byddai neb wedi eu clywed, beth bynnag: llenwyd y lle â cherddoriaeth salsa a chyfreithwyr siaradus.

– 'Neith rheina ddim o'r tro.

Nid oedd Juan yn un i wastraffu geiriau, hyd yn oed dros goffi.

– Beth?

– Galla i weld patrwm yn hwnna.

– Ond eu tynnu nhw mas o'r awyr 'nes i.

– Naddo, wir. Eu tynnu nhw mas o dy ben, rwyt ti'n meddwl. Ac os wyt ti'n gofyn cwestiwn i dy ben, ma fe'n cynnig i ti'r pethe ma fe'n gwbod ore gynta'. Nawr te… Fe gest ti dy eni ar y trydydd o Fai, mil naw chwech dau… Mm? Ac rwyt ti'n byw mewn tŷ rhif 71… Mm? Ydw i'n iawn?

Edrychodd Steffan yn syn ar ei ffrind.

– Ie, ro'n i'n amau 'ny. A'r ail gyfres wedyn… dyw honna'n ddim ond amrywiad ar y gynta, ychan! Smo'r rhain werth taten os wyt ti'n trial twyllo rhywun!

Tynnodd Juan lyfr o'i boced. *Groble's Random Numbers Vol. 27.*

– Dyma beth wyt ti moyn. Mae hwn yn rhoi rhifau i ti galli di fentro dy fywyd na allai'r un dyn byw weld patrwm ynddyn nhw. Compiwter sy' wedi'u dewis nhw. A 'set ti'n rhoi un o'r rhifau hyn, neu gant, neu fil ohonyn nhw i rywun, bydden nhw 'ma hyd Ddydd y Farn yn trial gweithio mas pa rif sy'n dod nesa'. Byddi di'n saff wedyn.

– Maddeuwch i mi…

Roedd Steffan ar fin diolch yn galonnog i Juan am achub ei groen pan ddaeth llais dyn o'r bwrdd nesaf.

– Mae'n wir ddrwg gen i dorri ar eich traws…

Trodd Steffan a gweld gŵr blonegog tua'r trigain oed, â llygaid macrell ganddo a napcyn mawr gwyn wedi'i stwffio yn nhop ei grys.

–… ond rydach chi'n cyfeiliorni yn fan'no, gyfaill.

– Be'ch chi'n feddwl, cyfeiliorni? gofynnodd Juan.

– Cyfeiliorni? Wel, cymryd cam gwag… mynd ar goll… gwneud camgymeriad…

– Wy'n gwbod beth mae'r gair yn feddwl! Pam ych chi'n 'weud e wy isie gwbod!

– O'r gora, ta. Fel y dywedais i, dwi'n ymddiheuro am fod mor hy, ond fel un sy'n digwydd gwybod rwbath bach am yr hen haprifa ma… a chitha, mae'n amlwg, yn awyddus i gael gafal ar set o'r cyfryw rifa, a hynny, decini, am resyma tra phwysig – rhesyma, carwn i bwysleisio, nad oes a wnelon nhw ddim oll â fi…

Roedd Juan erbyn hyn wedi anadlu llond paragraff hirwyntog o aroglau gwin a garlleg. Er mwyn ffrwyno tafod drycsawrus y dieithryn cododd ar ei draed a pharatoi at fynd. Ond roedd y gwin, beth bynnag am y garlleg, wedi caledu penderfyniad y dyn i rannu'i ddoethineb â dau gyfaill yr oedd cymaint o'i angen arnynt.

– Yr Athro Awberry Thomas… Mi fues i'n Athro Ffiseg yn y brifysgol 'ma ar un adeg, ond wedi symud i Brifysgol Caeredin erbyn hyn, ac wedi dod i lawr i weld y mab. Mae o'n gweithio yn y Cynulliad, wyddoch chi. Rhaid i ddyn ennill ei damad rywsut, decini, ha! A'r ferch-yng-nghyfraith hefyd, wrth gwrs, a'r wyres fach, mae hi'n dair…

Wedi cyrraedd pen ei dennyn byr, cododd Juan a chynnig ei law i'r Athro.

– Braf cwrdd â chi, syr… a thrueni mawr 'n bod ni'n gorffod mynd, a chithau wedi dod mor bell…

Cymerodd yr Athro law Juan, a'i siglo'n egnïol.

– O'r gora' hogia, o'r gora'! Ond gadewch i mi ddeud un peth bach…

Gan fod yr Athro bellach yn sefyll o flaen y brif fynedfa, nid peth hawdd oedd ei anwybyddu.

–… ac un peth bach yn unig. Mae ganddoch chi yn eich llaw, os nad ydw i'n camsynied, gyfrol o Groble's Random Numbers. Ydach chi'n tybio, gyfeillion, mai dyna'r unig gopi gafodd ei argraffu…? Mm…? Dichon eu bod nhw'n rhifau hap pan gafon nhw eu casglu, ond erbyn hyn…

– Ie? Erbyn hyn?

– Yr hyn dw i'n ceisio'i ddeud ydy hyn. Petai rhyw ddihiryn yn gwybod, fel yr ydw inna'n gwybod, er enghraifft, nid 'mod i'n ddihiryn, wrth reswm, 'dach chi'n saff yn fan'ma efo fi – er, wedi deud hynny, sut dach chi'n gwybod hynny, sut

```
77, 100, 1, 72, 87, 31, 5, 17, 42, 9, 17, 72, 25, 15, 15,
74, 60, 67, 71, 9, 72, 74, 16, 13, 71, 99, 99, 85, 25, 5, 36,
29, 72, 38, 33, 36, 55, 6, 76, 37, 40, 9, 82, 54, 69, 53, 54,
50, 99, 67, 18, 7, 93, 69, 97, 58, 76, 23, 39, 34, 94, 61,
27, 83, 20, 42, 7, 10, 74, 14, 87, 55, 84, 94, 4, 37, 81, 36,
30, 13, 71, 20, 35, 87, 48, 11, 43, 45, 21, 1, 78, 98, 29, 5,
85, 10, 80, 77, 63, 13, 5, 21, 77, 11, 67, 44, 22, 7, 92, 13,
41, 8, 46, 13, 55, 45, 41, 92, 13, 91, 58, 28, 86, 56, 43, 83,
50, 57, 83, 65, 20, 27, 68, 30, 9, 55, 19, 68, 22, 51, 94, 52,
91, 96, 7, 81, 66, 75, 18, 84, 46, 43, 51, 90, 43, 38, 41, 72,
100, 19, 92, 95, 14, 1, 85, 4, 53, 9, 35, 53, 54, 93, 73, 91,
42, 48, 66, 73, 20, 36, 71, 25, 94, 2, 73, 64, 32, 26, 2, 92,
83, 100, 79, 52, 66, 78, 55, 12, 88, 67
```

Ffig. 7   Detholiad o'r haprifau y bu Juan yn annog Steffan i'w defnyddio er mwyn dianc rhag Yr Eneiniog

dach chi'n gwybod? Mm…? Ond hidiwch befo… Tasa' rhyw ddyn felly yn gwybod mai allan o'r llyfr hwn yr oeddach chi'n tynnu'ch ffigura… wel, fasan nhw fawr o dro yn dod o hyd i'r gyfres o rifa roeddach chi wedi'i dewis… 'n enwedig os oedd hi'n gyfres hir… Mm…? Ond yn waeth na hynny…

Gwelwodd Steffan. Roedd ei gynlluniau manwl yn dymchwel o flaen ei lygaid a bygythiad yr ymgeisydd ga'dd ei wrthod yn pesgi ar bob brawddeg ddrewllyd a lifeiriai o enau'r Athro.

– Yn waeth na hynny?

– Wel, a'i deud hi'n blwmp ac yn blaen, gyfeillion, nid haprifa go iawn ydyn nhw, 'waeth be mae o'n ei ddeud ar y clawr.

– Nid haprifau go iawn?

Ochneidiodd y ddau.

– Dach chi'n gweld, hogia… fel y gwyddoch chi, cyfrifiadur ddaru greu'r rhifau sydd ganddoch chi yn y llyfryn bach yna, ynte…? Wel, er mwyn eu creu nhw, bu'n rhaid i'r cyfrifiadur hwnnw ddefnyddio be dan ni sydd yn gweithio yn y meysydd hyn yn ei alw'n algorithm. A. L. G. O. R. I. Th. M. Mm? Na hidiwch be 'di hwnnw… Ond mae'r algorithm yn ei gneud hi'n bosib i greu'r cyfresi welwch chi yn y llyfr ma. Ac maen nhw i gyd i'w gweld yn bethau hap a damwain llwyr, rhaid i chi gyfadda. A fedra neb brofi'n wahanol 'tasan nhw'n eu dadansoddi nhw o fore gwyn tan nos. Na fedran wir. Ond…

Crychodd yr Athro ei dalcen.

– Ac mae hyn yn ond go fawr, gyfeillion.

Ochenaid arall.

– Tasa' rhywun yn gwybod pa algorithm gafodd ei ddefnyddio… wel, efo rhyw faint o waith, medra fo greu'r un gyfres yn union. Mi fasa fo'n gwybod eich cyfrinach!

Roedd Juan fel plentyn oedd newydd ddarganfod mai lleidr pen-ffordd oedd Siôn Corn.

– Does dim shwt beth â rhif hap go iawn, te? Does dim ffigurau i'w cael heb fod rhyw fath o drefn neu batrwm yn perthyn iddyn nhw? Dyna rych chi'n 'ddweud?

Syrthiodd wyneb Steffan.

– Mae hi wedi canu arnon ni felly…

– Nac 'dy, nac 'dy, hogia.

Rhoes yr Athro law dadol ar ei ysgwydd.

– Peidiwch ag anobeithio. *Mae* anhrefn yn bod, credwch chi fi. Anhrefn go iawn hefyd. Ond os 'dach chi am gael hyd iddi, wiw i chi boetsian hefo petha fel hyn. Na, hogia, rhaid i chi droi oddi wrth waith dyn…

Bwrodd yr Athro olwg dirmygus ar y cyfreithwyr a sefai

o'u cwmpas ar bob llaw ac yna bwyntio ei fys at y ffenest.

–... ac edrych yn hytrach... draw fan'cw...

– Be? Ar yr awyr? cynigiodd Juan, yn ddiamynedd.

– Ar y ceir? awgrymodd Steffan.

–... draw fan'cw, hogia. Rhaid ichi edrych ar waith Natur...

– Gwaith Natur?

Roedd tinc amheus yn llais y cyfrifydd.

– Ia, 'ngwas i. Gwaith Natur. Dyna'r unig ffordd dowch chi o hyd i rywbath na fedrwch chi byth â deud amdano... A! Mi wn i be sy'n dŵad nesa'!

Ac yn awr ei angen, pa ddewis oedd gan Steffan ond mynd trwy'r unig ddrws a oedd yn agored iddo?

– Gwaith Natur. Iawn. Purion. Fel beth...?

Edrychodd yr Athro i fyw ei lygaid.

– Fel glaw yn disgyn. Fel llwch yn setlo. Fel sŵn y radio pan mae o rhwng dwy sianel. Fel... fel plant...

A dechreuodd wenu'n dadol.

–... fel plant yn chwarae. Sy'n f'atgoffa i, gyfeillion...

Trodd yr Athro at y drws.

–... Rhaid imi ei throi hi rŵan. Bydd Ioan yn aros amdana i...

– Ioan?

– Ia. Ioan, y mab... Ioan Thomas... Ydach chi'n ei nabod o?

8 Tatratattat

Disgybl ufudd fu Steffan erioed. Yn yr ysgol, ni fyddai byth yn esgeuluso ei waith cartref, na thorri'r gorchmynion a'r gwaharddiadau niferus a ddyfeisiwyd i reoli gwisg, hyd gwallt, tramwyo'r coridorau, cael gwared ar bapurau losin, siarad â'r athrawon, ac yn y blaen. Ac yn ddi-ffael, byddai'n well ganddo ddilyn cyfarwyddiadau rhywun arall nac ymddiried yn ei reddfau ei hun. Dychwelodd i'w swyddfa felly ac eistedd yno gan ddisgwyl yn amyneddgar, yn unol â chyngor yr Athro, i Natur amlygu ei hanhrefn.

Ac ni fu'n disgwyl yn hir. O'r ystafell nesaf, lle'r oedd Audrey a 24 swyddog monitro arall yn gweithio, fe ddeuai i'w glustiau synau nid annhebyg i'r synau yr oedd yr Athro wedi sôn amdanynt – yn sŵn cyson, di-dor, ac eto'n llawn amrywiadau, a'r rheiny'n amrywiadau na ellid priodoli'r un patrwm nac ystyr iddynt. Gwrandawai'n astud… *Tap, tap, tap* o fyseddell cyfrifiadur. Menyw'n chwerthin. Dau ddyn yn siarad. *Tatataptapapartapratrap.* Sŵn tri neu bedwar cyfrifiadur erbyn hyn? Dodi cwpan ar fwrdd… Drâr cabinet yn agor… *Pumed o Ebrill wedoch chi? Saith…naw… does… Bet! Dewch â'r…* Ffenest yn cau… *Tatratatrtatat… Cais rhif un wyth dim naw… Coffi?… pump… pump…* Drws yn cau… Peswch… *Miaaaaoow..*

Miaaaaoow?

Ac fel hyn y bu Steffan wrthi am weddill y prynhawn, yn ceisio gwahaniaethu rhwng synau mawr a mân ei gymdogion, gan nodi hefyd eu gwahanol gyfuniadau a'u hamrywiadau

mewnol. Ai hyn tybed, oedd anhrefn pur? Yr anhrefn pur
a fyddai'n rhoi lloches iddo rhag ei elynion? Teimlodd ias o
lawenydd o feddwl ei fod, o'r diwedd, efallai, wedi cael hyd
i'r ateb i'w broblemau. Nid oedd eto, rhaid dweud, wedi
dyfeisio dull o droi'r anhrefn hwnnw yn ganllaw ymarferol
ar gyfer trefnu – maddeuwch i mi, anhrefnu – ei deithiau
dyddiol. Nid oedd chwaith wedi cael y cyfle i sylweddoli mai
cystal peth fyddai tynnu rhifau allan o het. Ac wrth lwc, ni
fu'n rhaid iddo. Oherwydd, diolch i'r Drefn, neu'r Anrhefn,
cafodd Steffan wybod y prynhawn hwnnw fod Yr Eneiniog
wedi cael ei ddal gan yr heddlu ar ôl ymosod ar fardd yr hipo
a oedd, mae'n debyg, yn darged amgenach yng ngolwg yr
ymgeisydd rhwystredig ac yn darged haws hefyd, gan ei fod ar
y pryd yn arwain cwrs ysgrifennu mewn canolfan nid nepell
o'i gartref.

A dyna pam yr oedd Steffan, y tro hwn, am fod yn
llawer mwy gwyliadwrus ynglŷn â'r cais yr oedd newydd ei
drosglwyddo i Audrey. Ac ymhen hir a hwyr, ar ôl mwy o
dynnu coes, cochi, ebychu, chwerthin a rhyfeddu troes hithau
ei sylw at y ffurflen o'i blaen a dechrau chwilio trwy hen ffeils
am enwau ymgeisydd Rhif 86 a'i chanolwyr.

– Mae 'na P.O. Brian ma…

– Mm…

–… ac mae 'na Hannah Demondez…

– Mm…

– Ac mae rhywun o'r enw George Cumber ga'th grant yn
1964 i sgrifennu llyfr taith ond welwyd mo'r dyn byth wedyn.
Ody'r rheiny o ryw werth? Ond ta beth, does dim rhaid 'u
bod nhw yn y system, o's e?

– Wel o's, glei…

Cofiodd Steffan y llythyr cwta y bu'n rhaid iddo anfon

at brifardd rywdro, a hwnnw'n bygwth atgyfodi'r hipo bondigrybwyll.

Annwyl Fardd
RE: DY AWDL FAITH

*A rown ni nawdd? Sori. Na. Nid wyt yn ein bas data.
etc.*

Ond bardd oedd hwnnw. Ac wedi'r cyfan, onid oedd enwau'n cael eu hychwanegu at y gronfa bob awr o bob dydd? Oedd, roedd gan awdur berffaith hawl i fod yn newydd ac yn anadnabyddus. Ond nid yr awdur oedd yr unig broblem. Beth am y canolwyr? Doedd y canolwyr ddim yn y system chwaith! Mm.

– Wel, nac oes, falle, Audrey, a bod yn hollol deg, ond wedyn, shwt mae gwybod nad yw'r fenyw ma jyst wedi mynd at ei… at ei mam-gu, neu'r dyn drws nesa, neu…

– Mae eisie cael canolwyr i'r canolwyr, felly?

– Ond shwt alla i, Audrey, a hithau heb anfon yr un cyfeiriad ata i?

– Pam na wnei di ofyn iddi te?

– Gofyn iddi fod yn ganolwr i'w chanolwyr ei hunan?

– Jyst gofyn am fwy o wybodaeth.

A dyna a wnaeth.

Annwyl Ms O'Brien

Diolch am eich cais am gymhorthdal. Yn anffodus, nid yw'n gyflawn. Er mwyn ei gofnodi'n gais dilys, ac yna ei werthuso, ac yn y pendraw, os byddwch yn llwyddiannus, ei ôl-werthuso, rhaid ateb pob cwestiwn ar y ffurflen gais.

Carwn dderbyn mwy o wybodaeth yn arbennig am natur y gwaith yr ydych yn bwriadu ei ysgrifennu ac am eich cymwysterau i gyflawni'r gwaith hwnnw, a chymwysterau eich canolwyr i'ch cymeradwyo. Rhaid imi dderbyn yr wybodaeth hon, ynghyd â sampl o'ch gwaith, erbyn...

etc.

9 Ebbways

ONDD ROEDD RHESWM ARALL pam yr oedd Steffan a Llio'n chwilio am dŷ newydd. Ac nid dim ond am dŷ newydd, ond am newid eu holl ffordd o fyw. Y rheswm hwnnw oedd bod Llio'n disgwyl babi. A hithau'n tynnu am ei 39 blwydd oed yr oedd y trobwynt hwn yn ei bywyd yn golygu llawer mwy iddi na phe buasai ddeng mlwydd yn iau. Yr oedd yr hir ddisgwyl, yr hir ddyheu, a'r aml siom wedi troi beichiogrwydd yn gyflwr gwyrthiol bron. Gellid ymdrechu i gael babi ond ni ellid ei gynllunio. Yn y pen draw, roedd y cwbl y tu hwnt i reolaeth dyn a menyw.

Ond roedd y wyrth wedi digwydd ac roedd ymwybod Llio â'i chyfrifoldeb, y cyfrifoldeb am roi i'r babi bob swcwr a sicrwydd er mwyn hwyluso'i ffordd trwy'r byd, bellach yn llenwi pob awr o'i dydd. Gwyddai fod hyn yn ystrydeb ond roedd yn ystrydeb a'i siwtiai i'r dim. Y cam nesaf yn y bywyd ystrydebol hwn oedd prynu tŷ yn ei hen gynefin. Canys nid cartref nad yw'n gynefin. Yna, byddai'i bywyd yn gyfan. Yno, pe teimlai'r awydd, gallai droi ei hystrydeb yn gelfyddyd. Ond ni allai greu heb fod ganddi'r deunydd crai i fwydo'i dychymyg. Roedd rhaid setlo ar dŷ, a hynny'n ddi-oed. Cytunodd Llio a Steffan, felly, fod angen mynd ati ar unwaith i drefnu ymweld â rhai o'r tai roedden nhw wedi'u gweld ar wefan Ebbways. Roedd hi'n bryd mynd â'r maen i'r wal.

Edrychodd Steffan ar ei wats. Doedd hi ddim wedi taro pump eto. Efallai y câi well lwc o ffonio'r tro hwn. Siawns na fyddai rhywun yno a allai ei helpu. Chwiliodd am wefan y cwmni er mwyn cael hyd i'r rhif ffôn. Dau fys. Chwe llythyren. E..b..b..w..a..y…

Clic!

Wel, mi oedd 'na gwmni o dan yr enw hwnnw a allai werthu carped iddo. Ond dim sôn am estate agents. Ceisiodd eto, gan ddefnyddio chwiliwr gwahanol. E..b..b..w..a..y…

Clic!

Na, yr un peth. Rhyfedd. Ond nid yn gwbl anghyffredin efallai. Cwmni bach. Rhyw nam ar eu system. Doedd dim amdani ond ffonio Llio yn ôl a gofyn am y manylion ganddi hi.

– Eu cael nhw oddi ar y wefan wnaethon ni, cariad, a sa i'n cofio i fi sgrifennu nhw lawr.

– Nage, nage, fi wnaeth, Llio. Gadewais i fe ar bwys y ffôn pan driais i 'u ffonio nhw'r wythnos diwetha'. Rwy'n cofio sgrifennu fe lawr gan feddwl trial 'to.

– Ar y pad?

– Ie, glei.

– Wel, mae sawl rhif fan hyn… 0268 fyddai'r côd, yntefe? Dyma un. 473226. Wedyn… 473974… Wedyn… 476602. Aros… mae un arall… 476043..

– Oes rhagor?

– 'Na i gyd galla i weld, ond mae dy sgrifen di mor anniben, Steffan… ac mae lot fawr o bethach 'ma wedi' croesi mas, wy'n ffaelu gwneud rhych na rhawn…

– Paid becso. Tria i rheina.

Ac yntau'n meddwl, yn ei ddiniweidrwydd, bod ei bryder am rifau wedi dod i ben am heddiw, dyma Steffan yn gorfod rhoi cynnig unwaith yn rhagor ar wastrodi'r cythreuliaid bach diafael.

– Nid yw Gwilym na Bethan gartref ar hyn o bryd…
Gadewch neges…

Dim hwnna, beth bynnag.

– Tywi Aggregates… Tywi Aggregates… Hello?… Can I help you?

Tywi Aggregates? Pwy…? Beth…? Trawodd y ffôn yn ôl yn ei chrud.

Roedd dau rif ar ôl. Deialodd y cyntaf. Canodd y ffôn am funud gyfan. Yna peidiodd y grwndi ond ddaeth dim ateb. Dim. Dim ond distawrwydd.

– Mr Ebbway? Ai chi sy 'na?

Dim ateb.

– Ebbways Estate Agents? Oes 'na bobl?

Ac yna, megis llais o'r bedd.

– Bydd rhaid i chi siarad yn uwch, Mr Ebwys, rwy i bach yn drwm 'yn glyw..

– Na, nage fi..

– Pardwn?

– NAGE FI YW MR EBBWAY!! CHI YW MR EBBWAY!!… FI YW… mm… ble o'n i… RWY MOYN PRYNU TŶ, MR EBBWAY!!

– Wel, odych chi, wir…? Pob lwc i chi, syr!… Oes 'da fe dŷ i'w werthu te?… Helô?… Mr Ebwys?

Un cynnig arall. 4..7..6..0..4..3

– Mr Ebbway?

– Yn siarad.

Bu Steffan yn dafotrwm am eiliad.

– Chi YW Mr Ebbway…?

– Ie. Beth alla i 'neud i chi?

– A! Wel, meddwl o'n i… efallai… ac wedyn… ta waeth… wel, na, a gweud y gwir… Fe ges i'n nrysu braidd pan ffaelais i fynd i mewn i'ch gwefan chi gynne fach…

– Y webseit ych chi'n 'feddwl?… Wedi 'bennu.

– Wedi 'bennu? Yn cael ei diwygio felly?

– Na, wedi 'bennu, wedi 'bennu am byth. Mae'r tai wedi'u gwerthu i gyd, ych chi'n gweld, a does dim diben 'i chadw hi'n fyw…

– Ei chadw hi'n fyw?

– Y webseit wy'n feddwl.

– A!

– A finne wedi hen daro'r ffender, os gwedan nhw ffor' hyn…

– Y ffender…

– Pases i oed yr addewid dros… w… faint sydd nawr…

– A'r tai i gyd wedi'u gwerthu ych chi'n gweud?

– Gwerthais i 'nhŷ cyntaf yn nineteen forty-eight…

– Dim un?

– Cyn i chi gael eich geni, siŵr o fod… Pethe wedi newid lot ychan…

– A dim ar ôl?

– Wedi gwerthu'r cwbl lot… wel, heblaw un…

– Mae un ar ôl?… Heb ei werthu?

– Neb eisiau 'i brynu fe… lot o waith i 'neud arno fe 'chwel… Ydych chi'n handi 'da'ch dwylo?

Edrychodd Steffan ar fysedd ei law dde. Roedden nhw'n llyfn ac yn binc.

– Yn weddol… Pa dŷ yw e'n gwmws?

– Doedd e ddim ar y seit. Fyddech chi ddim wedi'i weld e fyn'na. Wel, fe fuodd e… dwy flynedd a rhagor yn ôl… ond roedd e'n hala pobl bant o'r tai erill, wy'n credu, so tynnais i fe o'na... Trueni... ffaelu gwerthu'r tŷ ola, a finne'n rhoi'r ffidil yn y to.

– A ble mae'r tŷ, wedoch chi?

– Ac mae e'n dŷ hanesyddol, 'fyd, serch bod enw Seisnigaidd arno fe erbyn hyn...

– Ie, a'r tŷ ma... ble mae e 'to?

– Mae e rhyngt... ydych chi'n gyfarwydd â'r ardal 'ma?

– O ffor' 'na mae'r wraig yn dod..

– Ife? Wel, gwedwch wrthi ei fod e ar bwys...

– Ie...

– Wel, erbyn meddwl, 'falle bydde hi'n haws i fi ddodi'r manylion yn y post iddi gael gweld... Bydd hi'n siŵr o nabod y lle.

Rhoes Steffan ei gyfeiriad iddo. Roedd ar fin holi Mr Ebbway ynglŷn â pha mor dawel oedd lleoliad y tŷ, faint o awyrennau fyddai'n hedfan drosto bob dydd, ar gyfartaledd, a nifer o faterion eraill, ond roedd Mr Ebbway eisoes wedi rhoi'r ffôn i lawr. A'r enw. Damia, roedd e wedi anghofio holi am enw'r tŷ.

Erbyn hyn, oherwydd y chwilio hir am y rhif, y galwadau ofer, a lapan digyfeiriad Mr Ebbway, roedd swyddfa'r Weinyddiaeth wedi cau. Golygai hynny fod rhaid i Steffan osod y larwm cyn mentro allan: tasg anodd, er gwaethaf y blynyddoedd o ymarfer, am fod Steffan o hyd yn anghofio rhifau'r côd. Ar yr un pryd, ofnai eu sgrifennu i lawr, rhag iddynt lithro i ddwylo'r gelyn, chwedl yntau. O'r herwydd, ac nid am y tro cyntaf, bu'n rhaid iddo wneud rhywbeth a aeth yn gwbl groes i'w natur, sef caniatáu i'w fysedd geisio cofio'r patrwm a wnaed gan y rhifau, er na chofiai'r rhifau ei hunain. Gwasgodd y botymau, gan ddilyn y reddf honno, a seiniodd

y larwm ei rybudd cras. Roedd ganddo ugain eiliad i ymadael
â'r adeilad a chloi'r drws. Dechreuodd gyfri'... Un... wrth
ymbalfalu... Dau... trwy ei bocedi... Tri... am yr allweddi...
Pedwar... Pump... Chwech...

10 Yr Ail Becyn

FORE TRANNOETH, cyrhaeddodd ail becyn Rhif 86. Synnai
Steffan at brydlondeb ac effeithiolrwydd yr ymgeisydd. Yn
wir, roedd yn anodd deall sut yr oedd hi wedi derbyn y neges,
paratoi'r pecyn a'i gyrru trwy'r post mewn cwta pymtheg awr.
Os mai trwy'r post y daeth, hefyd. Doedd dim stampiau ar y
pecyn, na chyfeiriad ychwaith. Dim ond ei enw. Ond efallai
bod Twm yn y stafell bost wedi'i ddadbacio. Pam y byddai'n
gwneud hynny ni allai Steffan ddeall, gan fod Twm fel arfer
yn rhy bwdr i glymu ei lasys ei hunan, ond ni allai feddwl am
unrhyw esboniad arall.

Yn ffeil rhagbaratoadol pob ymgeisydd roedd rhaid
amgáu disgrifiad o'r holl eitemau a dderbyniwyd. Gwnaed
hynny er mwyn diogelu'r Weinyddiaeth rhag cyhuddiadau o
gamwri a diffyg tryloywder, cyhuddiadau y byddai ymgeiswyr
aflwyddiannus a phwdlyd yn eu dwyn yn ei herbyn o bryd
i'w gilydd. Rhyw drofa fach yn y Llwybr Papur, ond nid un
i'w dibrisio, serch hynny. A dyma sut y cofnododd Steffan yr
eitemau a dderbyniodd y bore hwnnw.

Cynllun 5/b Cymorthdaliadau i Awduron
Blwyddyn 2004/5
Swyddfa C/1 Adran HC/2/c Swyddog SM
Cam 6 Cymal 12

Ymgeisydd 86 Kate O'Brien

Darn o gerddoriaeth, dan y teitl:
Regis Iusto Cantio El Reliqua Canonica Arte Resoluta

Eitem 2

Taflen Hysbysebu 'Paulo's Royal Circus', Western Avenue, Mynachdy, Caerdydd Sadwrn 13 – Sadwrn 20 Mai. Dim blwyddyn.

Eitem 3

Llun o hen ddyn wrth yr organ. Ar y cefn. Thomas. 3 Mai. Dim blwyddyn. Ac mewn llawysgrif wahanol, y geiriau: Dim Handl!

Annwyl Mr Steffan

Amgaeaf ragor o wybodaeth am fy nghais. Maddeuwch i mi am fethu anfon digon y tro cyntaf. Rhyw feddwl oeddwn i y byddai crynodeb mathemategol o'm prosiect yn achosi llai o broblemau i'ch systemau chi.

Gobeithio y bydd yr eitemau newydd amgaeedig yn goleuo'r hyn sydd gennyf mewn golwg. Ond cofiwch mai amlinelliad yn unig sydd yma.

Cewch gysylltu â'm canolwyr trwof i.

etc.

Er iddo gofnodi pob eitem yn gywir ac yn gymen, teimlai Steffan yn bur annifyr, a hynny am reswm penodol. Casâi fod yn wrthrych i rywun arall. Pan âi i'r sinema neu i gyngerdd, yn y rhes gefn y byddai'n eistedd bob tro, fel na fyddai'n teimlo llygaid y gynulleidfa yn gwanu ei wegil. Yn grwtyn bach, byddai'n dyfeisio esgus newydd bob blwyddyn dros beidio â chael parti pen-blwydd. 'Sa i'n moyn parti Mam achos… mae hwn a hwn yn ffaelu dod.' Neu 'Fi 'di cwmpo mas 'da hon a hon.' Neu 'Wy'n dost, Mam.' Neu 'Wy'n mynd mas i 'ware ffwtbol yn lle ca'l parti, Mam.' Ond y gwir amdani oedd bod ar Steffan ofn. Nid ofn y plant eraill fel unigolion, nid ofn y bwyd a'r anrhegion a'r gêmau, nid ofn yr heneiddio hyd yn oed. Yr hyn a godai ofn arno oedd yr eiliad frawychus honno pan fyddai pawb yn troi ato, ie, yn troi ato fe, a'i hoelio â'u llygaid, a'u gwenau, a'u disgwyl. Disgwyl beth? Disgwyl iddo gymryd y brif ran yn eu drama nhw. Ond heb roi sgript iddo, na dweud wrtho mai fel hyn y dylai symud ei ddwylo, ei ben, ei goesau. Ac yntau'n ddoli bren, sut gallai wybod?

Daeth teimlad tebyg drosto'r funud honno, wrth ddarllen llythyr Kate O'Brien. Beth oedd ar droed fan hyn? Beth oedd bwriad yr ymgeisydd? Ai rhoi ei amynedd ar brawf? Neu ei ddyfeisgarwch? Ynte, ai rhyw gêm seicolegol oedd y cyfan, a hithau eisoes wedi cynllunio'r symudiadau i gyd? Neu, yn waeth byth, o gofio helynt Yr Eneiniog y flwyddyn gynt, tybed... ha!... tybed ai rheolwyr y Weinyddiaeth oedd yn ei roi ar brawf, gan ddefnyddio enw ffug a phob math o gastiau er mwyn gweld a allai ymdopi, a allai gyflawni ei ddyletswyddau fel swyddog cyflogedig syber, cyfrifol? Ac roedd yn rhaid iddo ymateb. Roedd yn rhaid iddo chwarae ei ran. Ond yn wahanol i'r parti pen-blwydd yr oedd yn ei ffieiddio gymaint pan oedd yn fach, ni wyddai'r tro hwn hyd yn oed pwy oedd awdur y ddrama yr oedd disgwyl iddo gymryd rhan ynddi. A phwy allai roi gwybod iddo?

Ac efallai, wedi'r cyfan, mai ffigwr truenus, ansefydlog oedd 86: yn ffigwr a haeddai gydymdeimlad, nid dicter. Ac efallai, hefyd, mai ar gyrion eithaf ei drama hi y gwnâi Steffan ei gyfraniad bach ef. Os mai dim ond *extra* roedd e i fod, yn rhan o'r gefnlen yn hytrach na'r plot, doedd dim lle i boeni. Ond wedyn, onid oedd disgwyl hyd yn oed i *extra* symud weithiau, a symud yn unol â gofynion y stori a gweithredoedd ac emosiynau'r prif gymeriadau? Oedd, roedd Steffan yn casáu bod yn wrthrych i rywun arall.

11 Cnoi Bresych

ROEDD UN PETH yn sicr. Ni allai Steffan fforddio aros am
atebion i'r cwestiynau hyn. Ymhen deng niwrnod byddai
disgwyl iddo fod wedi cofnodi pob un o'r 172 cais a'u didoli
yn ôl eu haeddiant. Neu, yn hytrach, yn ôl eu haeddiant
yn ei farn ef. Câi'r farn honno ei hystyried, wrth gwrs, ond
byddai'r penderfyniad terfynol, fel y gwelsom yn barod, yn
dibynnu ar nifer fawr o ffactorau nad oedd eu goleuadau
llachar erioed wedi gwawrio ar ddyffrynnoedd cul, cysgodol
meddwl Steffan. Wedi mynd â'r ceisiadau at ffin eithaf ei
diriogaeth ef, byddai Steffan yn eu trosglwyddo i ofal ceidwaid
y tiroedd uchel, yr Adran Draws-werthuso Strategol. Neu, a'i
ddweud yn symlach, byddai'r gwaith yn cael ei symud o Gam
8 i Gam 9 o'r Trigain Cam Namyn Un. Ac er na fyddai'r
dyfarniad olaf yn cael ei wneud am dri mis arall, roedd yr
amserlen yn dynn. Roedd yn rhaid i'r Adran Draws-werthuso
Strategol yn gyntaf dderbyn barn y Swyddogion a'r Is-baneli
Rhanbarthol (Camau 10 i 16) ar bob un o'r ceisiadau, yna
barn y Swyddogion Cyfleoedd Cyfartal (Camau 12 i 14) a'r
Cynghorwyr Annibynnol (Cam 15). Yna, byddai gofyn iddynt
ymgorffori cyfraniad pob un o'r Swyddogion a'r Cynghorwyr
hyn mewn adroddiad (Cam 16) a gâi ei anfon (Cam 17), yn ei
dro, at aelodau Pwyllgor y Weledigaeth gogyfer â'u cyfarfod
ar 26 Gorffennaf (Camau 18 i 21).

Rhygnai Steffan ymlaen, a chloc mawr y Weinyddiaeth, y
cloc bythol effeithiol, diduedd, yn tician yn ei enaid.

88... Na... 89... Na... 90... Na..91... mm... Na...
92... Na... 93... pa!... 94... Nid oedd angen llawer o

bwyso a mesur cyn dedfrydu'r rhan fwyaf o'r ceisiadau i bentwr y colledig. Gyda'r blynyddoedd, daethai Steffan i adnabod nodweddion y rhain, weithiau heb agor y pecyn: gor-daclusrwydd y sgrifen ar yr amlen, er enghraifft, neu label cyfeiriad yr ymgeisydd *wedi'i deipio fel hyn ar y cefn*, ac unwaith, rhuban coch wedi'i glymu'n dwt am y cyfan. Gydag eraill, roedd brawddeg neu ddwy yn ddigon. 94...

Y mae yn rhaid cael newidiad cyflwr, newid yr anian, bod yn feddiannol o ras cadwedig yma cyn meddiannu gogoniant tragwyddol wedi myned oddi yma...

Crefyddoldeb eildwym... 95...

Cusanodd y ddau gariad, gusan winglyd hir yn cordeddu eu gwefusau, dringodd y miwsig i ddiweddglo buddugoliaethus a chyfododd y gynulleidfa o'u seddau plwsh

Plwsh! 96...

Pan y daethom i'r man y saif y dref, gwnaethom wersyll, gan mai dyma derfyn ein taith ar y pryd, ac ym mhen rhai dyddiau tra yn gwneud ein hunain yn drefnus a chysurus, am yr ysbaid yr oeddym i aros yno, gwelem un nawngwaith wrth ein hymyl rhyw fwystfil yn rhedeg trwy y gwellt i gyfeiriad y behemoth..

Y behemoth. Duw a'n helpo! Na, cyd-ddigwyddiad, mae'n rhaid. *Pastiche arall,* nododd Steffan yn ei gofnod ar rif 96, *ymgais i ddynwared arddull glogyrnaidd a safbwynt naïf yr awdur o anturiaethwr yn oes...* yn oes Fictoria, mae'n debyg... *yn oes Fictoria.* A yw dynwarediad clyfar yn llai diflas na'r gwreiddiol diflas? Nac ydy.

97... ach-y-fi!... 98... Pam mae'r safon... 99... yn dirywio... 100... blwyddyn... 101... ar ôl blwyddyn... 102... a'r rhygnu ymlaen am... 103... yr un pethau o hyd... 104... ac o hyd... 105... ac o hyd... 106... fel gwylio... 107... 'r... 108... meirw'n... 109... cogio... 110... byw... 111... yn cogio… mm… dynwarediad eto fyth… ond…

*Cyn gynted ag y traddodwyd y ddedfryd, dechreuodd pawb
yn y llys ymwau drwy'i gilydd fel gwenyn; ac yr oedd
cynnwrf y bobl yn myned allan ac yn siarad, mor uchel fel
mai prin y gallwn fy nghlywed fy hunan yn crio, yr hyn a
wneuthum i bwrpas…*

Ond dynwarediad o bwy?

*… ac yna daeth tincial y llestri swper o'r gegin fach. Diar,
yr oedd rhywbeth cyfeillgar ac agos atoch mewn sŵn llestri,
onid oedd? Eisteddodd drachefn, a'r tro hwn edrychodd ar ei
dodrefn o'r ochr arall i'r bedd. Rhyw ddiwrnod fe'i didolid i
gyd…*

Llestri? Mewn llys? Dryslyd, braidd… Ond yn well na'r
behemoth. Doedd y clytiau ddim yn perthyn i'w gilydd
rywsut. Ai gwaith gwreiddiol oedd hwn? Neu, os nad oedd
pob rhan ohono'n wreiddiol, a oedd y ffordd y cawsai'r
clytiau eu gweu trwy ei gilydd yn wreiddiol? Ac i ba ddiben
wedyn? Heb ddarllen y cwbl lot, ni fedrai ddweud. A doedd
amser ddim yn caniatau. 107… 108…

Rywsut neu'i gilydd, cafwyd y cyfan i fwcwl erbyn
amser cinio. Erbyn hyn, roedd Steffan yn eistedd ar un o
feinciau pren bar salad *Nuts*, yn cnoi bresych. Roedd arafwch
angenrheidiol y bwyta yn gysur i'w feddwl. A rhwng y
bresych a'i ddiwydrwydd y bore hwnnw, byddai Steffan
wedi bod yn ddigon bodlon ei fyd, oni bai am un peth. Rhif
86. Oherwydd, fel y dywedai'r rheol berthnasol yn y Llyfr
Canllawiau, *Rhaid i swyddog gwblhau'r gwaith a chau'r ffeil ar
yr holl geisiadau, a chael llofnod y Cyfarwyddwr perthnasol i dystio
bod hynny wedi'i wneud yn foddhaol, cyn trosglwyddo'r ffeil i adran
arall.* Sgrwnsh.

Yr oedd y cais yn bod. Sgrwnsh. Ni allai ei ddadfodoli.
Sgrwnsh. O leiaf, yr oedd y cais yn bod yn yr ystyr bod

ganddo rif sgrwnsh a bod y rhif hwnnw wedi'i gofnodi yn
annileadwy, hyd dragwyddoldeb, yng nghyfrifsgrwnshiadur
y Weinyddiaeth. Ond prin ei fod yn bodoli mewn unrhyw
ystyr arall… Sgrwnsh… Sgrwnsh… Yr oedd, meddyliai
Steffan wrtho'i hun, yn Lefftenant Kije o gais. Roedd olion
sgrwnsh ei draed eisoes i'w gweld ar y Llwybr Papur sgrwnsh
ac ni wyddai neb ond efe mai olion esgidiau yn unig oeddent
sgrwnsh, ac nad oedd na thraed na choesau na'r un aelod arall
sgrwnsh o'r corff ar gyfyl y lle. Rhywsut sgrwnsh byddai'n
rhaid llenwi'r sgidiau. Sgrwnsh.

Cerddodd yn ôl i'r swyddfa. Aeth heibio'r hen fynwent, ei
cherrig bellach wedi'u troi'n forder i ardd flodau ac yn dynodi
dim i neb. Wel, yn dynodi, o bosibl, farwoldeb mynwentydd.
Mynwent cerrig beddau oedd hon. Aeth heibio caffe *Ho! Ho!*
a fu, gynt, yn siop lyfrau, yn fanc, yn ddim byd o bwys ac,
ymhell bell cyn hynny, yn fynachlog. Nid bod olion o hynny
i'w gweld mwyach. Nid ar yr wyneb, beth bynnag. Doedd
dim byd eglwysig am y bensaernïaeth hon, er mor agos oedd
i'r entrychion. Troes y gornel a cherdded i ddannedd y gwynt
meinaf yn y ddinas: gwynt oer a chwythai lwch i'r llygaid. Yr
un gwynt a fu'n chwythu fan hyn erioed ers bod gwyntoedd
yn bod. Ond efallai ddim. Efallai mai llwch yn y llygaid oedd
hynny hefyd. Efallai mai plentyn y bensaernïaeth oedd e,
wedi'r cyfan, yn newydd-ddyfodiad o wynt, yn fabi bach
piwis i'w fam goncrid.

Wedi dychwelyd i'w stafell, edrychodd Steffan ar y
defnyddiau roedd 86 wedi eu hanfon ato. A oedd rhywbeth
ynghudd ynddynt? Ynte, ai twyll oedd y cyfan? Ai twyll,
yn wir, oedd y peth cudd? Y llun… Dyn yn eistedd
wrth yr organ. Mr Thomas mae'n debyg… Neu Thomas
rhywbeth, efallai… Beth oedd e'n ei chwarae…? Doedd dim
cerddoriaeth yn unman. Ond efallai nad llun o rywun yn
chwarae'r organ oedd hwn ond, yn hytrach, llun a dynnwyd i

gofnodi rhyw achlysur neu'i gilydd.

A'r gerddoriaeth wedyn. Roedd synnwyr cyffredin, tybiai Steffan, yn awgrymu bod a wnelo'r gerddoriaeth rywbeth â'r dyn wrth yr organ. Ond doedd ganddo ddim tystiolaeth i brofi hynny. Doedd dim enw cyfansoddwr wrth y darn, na dyddiad. Doedd y teitl, os teitl ydoedd, yn golygu dim iddo. *Regis Iusto Cantio El Reliqua Canonica Arte Resoluta... Regis...* Roedd mewn Lladin, fe wyddai hynny, ac yn sôn am ryw frenin, efallai. Oedd gan y brenin hwn rywbeth i'w wneud â'r achlysur 'roedd yr organydd yn cymryd rhan ynddo? Ac os felly, pa frenin? Wel, y brenin nefol, fwy na thebyg. Golygai hynny, meddai greddfau duaf Steffan, mai cranc crefyddol arall oedd 86. Ie, cranc arall, doedd dim amheuaeth. Ond na, cymer bwyll, meddai greddf oleuach, â rhyw dinc o gerydd yn ei lais: siawns nad oes yna ryw esboniad rhesymol am hyn i gyd. Rhaid peidio â rhuthro i gasgliadau, meddai, dim ond am eu bod nhw'n gasgliadau cyfleus ac yn ategu rhyw syniadau afiach sydd gen ti ynglŷn â sut mae'r byd yn gweithio.

Paulo's Royal Circus. Dim ond y daflen oedd ar ôl. Yr oedd hon, o leiaf, yn ddigon pendant o ran lle, os nad o ran amser. Heblaw am y dyddiad: 13 – 20 Mai. Oedd y manylyn hwn yn arwyddocaol, tybed? Neu a oedd yn *ymddangos* yn arwyddocaol dim ond am ei fod yno, ac am fod cyn lleied o wybodaeth arall ar gael? A beth am y 'Royal', wedyn? A oedd cysylltiad rhwng y 'Royal' yma a'r brenin yn y gerddoriaeth? Roedd angen gochel rhag mynd i gors fan hyn. Os mai twyll oedd y cyfan, dyma'r union fath o fwydyn, o gynrhonyn yn hytrach, a gâi ei ddefnyddio i'w hudo at y bachyn.

12 Cranc yn y Bae

– FE WELA' I DI TU FAS, LLIO.

Doedd Steffan ddim wedi cael cyfle i ystyried dirgelion Rhif 86 ymhellach. Roedd y ffurflen gais a'r deunydd ategol wedi gorwedd yn ei *briefcase* trwy'r dydd, yn dwyllodrus o ddof a llonydd, tra bu Steffan yn cynnal Archwiliad Effeithiolrwydd Misol gydag un o brif gleientiaid y Weinyddiaeth.

– Wy lawr yn y Bae yn barod, Llio… Cerdda' i draw 'nawr…

Roedd yr Archwiliad wedi para'n hwy na'r disgwyl am resymau sy'n rhy ddiflas i'w cofnodi'n llawn, ond a darddai o un cwestiwn canolog, sef: ai un profiad llenyddol i saith deg o bobl oedd cynhadledd ddiweddaraf y sefydliad; ynte, o ddidoli'r achlysur yn sesiynau unigol, ai un ar ddeg o brofiadau a gafwyd gan saith cant a deg o bobl? (Ac eithrio'r hen gwpl a aeth i'r gwely'n gynnar ar y nos Sadwrn, wrth gwrs, a'r bardd adnabyddus na chododd tan dri o'r gloch brynhawn trannoeth.) Neu, fel arall, o gyfrif pob cerdd a drafodwyd neu a ddarllenwyd fel y byddwn yn cyfri'r *Coco Pops* a'r *Chianti* a'r *Peach Melba Yoghurt* etc. etc. a brynwn yn Tesco, fel nwyddau gwahanol, onid 220 a 1540 yw'r gwir ffigurau, a hynny, wedyn, yn gwneud cyfanswm o 338,800 o berson-brofiadau llenyddol dros gyfnod y gynhadledd. A'r cyfan yn profi bod y corff, yn ei dyb ef ei hun, yn haeddu cynnydd o 186% yn ei gymhorthdal, a hynny, yn bennaf, er mwyn cyflogi staff ychwanegol i baratoi'r adroddiadau hunanasesu angenrheidiol ar bob un o'r person-brofiadau hynny. Ac yn y blaen.

Ta waeth, erbyn diwedd y cyfarfod maith a dyrys hwn, doedd gan Steffan ddim dewis ond llusgo ei holl bapurau draw i'r Ganolfan, lle'r oedd wedi trefnu cwrdd â Llio. Papurau'r Archwiliad, wrth reswm; hefyd y rhai a oedd yn ymwneud â chais Rhif 86, fel y nodwyd eisoes, am fod Steffan yn gobeithio bwrw golwg drostynt yn nes ymlaen ar ôl mynd adref; ond, yn ogystal â'r rheiny hefyd, llwyth nid ansylweddol o bapurau eraill a gadwai Steffan wrth law bob amser 'jyst rhag ofn', dogfennau megis y Llawlyfr Asesu a fersiwn cryno o Siart y Trigain Cam Namyn Un Tuag At Lwyr Effeithiolrwydd Gweinyddol. Na, doedd ganddo ddim dewis ond eu llusgo draw i'r Ganolfan, a'u cadw nhw'n dynn rhwng ei goesau gydol y cyngerdd, yr egwyl a'r cwbl. Roedd un peth yn sicr: ni feiddiai eu gollwng o'i afael. Os oedd y papurau'n gorwedd yn oeraidd ddigynnwrf yn y *briefcase*, roedd peth o'u cynnwys yn dal i fudlosgi yn ei feddwl.

– Wy'n ffaelu dy weld di, Llio...

Pan gyrhaeddodd Steffan y Ganolfan, roedd torf o rai cannoedd yn sefyll y tu allan. Yn eu plith, gwelodd griw o ferched y Weinyddiaeth yn ymochel dan ymbarél mawr melyn. Roedd y merched wedi sylwi arno yntau, hefyd, er iddo dynnu cwffwl ei anorac dros ei ben. A fuon nhw'n ei wylio'n dod o bell? Roedd Steffan yn ofni hynny. Os nad oedd yn camsynied yn fawr, roedd rhyw olwg ddireidus, watwarus hyd yn oed, ar eu hwynebau. Cafwyd codi llaw a gwaeddi cyfarchion diangen o frwd, ym marn Steffan, ac yna rhyw chwerthin a sibrwd, y naill wrth y llall, a'r cwbl yn achosi cryn embaras iddo. Roedd Audrey'n sefyll y tu ôl iddynt, dan ei hymbarél bach coch ei hunan, yn rhannu'r jôc. Yn eu procio, o bosib. Cododd Steffan yntau ei law a gwenu'n ôl arnynt, heb argyhoeddiad. Nid oedd Llio i'w gweld yn unman. Penderfynodd Steffan mai cam gwag fyddai paredio i fyny ac i lawr y rhesi hir er mwyn chwilio amdani.

Byddai'n edrych yn rhy debyg i gadfridog yn archwilio ei filwyr. Deuai mwy o wenu a sibrwd yn ei sgil. Trodd ar ei sawdl a sibrwd i mewn i'w ffôn bach.

– Ble rwyt ti, Llio?

A chael gwybod ei bod hi eisoes wedi mynd i mewn, ar ôl syrffedu ar sefyll cyhyd yn yr oerfel, ac yn disgwyl amdano, bellach, yn y dderbynfa.

– 'Sa di f'yna, cariad! Bydda i 'da ti yn y man!

Ond un araf ar y naw oedd y ciw, a'r rheswm am hynny, fel y gallai Steffan ganfod erbyn hyn, wrth falwenna ei ffordd tuag at y fynedfa, oedd bod mesurau diogelwch llym ar waith: mesurau hynod drwyadl ac effeithiol, doedd Steffan ddim yn amau, mesurau y byddai ef, o dan amgylchiadau gwahanol, wedi'u hedmygu a'u canmol, ond heddiw doedd ganddo mo'r amser na'r amynedd i ddygymod â thrylwyredd o'r fath.

– Byddwch yn barod i wagio eich *briefcase*, syr.

Na, nid ar chwarae bach mae dyn yn dangos ei ddillad isaf i'r byd, na lliw ei ofnau chwaith, a phethau go debyg i'w gilydd ydyn nhw hefyd, meddyliai Steffan, pethau preifat, pethau i'w cadw dan glo, o'r golwg. O'r tu mewn iddo, clywai lais bach yn gofyn: os gall ychydig o bapurau beri'r fath ofid iddo fe, Steffan, ac yntau'n hen law ar drafod awduron piwis, bygythiol, sut byddai swyddog diogelwch drwgdybus, ond anwybodus, yn ymateb i'w gasgliad o luniau a deiagramau a hafaliadau a negeseuon amwys ac awgrymog? Yn waeth na hynny, ychwanegodd y llais bach, beth petai'r swyddog, wedyn, yn cysylltu'r pethau hyn â phethau eraill, gwirioneddol frawychus, megis hanes y *ffatwah* a'r Eneiniog? A oedd yno lwybr papur i'w gael yn y *briefcase*, tybed, heb iddo wybod? Llwybr a arweiniai o'r naill i'r llall? Cafodd Steffan haint o feddwl am y fath bosibilrwydd hunllefus. Ond yn y cyfamser...

– Na, chewch chi ddim dod â hwnna i mewn i'r Ganolfan, madam…

Rhyw ddeg llath o'i flaen, gwelai Steffan neb llai na Haf Griffin, pennaeth y mudiad ieuenctid, a babi ar un fraich a fflasg o de, yn ôl pob golwg, o dan y llall, yn cael ei hebrwng o'r Ganolfan gan un o'r swyddogion diogelwch.

– Mae'n rhaid i fi dwymo bwyd y babi, w!

– Wel chewch chi ddim! Mae'r Frenhines yn dod fory… !

– Beth…?

– Shwt ydw i'n gwybod nag ych chi ddim wedi cwato dou bownd o ffrwydron ynddi…?

Doedd Steffan ddim yn siŵr p'un ai yn y babi neu yn y fflasg yr oedd y ffrwydron i fod wedi'u cuddio, yn ôl dychymyg lliwgar y swyddog, ond cawsant i gyd – y fam, y babi, a'r fflasg – eu gwthio'n ddiseremoni i'r gwynt a'r glaw mân a'r lluoedd cegrwth.

Ac os gallai babi gael ei ddiarddel am fygwth cyfraith a threfn – a babi mor uniongred o sefydliadol, hefyd – faint o obaith oedd gan Steffan i gael mynediad? Na, doedd ganddo ond un dewis. Byddai'n rhaid sleifio o'r golwg ar y cyfle cyntaf, agor y cês, tynnu allan y dogfennau mwyaf amheus a'u rhoi nhw naill ochr rhywle tan ddiwedd y cyngerdd. Ni wyddai ble. Doedd wiw iddo geisio eu cuddio yn ei bocedi, neu hyd yn oed yn leinin ei siaced, rhag ofn byddai rhyw swyddog mwy cydwybodol na'r rhelyw yn mynnu gwacáu'r rheiny hefyd. Câi chwilio yn y man.

– Rhowch eich allweddi a'ch arian mân yn y bagiau hyn, os gwelwch yn dda.

Ond doedd Steffan ddim am dynnu sylw ato ei hun, chwaith, wrth ymadael â'r ciw. Byddai'n rhaid disgwyl ei gyfle, gan ymdrechu, orau y gallai, i ymddangos yn hamddenol a di-hid. Clywodd rywrai y tu ôl iddo'n canu *Singing' in the Rain*, a dechreuodd chwibanu'n dawel. Ond nesáu roedd y drws o hyd: y drws i *ffwrnais yr awen*, ys dywedai'r geiriau anferth uwch ei ben. *Ffwrnais* oedd hon, yn wir, meddyliai Steffan, a'r diferion chwys yn plorynnu ei dalcen, er gwaethaf yr oerfel. Yna, wrth glosio at ddibyn anobaith, dyma'r swyddog yn claddu ei ben yng nghynhwysion bag rhyw anffodusyn arall. Rhuthrodd Steffan i ddiogelwch y gwyll.

Wrth lwc, a heb i Steffan sylweddoli, gan gymaint ei bryder am gynnwys y cês, yr oedd torf fawr arall wedi ymgynnull ryw ganllath o'r Ganolfan, lle'r oedd un o'r hen ddociau wedi cael ei droi'n amffitheatr awyr agored. Yn wir, yr oedd y dorf hon yn fwy o lawer na'r un y bu'n rhan ohoni, a chynigiai loches berffaith iddo. Cerddodd draw, yn sioncach ei gam erbyn hyn, i ymuno â'r lluoedd. A gyda hynny, dechreuodd y canu. Canu ychydig yn betrus ac ansicr i ddechrau, cyn magu hyder o linell i linell. Y rhain, felly, fu'n canu *Singing' in the Rain*, meddyliai Steffan, gan ddiolch i Ffawd am ei drugaredd. Aeth y canu o nerth i nerth, a band pres yn gyfeiliant hwyliog iddo.

Pe dymunwn olud bydol
chwim adenydd iddo sydd...

Cymerodd Steffan daflen geiriau gan un o'r merched a oedd yn eu dosbarthu ymhlith y dorf. Anadlodd yn ddwfn. Gallai ymlacio am ychydig. Yr oedd ei alibi'n gyflawn. Yr oedd bellach yn ganwr dilys, trwyddedig a châi dwrio ym mherfeddion ei gês dan gochl ei statws newydd, heb dynnu sylw na cherydd gan neb. Yn wir, ac er gwaethaf yr awel fain a'r briwlan cyson, yr oedd Steffan erbyn hyn yn cael cymaint o flas ar y canu ac ar ymgolli yn y dorf, bu ond y dim iddo anghofio am ei dasg.

... golud calon lân, rinweddol

yn dwyn bythol elw fydd.

Neges destun gan Llio a'i dihunodd o'i anhysbysrwydd clyd.

– rhd mnd mn bl diawl wt ti?

Dechreuodd Steffan dapo ei ateb â'i law dde.

– ar fy ff

Yn anffodus, ni orffennodd y neges, er mai dim ond tair llythyren oedd yn weddill: llai, o ddefnyddio iaith destun. Oherwydd, yn ei frys i ddychwelyd at ei gymar ac i ail-orseddu rhyw rith o drefn yn eu bywyd, fe agorodd ddrws, yn hytrach, i anhrefn: nid rhyw annibendod dibwys, chwaith, rhyw flerwch, rhyw frycheuyn bach ar wyneb Trefn, ond Anhrefn yn llawn ystyr y gair, yr ystyr yr oedd yr Athro Awberry Thomas wedi'i phriodoli iddo. Wrth dapo'r neges â'i law dde, a dal y daflen ganeuon rhwng ei ddannedd, fe geisiodd Steffan, ar yr un pryd, ddatgloi ei *briefcase*, gan feddwl tynnu allan y papurau tramgwyddus a'u saco nhw i lawr ei drywser, yn niffyg unrhyw gynhwysydd arall, a chan hyderu na fyddai'r swyddog diogelwch mwyaf pybyr am chwilmentan yn ormodol yn y parthau hynny.

Dyna oedd y bwriad. Dyma'r hyn a ddigwyddodd. Yn gyntaf, fe gwympodd ffôn bach Steffan o'i law wrth iddo geisio gwasgu'r llythyren 'o'. Yna, bu'r dyn mawr barfog a fu'n sefyll wrth ei ochr mor garedig â chodi'r ffôn bach a'i roi yn ôl i Steffan.

– Dyna chi. Yn dal i weithio, gobeithio!

Ond mor fawr yw'r bwlch rhwng bwriad a chanlyniad! Wrth ynganu ei ddiolch i'r dieithryn, anghofiodd Steffan fod y daflen ganeuon yn dal yn sownd yn ei geg. Cipiwyd y daflen gan wth o wynt. Ac unwaith eto, cafodd greddf y blaen ar reswm. Mewn ymdrech wyllt i fachu'r papur, hyrddiodd Steffan ei ddwy law i'r awyr.

– Brafo! gwaeddai'r dyn barfog.

Ond byr fu'r gorfoleddu. Wrth achub y daflen, gollyngodd Steffan ei afael yn y cês. Cwympodd hwnnw, gan daro'r llawr a'r fath ergyd, beth allai ei wneud ond agor ei fol a chwydu ei gynnwys i drugaredd y gwynt? A dyna a wnaeth.

Morio canu roedd y dorf o hyd.

And in the evening by the firelight's glow

You'll hold me close and never let me go…

Ac o weld papurach Steffan yn cael eu chwythu dros eu pennau, a chan feddwl mai taflenni caneuon oeddent, yn hytrach nag adroddiadau archwilio a ffurflenni cais a llawlyfrau rheoli ac ystadegau, dyma nifer ohonynt yn ymuno yn y rhialtwch a thaflu eu papurau hwythau i'r awyr, fel haid o golomennod gwynion.

We'll gather lilacs in the spring again…

★ ★ ★

Awr yn ddiweddarach, eisteddai Steffan ar ris isaf yr arena. Roedd y canu wedi dod i ben a'r dorf wedi teneuo. Er gwaethaf blerwch eithriadol ei ddillad a'i wallt, roedd sylw Steffan wedi ei hoelio ar ei dasg gyda'r unplygrwydd dwys a gysylltir, yn amlach, â gwaith trwsio watsys neu dorri diemyntau. Roedd wedi dodi ei *briefcase* o'i flaen ac, ar y ddesg honno, roedd yn didoli a chymhennu'r holl bapurau roedd wedi eu casglu ers y ddamwain anffodus. Hyd y gallai weld, roedd wedi cywain y rhan fwyaf ohonynt, yn ogystal â rhai ugeiniau o daflenni caneuon, amryw fwydlenni bwytai cyfagos, a rhyw fflwcs a sbarion eraill yr oedd pobl wedi'u rhoi iddo – yn bacedi creision a blychau byrgyrs a hen facynon papur a phetheuach ffiaidd eraill – gan feddwl mai dyn casglu sbwriel ydoedd.

– Ach-y-fi!

Cyrhaeddodd waelod y pentwr, a sychu olion y saws coch oddi ar ei fysedd. Ie, achubwyd y rhan fwyaf o'r papurau, er bod hanner y rheiny wedi'u gwlychu a'u malu a'u stwnsho dan draed a byddai angen eu rhoi nhw yn y cwpwrdd crasu dros nos cyn y byddent o unrhyw werth eto. Y rhan fwyaf. Ond yr oedd un neu ddau ar goll o hyd.

Cododd Steffan ei ben ac edrych o'i gwmpas. Roedd y dorf wedi symud i dir uwch erbyn hyn, er mwyn cael gwell golwg ar y tân gwyllt. Gadawodd wacter mawr ar ei hôl. A fentrai chwilio eto trwy'r mochyndra afiach? Byddai'n haws gweld erbyn hyn, meddyliai Steffan, a'r lle yn wag. Ond nid yn hollol wag, chwaith, oherwydd, o gysgodion pellaf yr arena, yn ymyl y llwyfan, fe welodd Steffan ddyn yn dod i'r golwg, a hwnnw'n ddyn anghyffredin o drwsiadus, mewn siwt ddu a dici-bo gwyn. Dyn tal, golygus, a chanddo osgo urddasol i'w ryfeddu. Dyn, os nad oedd Steffan yn camsynied, a oedd bellach yn brasgamu tuag ato ef, gan ddal darn o bapur yn ei law.

– Nene 'dech chi'n edrych amdano, mae'n debyg?

Estynnodd y papur i Steffan.

– Cael ei chwythu acw gan y gwynt wna'th o, siŵr iawn.

– Acw?

– I ganol y llwyfan… Achosi tipyn o ddryswch, hefyd… A finne'n meddwl bod a wnelo fo rywbeth efo'r canu… A neidio arno fo wedyn… Er mwyn ei achub o… Fel ene…

A dyma'r dyn urddasol yn dangos i Steffan sut yr achubwyd y papur gan un naid chwim a phenderfynol.

– Fel ene… Gobeithio nad ydy o wedi'i faeddu gormod…

– Na, na… Diolch o'r galon… Mae'n berffaith iawn…
Cymrodd Steffan y papur o law'r dyn.

– Ydy, ydy… yn berffaith iawn, mil diolch ichi… bach yn
fudur, efallai, rhyw fymryn bach, 'na'i gyd…

– Ha! Go dda, was! Go dda!

– Mm…?

– Bach yn fudur. Mae o, hefyd!

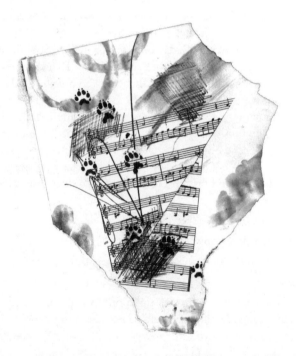

Ffig. 8   Y RICECAR ar ôl iddi gael ei rwygo a'i sathru dan draed

Mm…?

– Wel, Bach, wrth gwrs. Johann Sebastian. Neb llai. Y fo sgwennodd y darn ene, fel 'dech chi'n gwybod, mae'n siŵr…

Ar hynny, gollyngwyd y tân gwyllt cyntaf i'r awyr, yn ffynhonnau melyn a choch a gwyrdd.

– Ond ma 'ne un peth sy'n fwy doniol na 'nene…

– Mm…?

Ac ar ôl y ffynhonnau, ffrwydrodd llond gardd o flodau amryliw, gan wasgaru eu petalau dros ddyfroedd llonydd y bae.

– Mae o'n deud 'thych chi am fynd i chwilio amdano… Ha!… A dene 'dech chi wedi'i wneud, ynte?

– Mm…? Chwilio?

– Chwilio! Dyna 'di RICECAR. Dech chi'n dallt? Dech chi fod chwilio am yr alaw gudd.

– Eh…?

– Yr alaw gudd. Alaw'r brenin. Rhaid ichi chwilio am alaw'r brenin. Dene mae o'n 'i ddeud…

Ac o'r bae, fe gododd elyrch tanllyd, saith ohonynt, un ar ôl y llall, yn hyrddiadau penderfynol o liw trwy'r nos ddudew, cyn plygu eu gyddfau'n ôl, ac aros, ac aros, gan wrthod ildio i rym disgyrchiant.

– A gyda llaw…

– Ie…?

– Cranc maen nhw'n galw 'nene, am fod…

Ond collwyd gweddill ei frawddeg yn sŵn y rhaeadrau tân. Beth bynnag, roedd Steffan wedi clywed digon am un noson.

13 Chwilio yn y Llyfrgell

DECHRAU GYDA'R HYN sydd agosaf atoch chi. Dyna fu'r
cyngor gorau i awduron erioed. Yr oedd yn gyngor digon
addas i'r biwrocrat, hefyd, wrth iddo geisio dilyn y Llwybr
Papur yn ôl i'w darddiad. Gwyddai Steffan y byddai gofyn
iddo fod yn rhywbeth amgenach nag awdur na biwrocrat er
mwyn datrys y benbleth bresennol, ond nid oedd hynny'n
siglo ei ffydd yn y dulliau rhesymegol o weithio yr oedd
blynyddoedd o wasanaeth yn y Weinyddiaeth, yn ogystal â'i
fagwraeth ddisentiment, wedi eu gwreiddio'n ddwfn yn ei
gyfansoddiad. Ystyr y ffydd honno i Steffan oedd mai adar o'r
un lliw a ddôi o'r un nyth, yn ddi-ffael; neu, petai'n gollwng
cyfrifiadur o ffenest ei swyddfa ar y trydydd llawr byddai'n
taro'r pafin yn yr un lle ac ar yr un cyflymder heddiw, yfory,
drennydd a thradwy; neu, petai ei frawd, brawd nad oedd
wedi'i weld ers deng mlynedd, yn taro'r bêl wen yn erbyn y
bêl goch ar yr un bwrdd snwcer, fel yr arferai ei wneud yn
llawer rhy aml, a'i tharo gyda'r un nerth, o'r un ongl, yna
byddai'r bêl yn cwympo – ploc! – i'r boced bob tro. Ac felly,
yn yr achos hwn, a bwrw nad twyll oedd y cyfan, nid oedd
ond angen dilyn llinyn y dystiolaeth ac fe gâi hyd, yn y pen
draw, i'w ffynhonnell. Onid dyna neges y gerddoriaeth?.
Roedd ymgeisydd 86 ei hun wedi dweud mai amlinelliad
oedd y darnau i gyd. Onid oedd hynny'n cadarnhau bod yna
batrwm o ryw fath i'w ddirnad, o chwilio'n ddyfal amdano?
Chwilio. Dyna'r gorchymyn, fel roedd y dyn gwybodus
yn y dici-bo wedi dweud yn ddigon plaen. Ac o dderbyn
gorchymyn, ymateb greddfol Steffan oedd ufuddhau.

A dyna pam aeth Steffan i Adran Astudiaethau Lleol
Llyfrgell y Ddinas ar fore dydd Sadwrn, drannoeth y tân
gwyllt, ac wyth niwrnod yn unig cyn y byddai'n rhaid iddo
drosglwyddo'r ffeils i'r Adran Draws-werthuso Strategol.

Paulo's Royal Circus. Dyna'r eitem agosaf ato, yn
ddaearyddol, beth bynnag. Roedd ganddo ddyddiad. Cael hyd
i'r flwyddyn fyddai'r cam nesaf. Eglurodd wrth y llyfrgellydd
natur ei gwest a chael golwg hurt yn ôl.

– Does gyda chi ddim byd ar hanes y syrcas ffor' hyn te?

Gŵr tal, tenau, llwyd ei groen oedd y llyfrgellydd. Siglodd ei
ben yn araf.

– Nac oes.

A phetasai Steffan yn ddieithryn, diau y byddai ei ymchwil
wedi dod i ben yn y fan a'r lle. Ond gwyddai Steffan, ar
sail ei adnabyddiaeth hir o'r dyn dwys a phwyllog hwn, nad
dyna'r cyfan oedd ganddo i'w gynnig. Gwyddai hefyd fod
yr adran hon o'r llyfrgell, a neilltuwyd ymhell o brysurdeb y
lloriau islaw, yn gweithredu yn ôl clociau arafach na'r rhelyw
o glociau'r sefydliad. Bu rhai wrthi fan hyn am wythnosau,
am fisoedd, am flynyddoedd hyd yn oed, yn mân gribo eu
ffordd trwy hen bapurau, archifau, cofnodion, cofrestrau ac
adroddiadau, er mwyn darganfod eu Greal personol.

– Ond chi yw'r ail i holi amdano yn ystod y pythefnos
diwethaf.

Saib arall. Mwy o ymbwyllo.

– A dim ond chi'ch dau sydd wedi holi amdano erioed, am
wn i.

– Yr ail mewn pythefnos? Ydych chi'n cael dweud pwy oedd
y cyntaf?

Saib hir.

– Dwi ddim yn siŵr ydw i'n cael dweud, ond alla i ddim

dweud… oherwydd chymerais mo'i henw hi.

– Menyw oedd hi, felly?

– Ie. Menyw oedd hi. Ond do'n i ddim yn ei nabod hi.

– Ydych chi'n cael dweud a ffindiodd hi rywbeth?

– Dim ond beth roies i iddi… P'un a ffindiodd hi rywbeth yn hwnna, alla i byth â dweud…

– A beth oedd hwnna, os ca i ofyn?

– Gofynnodd hi am gael gweld hen rifynnau o'r *Western Mail*.

– Ife? Mm… A pha mor hen yn gwmws… Ydych chi'n cofio?

Edrychodd y llyfrgellydd ar y nenfwd. Nid oedd yn ddyn i'w frysio.

– Nac ydw. Ddim yn fanwl. Rywbryd wedi'r rhyfel diwethaf… mm…

– Ie?

– Na… dyna i gyd galla' i gofio.

Dechrau yn y dechrau, felly. Pryd daeth y rhyfel i ben? Medi 1945? Tachwedd? Hydref amdani.

Roedd yr hen bapurau wedi'u rhwymo'n gyfrolau mawr, trwm, a hawdd deall diffyg brwdfrydedd y llyfrgellydd ynglŷn â'r ymholiad o'i weld yn pwlffacan dan eu pwysau. Serch iddynt gael eu rhwymo'n gadarn, roeddent wedi melynu a breuo'n ddifrifol. Cyn bo hir, meddyliai Steffan, dwst mân fyddai'r cyfan. Oni bai, wrth gwrs, bod rhywun yn eu trosglwyddo i ficroffilm, a hynny ar fyrder. Roedd yn siŵr y byddai'r llyfrgellydd, druan ohono, yn cytuno; ond barnai nad hwn oedd yr amser gorau i grybwyll y mater.

Agorodd y gyfrol gyntaf a llanwyd ei ffroenau â llwch yr oesoedd. Pesychodd. I Hydref 1945. Dim byd. 2 Hydref. Dim. 3 Hydref… Dim. Roedd rhaid troi pob tudalen yn ofalus iawn, gan ddefnyddio cledr y llaw dde, ac yna'r llaw chwith, i sicrhau na rwygai dan ei phwysau ei hun. Fel dadrwymo mymi o'r Aifft, y gallai'i ddirgelion hynafol ddadfeilio'n ddim o dan y llygedyn cyntaf o olau. 9 Tachwedd. Pesychodd eto. A thisian. 11 Tachwedd. Bette Davies yn *The Corn is Green*. A pha ots am hynny? Heb yr un cliw na chanllaw, doedd gan Steffan ddim dewis ond sganio pob colofn o bob tudalen o bob rhifyn, gan geisio anwybyddu, orau y gallai, y sgwarnogod dirifedi hynny a arferai, trwy bipo'n slei trwy fwlch yn y clawdd a siglo eu clustiau bach ciwt, ei arwain ar ddisberod. 15 Tachwedd. *'Join the Crusade Against Waste of Bread'*. A cheisio ymatal rhag gwneud y peth a ddaeth mor naturiol iddo ag anadlu a breuddwydio, sef, yn hytrach na bwrw ati i gyflawni'r dasg a'i hwynebai, ei fesur. 6 Rhagfyr. Richard Tauber yn canu… a rygbi. 16 Tachwedd. Mwy o rygbi. 17 Tachwedd. Dim.

Serch hynny, ei fesur a wnaeth. A chael, o wneud y syms yn fras yn ei feddwl, bod angen sganio 1176 tudalen ar gyfer pob mis. 22 Tachwedd. Dim… A phetai'n treulio… 23 Tachwedd. Dim byd. 24 Tachwedd… a phetai'n treulio deg eiliad i sganio pob un… 25 Tachwedd. Dim… byddai'n cymryd… 26 Tachwedd… cartŵn… *'Do you think there can be a secret hidden in it somewhere?'* Wel, gwir y gair! 29 Tachwedd… byddai'n cymryd… byddai'n cymryd dros dair awr i fynd trwy fis o bapurau!

A doedd syms Steffan ddim yn bell ohoni. Erbyn hyn, roedd wedi bod wrth ei dasg ers pum awr a deugain munud, heb ddim llwyddiant ond gan achosi, tybiai Steffan, gryn niwed i'w ysgyfaint a'i lygaid. Caeodd gloriau coch y gyfrol gyntaf a'i symud o'r neilltu. Agorodd yr ail, gan roi ei law dros ei drwyn a'i geg er mwyn arbed anadlu'r llwch. Ond, am ryw reswm, ni chafwyd cymaint o lwch y tro hwn. Doedd y papur ddim wedi melynu cymaint, chwaith: testun syndod i Steffan,

gan nad oedd y gyfrol hon ond ychydig wythnosau'n iau na'r
llall. Yn wir, parai gymaint o syndod iddo, fel na allai ymatal
rhag ymchwilio ymhellach i'r mater, a threuliodd yr awr
nesaf yn adran wyddoniaeth y llyfrgell, yn astudio sut a pham
mae papur yn dirywio. Dysgodd am effeithiau'r awyr ac asid.
Cafodd wybod, hefyd, bod yr effeithiau hyn yn gyfyngedig,
at ei gilydd, i bapur a wnaed o goed. A bu'n pendroni wedyn
ynglŷn â sut i gymhwyso'r ffeithiau hyn at yr achos dan sylw.
A fu newid yng ngwneuthuriad y papur yr argraffwyd y
Western Mail arno yn ystod y cyfnod dan sylw, tybed? Ynte, a
oedd rhyw lyfrgellydd mentrus, rhywun o flaen ei amser ym
materion cadwraeth ac ati, wedi trin y papurau â rhyw alcali
pwrpasol? Calsiwm carbonad, efallai. Neu bariwm hydrocsid.
Tybed.

  A'r tu hwnt i'r cwestiynau hyn, wrth gwrs, y gorweddai'r
cwestiwn sylfaenol, tragwyddol, mwyaf treiddgar o'r cwbl,
sef: a oedd Steffan un tamaid yn gallach o ddarganfod yr
esboniadau posibl hyn? Ac am ryw hyd, tra bu'n ymdrochi
yn y mwydion coed a'r cemegau, doedd dim gwahaniaeth
ganddo. Y chwilio oedd ei wobr ei hun. Ond am ryw hyd yn
unig. Am hanner awr wedi pedwar, canodd cloch i rybuddio
pawb bod y llyfrgell am gau ymhen hanner awr. Trawodd
Steffan y sgwarnog ddiweddaraf ar ei phen a dychwelyd at ei
ddesg.

  Ie, waeth beth oedd y rheswm, roedd papur yr ail gyfrol
yn rhyfeddol o glaerwyn am ei oedran. Dechreuodd droi'r
tudalennau, yn gynt y tro hwn, am eu bod nhw'n llawer llai
bregus na'r rhai blaenorol. 8 Ionawr 1946. Dim. 9 Ionawr.
Dynion yn chwarae snwcer. Biliards, efallai. G. O'Brien ar y
chwith. O'Brien. Tybed? Enw digon cyffredin. 10 Ionawr…
Ac yna gwelodd Steffan rywbeth nad oedd wedi sylwi arno
o'r blaen. Darn papur. Ac yn wir fe gafodd gryn anhawster i

ddeall sut roedd e wedi'i golli. Oherwydd roedd y darn papur hwn, er ei deneued, yn goch llachar ac yn ymestyn rhyw chwe modfedd o ymyl y gyfrol. Bwrodd gip draw at ddesg y llyfrgellydd. A wyddai ef rywbeth am hyn, tybed? Ond doedd y gŵr tawel a dwys ddim yno.

Aeth Steffan at y papur coch, gan anwybyddu mis Chwefror a mis Mawrth a mis Ebrill, a'r holl ddirgelion a allasai fod yng nghudd ym mhlygion eu dyddiau, a'i gael ei hun ar ddechrau mis Mai. Y cyntaf o Fai. Roedd y darn papur yn nodi'r cyntaf o Fai. Ac i beth? Sganiodd y colofnau. Dim. Aeth i'r dudalen nesaf. Dim! A'r nesaf. *How to get into a Crack Force and earn 320 a month and all found! Decide on the Palestine Police Force. A MAN'S JOB!*... Neidiodd Steffan i'r ail o Fai. Dim byd ar y dudalen gyntaf. Canodd y gloch eto. Roedd ganddo chwarter awr. Aeth i'r ail dudalen. Cyhoeddi *Trwm ac Ysgafn* T. J. Morgan... Edrychodd draw tuag at ddesg y llyfrgellydd eto. Roedd wedi dychwelyd. Ac roedd yn edrych arno. Ddim heb reswm, chwaith, meddyliai Steffan, gan mai ef oedd ei gwsmer olaf. Neidiodd eto, i 3 Mai... Nuremburg... a...

A dyna hi! Dyna hi! Dyna'r syrcas. Western Avenue, Caerdydd, o'r trydydd ar ddeg hyd yr ugeinfed o Fai. Yr un dyddiadau'n union. Ffaith i chi! Nid oedd yn cael ei dwyllo wedi'r cyfan. A 1946... 1946 oedd y flwyddyn. Ffaith arall i chi! Ildiodd Steffan i donnau cynnes ei fuddugoliaeth. A chan nad oedd neb arall wrth law i'w longyfarch, fe longyfarchodd ei hun. A sut, meddai llais bach y tu mewn iddo, *sut yn gwmws* y tynnoch chi'r ffeithiau hyn i'r wyneb, yn dalpau mor fawr a llachar? Wel, daeth yr ateb parod a brwd, trwy gloddio dyfal a rhesymegol, wrth gwrs! Ac wrth ymfalchïo yn ei lwyddiant, aeth y papur coch yn angof. Doedd e ddim yn y lle cywir, beth bynnag, nac oedd? Felly, hefyd, gwynder y papur. Roedd wedi darganfod sawl esboniad posibl am y rhyfeddod hwnnw. A'r fenyw a fu'n holi am y syrcas? Roedd yn fwy cyfleus peidio â meddwl amdani hi. Ar hyn o bryd, o leiaf. Oherwydd, am y tro, ffeithiau a lenwai feddwl Steffan. Ffeithiau diymwad. O'r diwedd, roedd wedi cael gafael ar ffeithiau.

14 Colton

GARTREF, YN CARNFORTH, roedd Llio wedi dechrau pacio. Dychwelodd Steffan o'r llyfrgell, yn ddigon bodlon ei fyd ar ôl ei ddarganfyddiad diweddaraf, a chael y byd hwnnw wedi'i droi tu chwith allan unwaith yn rhagor.

– Alla' i ddim aros nes bod y babi'n dod, Steffan. Mae cymaint o bethau i'w gwneud.

– Ond does gyda ni unman i fyw...

– Maen nhw wedi dod.

– A dwi ddim wedi rhoi fy notis i mewn eto... Beth sy wedi dod?

– Y tŷ... Manylion am y tŷ ry'n ni'n mynd i'w brynu.

– Ond shwt...

Aeth Llio at fwrdd y gegin a chodi darn o bapur. Cymerodd Steffan olwg sydyn drosto.

– Ond dwedodd Ebbways fod angen lot fawr o waith ar y tŷ hwn...

– Does dim ofan gwaith arnon ni, Steffan bach, oes e? Dyna pam ryn ni'n prynu tŷ, yntefe? Fel 'yn bod ni'n dau yn gallu gweithio fel ryn ni'n moyn, a magu'r babi mewn lle mwy addas na rhyw dŷ teras sy'n rhy fach i fagu llygoden.

– Wel...

– Dwyt ti ddim wedi newid dy feddwl, wyt ti? A finne wedi pacio fy offer paentio i gyd, 'yn llyfre i i gyd, a'r llestri...

– Y llestri?

– Dim ond y llestri gore. Nage'r stwff bob dydd…

– Ond y…

–… a dwi ddim eisie mynd trwy'r strach 'na eto.

– Ond wyt ti'n siŵr taw dyma'r lle rwyt ti eisie byw…? Mae e i gyd i'w weld mor sydyn…

Edrychodd eto ar y manylion. Doedd dim llun. Doedd dim cyfeiriad, hyd yn oed. Dim ond yr enw. Colton. A chyfeirnod map. Pa fath o dŷ allai hwnna fod, heb gyfeiriad? Sut roedd disgwyl i ddyn gael hyd i dŷ heb gyfeiriad?

– Llio, smo ni'n gwybod dim am y lle…

– Steffan, rwy'n disgwyl babi ymhen tri mis, a dwi ddim eisiau magu babi mewn cwt ieir. 'Drycha… mae tair stafell wely yn y lle Colton ma, dim bo' fi i am gadw'r enw, mae'n hen enw gwirion ar dŷ yng nghefen gwlad Cymru, cawn ni wared ar hwnna, ac wedyn, mae dwy stafell lawr stâr, ac mae un o'r rheiny'n ddigon o faint i fod yn stiwdio i fi. Ac mae hynny heb gyfri'r gegin… Mae gardd fawr… A thipyn o nant yn rhedeg heibio… Sa i'n gweld dim byd yn bod ar y lle.

– Ond dwedodd y dyn…

– Dwedodd y dyn! Dwedodd y dyn! Pam nad yw e'n dweud fan hyn, te, os oes angen lot o waith ar y tŷ…? A ta beth… rwy'n moyn bod ar bwys Mam a Dad tra bod y babi'n fach… iddo fe gael bod yn rhan o deulu…

– Ond…

–… Ac mae teulu'n bwysig, a bydde fe'n bwysig i tithe hefyd, Steffan, 'set ti wedi cael magwraeth hanner normal… A dwi ddim eisie gweld yr un bach yn tyfu lan heb fod 'da fe deulu'n gefen iddo fe, a chymdogaeth o ryw fath, nage dim ond rhesi ar ôl rhesi o dai yn gwmws yr un…

– Ond shwt…

−…yn gwmws yr un peth, a'r bobol yn symud mewn a mas o hyd, a neb yn sefyll yn ddigon hir i ddala annwyd bron, a ble elai fe i 'ware wedyn, a…

– Ond shwt wyt…

−… a dim parc yn agos aton ni, a'r heolydd yn llawn ceir a baw cŵn, a drygis tu fas i…

– Ond shwt wyt ti'n…

−…tu fas i'r ysgol, a phlant bach… wyt ti'n gwrando, Steffan?

– Ond shwt wyt ti'n gwybod 'i fod e…

−…hyd yn oed plant bach… yn gaeth i…

    …Ond shwt wyt ti'n gwybod 'i fod e'n agos atyn nhw? Does…

−… plant bach yn gaeth i…

−… does dim cyfeiriad, Llio, does dim cyfeiriad yn y manylion, mae hynny'n meddwl 'i fod e 'nghanol unman, ym mhen draw…

−… yn gaeth i…

−… heb siop, heb dafarn, heb ysgol, heb fysus, heb signal i dy fobeil fwy na thebyg, heb fywyd, heb ddim… ond defaid! Cawn ni edrych ar y map os nad…

−… yn… Beth?

– Map!

– Map?

– Ie, mae gyda ni fap… Mae cyfeirnod map yn y manylion… 'Drycha…

    Tynnodd Steffan fap OS 146 o ddrâr y dresel.

– Beth yw'r rhif eto?

– Saith chwech saith dau wyth saith

– Saith chwech saith… ar draws… Dau wyth saith… Dyma fe… Mae 'na dŷ f'yna.

– Wrth gwrs bod 'na dŷ f'yna, Steffan. Dyna'n tŷ newydd ni.

15 Syrffio

YR ORGANYDD OEDD NESAF. Roedd Steffan yn cymryd
yn ganiataol, er nad oedd dim tystiolaeth i'w brofi, fod
cysylltiad rhwng llun yr organydd a'r gerddoriaeth. Bach a
gyfansoddodd y darn, yn ôl y dyn â'r dici-bo. Ar gefn y llun
cafwyd y geiriau 'Nid Handl!' Wel, er gwaethaf y camsillafu,
nid Handel oedd Bach, yn bendant. Ond roedden nhw'n
gyfoeswyr. Awgrymodd yr ebychnod fod rhyw jôc breifat ar
waith yma nad oedd gan Steffan yr allwedd iddi. Ddim eto.
Ond erbyn hyn, yr oedd wedi canfod cysylltiad diamheuol
rhwng yr hen ddyn a'r syrcas. 3 Mai oedd ar gefn y llun, ac ar
yr un dyddiad yn union y gwelwyd hysbyseb am y syrcas yn y
Western Mail. Ac er na allai ddirnad yr ystyr a gelai y tu ôl i'r
cysylltiad, yr oedd ei ymdrechion yn dechrau dwyn ffrwyth.
Dyfal donc.

Fyddai Steffan ddim yn hoff o unrhyw fath o syrffio. Ni
hoffai fod ar drugaredd tonnau na thechnoleg na'r un grym a
allai siglo ei gydbwysedd brau. Gwell ganddo droedle sicr ar
dir cadarn, gan symud yn bwyllog, gam wrth gam, a hynny
dim ond pan fyddai'n rhaid a phan fyddai'r nod o fewn golwg.
Anfynych iawn, felly, y byddai'n taflu ei hun i ganol y we
fyd-eang heb wybod i sicrwydd mai ef oedd y corryn ac nid
y gwybedyn. Ond y tro hwn, roedd pethau'n wahanol. Ni
welai'r un nod, ac roedd y cliwiau a gawsai gan 86 wedi'u
gwasgaru yma a thraw fel nad oedd gobaith yn y byd symud
o'r naill i'r llall heb gymryd camau brasach o lawer na'i rai
bach carcus arferol. Yn waeth na hynny, yr oedd amser
bellach wedi magu dannedd ac roedd y dannedd hynny'n

dechrau cnoi ei sodlau. Doedd dim unman arall i droi.

Beth i'w roi'n gyntaf? Y syrcas? Bach? Organ? Thomas? 1946? 3 Mai? A wnâi unrhyw wahaniaeth?

Paulo's Royal Circus + Bach Clic!

Clywodd sŵn crensian ym mherfeddion y peiriant… Dim canlyniadau.

Dim!

Thomas + 1946 Clic!

Mm. Saith mil o bosibiliadau ond… Yn ôl y ddwy dudalen gyntaf… Dim byd o werth.

Thomas + Bach Clic!

435,000 o bosibiliadau. Wel, wrth reswm, meddyliai Steffan, roedd sawl un wedi cael ei fedyddio'n Thomas Bach. Gellid diystyru'r rheiny ar unwaith. Serch hynny, cafodd ei hudo am dros hanner awr gan atgofion rhyw hen Americanwr hynaws a arferai ymddiddori mewn blodau a threnau model. Pan ddychwelodd i'w orchwyl, gwelodd fod amryw byd o Thomasiaid wedi arwain, canu, chwarae, recordio, golygu a chyhoeddi gweithiau Bach, a bod nifer sylweddol wedi ysgrifennu am y teulu: y tad, y fam a'r meibion, ar wahân neu gyda'i gilydd, heb sôn am yr holl blant a fu farw'n ifanc, yr holl fynd a dod, o dre i dre, o dŷ i dŷ, amgylchiadau cyfansoddi'r cantata hwn a'r tocata arall, ac yn y blaen. Bu Steffan wrthi am awr a hanner arall yn hidlo'r rhain. Ac roedd hefyd ambell Bach nodedig arall (yn ôl nifer y gwefannau a gysegrwyd iddynt) nad oedd yn gyfarwydd i Steffan na hyd yn oed yn ymhél â cherddoriaeth, gan gynnwys un a fu'n troi ym myd meddyginiaethau llysieuol ac yn llythyra â rhyw Thomas… ond, mewn difrif, oedd disgwyl iddo gwrso sgwarnogod fel 'na?

Ac yna, ynghanol y cwbl, gwelodd Steffan yr hyn yr oedd

yn chwilio amdano. Neu'r hyn yr oedd yn barod iawn i gredu oedd nod ei ymchwil. Nid Thomas oedd fel y cyfryw, ond yn hytrach rhyw *Thomaskirche*. Ond ni waeth am hynny. Dyma, o'r diwedd, ryw fath o Thomas yr oedd ganddo gysylltiad uniongyrchol â Bach, cyfansoddwr y *Ricecar*.

Clic!

Ni fedrai Steffan lawer o Almaeneg ond fe wyddai, ar sail gwyliau dyddiau ysgol, mai *eglwys* oedd ystyr *kirche*. A fan hyn, yn yr eglwys hon, y *Thomaskirche*, yn ôl y wefan, y bu'r cyfansoddwr yn gweithio am ran helaeth o'i oes: am saith mlynedd ar hugain, i fod yn fanwl. Fan hyn yr oedd ei weddillion corfforol. Er nid fan hyn, mae'n debyg, y cawsai ei gladdu gyntaf. Thomaskirche, Leipzig. Eglwys Thomas yn Leipzig. Ai Leipzig oedd y ddolen gyswllt, tybed? Ac os felly, a oedd cysylltiadau pellach i'w canfod? Gyda'r dyn wrth yr organ, efallai? Organ eglwys oedd yn y llun, mae'n rhaid. Ai organ Thomaskirche? Ai yn 1946 y tynnwyd y llun? Bu Steffan yn ystyried y cwestiynau hyn, ac eraill, dros goffi; ac ystyried, hefyd, oblygiadau'r gwahanol atebion posibl iddynt. Ac o farnu, dros goffi arall, nad oedd ei fyfyrdodau'n arwain i unman fe benderfynodd, yn gam neu'n gymwys, ei fod wedi dihysbyddu'r wybodaeth y gallai Thomas + Bach ei chynnig iddo. Yr oedd yn bryd archwilio cyfuniadau eraill. Bwrodd Steffan ati o'r newydd.

Leipzig + 1946 Clic!

Tipyn llai y tro hwn. Y brifysgol yn ail-agor. Lluniau o'r difrod a wnaed gan y cyrchoedd bomio, a'r gwaith clirio'n dechrau. Llun o'r miloedd yn Augustusplatz yn galw am sefydlu gweriniaeth sosialaidd. 1946. Blwyddyn geni Anton, Leipzig. Anton? Aeth Steffan i'r dudalen nesaf. Ac yna, dan bwysau ei chwilfrydedd, camodd yn ôl. Anton pwy? Agorodd y wefan. A theimlo'r chwys yn ddrain bach ar ei gorun.

Anadlodd yn ddwfn a chydio'n dynn yn ochrau ei gadair. *Ganed 7 Awst 1946 yn Sŵ Leipzig.* Yn gwbl ddamweiniol ac eto, fe deimlai, yn gwbl anochel, yr oedd yn ôl ym myd yr hipo. *Hippopotamus Amphibius.* Y Behemoth. Oherwydd dyna pwy oedd Anton. Yr hipo cyntaf i gael ei eni yn Leipzig wedi'r rhyfel. Aeth Steffan ati i ddarllen. Wedi iddo gael ei ddiddyfnu, ei symud i Kiev. Yn rhan… yn rhan o iawndal Yr Almaen i'r Undeb Sofietaidd! Babi o hipo'n iawndal! Ystyriai Steffan am ychydig eiliadau sut y byddai terfysgoedd daear yn cyffwrdd â phawb a phopeth yn ddiwahân, y da a'r drwg fel ei gilydd, gan chwalu hyd yn oed y teulu hwn o hipos diniwed. Treuliodd funud neu ddwy hefyd yn rhyfeddu at ffrwythlondeb y fam, Grete, a fu'n planta am 35 mlynedd; ac at greulondeb Natur, a laddodd dros hanner ei hepil yn eu babandod. Ai dyna arwyddocâd y cysylltiad, o bosibl? Ai myfyrdod ar erchylltra hanes, ar ddallineb ffawd, oedd mewn golwg gan ymgeisydd 86? Ac o ymgolli yn y posibilrwydd hwn, ciliodd ofnau Steffan rywfaint ac fe'i lapiodd ei hun mewn cot fawr o ddoethineb clyd.

Ond sut roedd y gerddoriaeth a'r organ a'r eglwys yn berthnasol i hyn i gyd? Chwiliodd eto.

Bach + hipo Clic!

Dim ond 3561 o bosibiliadau'r tro hwn.

Hippo provides abundance of sensory stimulation as baby develops hand-eye…Bach for babies…

Go brin. Ac eto roedd hyn yn rhywbeth i'w gadw mewn cof ar gyfer y babi. Nododd Steffan y cyfeiriad yn ei ddyddiadur a throi at y nesaf… A darganfod gwefan rhywun neu rywbeth o'r enw *elwe*… Elwe…? Elw+gwe, efallai…? Ac yna *augustine of hippo, bach*…

Clic!… a neidio i'r dwfn eto…

ala miscellanea, anglicans, anna wierzbicka, anti-
postmodernism, audhumla, augustine of hippo, bach, beowulf,
bible, biblical criticism, bookbinding

Aeth ymlaen…

… hawley-cooke, head-driven phrase structure grammar,
hebrew, historical linguistics, history, humanities computing,
indian food, inflectional morphology, inklings

ac ymlaen…

… natural semantic metalanguage, neil gaiman, nerd girls,
nerds, noam chomsky, npr, office products, old english,
optimality theory, origen, oxford, paradigm function
morphology, perl, philosophy, physicist!, philosophers,
physics, poetry, politics, pragmatics, pragmatism,
pseudepigraphia, quenya

ac ymlaen…

rachmaninoff, raspberries, roleplaying, sanskrit, scarves, science
fiction, science olympiad, semantics, simplicity, socrates,
solzhenitsyn, spanish, stanley ned rosenbaum, stravinsky,
syntax, t.s. eliot, tex, the great divorce, the odyssey, the
sparrow, theology, this american life, till we have faces,
tolkien, topology, typology, universalism, university of
kentucky, ursula leguin, used book stores, vladmir nabokov,
vonnegut, water, welsh, william james, wizard people, xml,
zubrin, zwitterions

Welsh…? Zwitterions?

Edrychodd Steffan yn ei eiriadur Saesneg.

zwitterion, n. an ion carrying both a postive and a negative
charge (Ger. zwitter, a hybrid, and ion)

Edrychodd yng Ngeiriadur Bruce.

zwitterions, *n: Ph. switerion(-au) m.*

Trodd y gair o gwmpas yn ei geg. *Switerion*. Ei ddeintio'n dawel. *Swit-er-i-on*. Ei ddatgan yn uchel. *Swit-ER-ion!!!!* Ac o gael gwir flas ar y gair, fe'i hychwanegodd at y rhestr o eiriau arwyddocaol a mynd ati o'r newydd i archwilio'r cysylltiadau rhyngddynt.

16 Ynglŷn â Switerionau, Bach, Leipzig a'r Syrcas Frenhinol

17 Egwyl Fach Dyngedfennol

YN YSTOD YR AWR NESAF, cafwyd rhai cannoedd – miloedd, efallai – o ddigwyddiadau o fewn cylch profiad Steffan. Ar un adeg, byddai Steffan wedi mentro eu hamcangyfrif, petasai rhaid. Gwnaeth hynny unwaith pan ofynnwyd, mewn cwis tafarn, sawl disg llipa y byddai'i angen i storio pob profiad mewn bywyd, a bwrw bod y bywyd dan sylw'n para am 80 mlynedd. Tri, meddai Steffan, o gofio natur ddigynnwrf ei fywyd ef a hefyd cof rhyfeddol y disg llipa. (Pethau cymharol newydd a syfrdanol oedd cyfrifiaduron bryd hynny.) '7,142,857,142,860,000,' oedd yr ateb. '… yn fras.' Erbyn hyn, barnai Steffan nad oedd damaid callach o roi sylw i'r fath ddwli. Dibwys oedd y rhan fwyaf o'r digwyddiadau hyn, beth bynnag.

A dibwys, bid siŵr, oedd y rhan fwyaf o'r pethau a ddigwyddodd iddo yn ystod yr awr nesaf. Mynd i'r tŷ bach, er enghraifft, a'r symudiadau corfforol niferus y bu'n rhaid eu cyflawni er mwyn gwneud hynny'n foddhaol – o agor drws ei stafell, i gau ei gopis; o gamu'n ofalus dros y wifren a ymdroellai o gefn ei gyfrifiadur, i anelu'n gywir i'r lle piso. Ond yr oedd pethau eraill, er eu bod nhw'n ddibwys ar yr olwg gyntaf, yn magu ystyr o'u gweld yn eu cyd-destun, mewn perthynas â'r pethau a'u hamgylchynai, mewn gofod neu mewn amser.

Cafwyd tri digwyddiad o'r math hwn yn ystod yr awr hon. Yn gyntaf, daeth piffgynen i mewn i'r stafell trwy'r ffenest fach uwchben desg Steffan. Ymweliad annhymhorol, meddyliai. Ac eto, onid oedd y tywydd wedi bod yn

annhymhorol hefyd? Dyna pam yr oedd Steffan wedi agor
y ffenest fach yn y lle cyntaf. A dyna pam mai honno, o
bosibl, oedd gweithred bwysicaf y bore, sef agor y ffenest
a rhoi mynediad rhwydd i dresmaswyr adennog. Ta waeth
am hynny. Piffgynen flin a checrus oedd hon, yn ôl ei sŵn.
Hedfanai ar hap o gwmpas y stafell, gan fwrw'n fyzlyd yn
erbyn y cwpwrdd, y golau, y planhigyn deiliog, a Siart y
Trigain Cam Namyn Un, cyn glanio'n benchwiban feddw ar
y silff lyfrau yn ymyl y drws.

Daeth yr ail ddigwyddiad yn dynn ar sodlau'r cyntaf ac
roedd yr un mor annisgwyl. Gyda chlec sydyn, agorwyd drws
ystafell Steffan gan neb llai na Phennaeth y Weinyddiaeth ei
hun, Dr F. MacCumhail. Dihunwyd Steffan o'i fyfyrdod a
gwenodd y Gwyddel mwyn arno.

– A! Steffan…! Sut mae hi ym myd y llyfrau…?

– Y…

– Diwrnod braf bla bla bla Haul bla Adar bla bla Coed bla…

– Ie, hyfryd iawn…

Troes y Pennaeth lifolau ei wên ar gynnwys anniben y stafell,
gan ymestyn ei freichiau fel petai'n ceisio cofleidio'r- cyfan.

– A…! Llyfrau bla bla Barddoniaeth bla bla bla Diwylliant
bla bla bla'r Artist bla'r Llenor bla bla… Ond bla .7% bla a
£3,567,450 bla bla ac yn y byd go iawn, Steffan, Y BYD GO
IAWN, A!… bla bla Llew Smwt a bla bla bla Paul Drudwen
bla bla… A sawl cais ydych chi wedi ei dderbyn o ardal bost
SA dau ddeg…?

– Y?

– Ardal Bost SA dau ddeg… Sawl cais?

– A!… Dwi ddim yn siŵr… Gadewch i mi weld…

Teipiodd y côd post i mewn i'w gyfrifiadur a chael yr ateb
swta: dim.

– Dim, mae arna i ofn…

– Dim…? Dim!?

Pwysodd y Pennaeth ei law ar y silff lyfrau er mwyn sadio ei hun. Nid nepell oddi wrtho, yr oedd y biffgynen – sef, y biffgynen ddig a fu'n ffigwr canolog yn y cyntaf o'r tri digwyddiad a ddisgrifir yma – wedi dechrau dadebru ar ôl ei ymdrechion blinderus ac ofer i ddianc. Roedd wedi sythu ei hadenydd a glanhau ei chwe choes, a phe caniatasid iddi'r amser, dichon y byddai wedi dod i werthfawrogi manteision ei hamgylchfyd newydd, gan mai papur yw deunydd crai pob nyth piffgwn. Ond yn lle hynny, fe darfwyd ar ei myfyrdod gan fawd ymwthiol y Pennaeth. Ac yn ddigon rhesymol o'i safbwynt amddiffynnol hi, a hithau erbyn hyn, o bosibl, yn feichiog â'r genhedlaeth nesaf o biffgwns bach, fe blannodd ei cholyn yn ddwfn yng nghnawd y bawd hwnnw.

– Aaaaaaaaaaaaaaaaaaaaaaaaaaaaaaaaaaaaaaa!

Yn llai rhesymol, ar ôl i bangfeydd y gwewyr cyntaf gilio, troes y Pennaeth ei lid ar Steffan.

– Wel mi FYDD… mm…

Sugnodd fawd ei law dde. Ar yr un pryd, mewn ymgais i'w amddiffyn ei hun, chwifiai'r llaw chwith o flaen ei wyneb: gweithred resymol, eto, o'i safbwynt amddiffynnol yntau, ond bygythiad newydd i'r biffgynen, a oedd erbyn hyn yn anelu am y bawd arall.

– Blydi… bastard… ffycin… mi FYDD cais o Ardal Bost SA dau ddeg…

– Ond…

– MI FYDD CAIS…

Sugnodd a chwifiodd.

– Ond beth am Gam 29… a Cham 32… a…

– AC MI FYDD YN LLWYDDIANNUS!!!!!

Ymadawodd y Pennaeth, dan sugno'i fawd. A dyna ddiwedd yr ail ddigwyddiad.

Daeth y trydydd digwyddiad o fewn ychydig eiliadau, pan gafodd Steffan alwad ffôn gan Llio.

– Steffan, rwy wedi trefnu mynd draw i weld y tŷ Colton ma...

– Ond, Llio... pryd?

– Dydd Mercher.

– Dydd Mercher? Drennydd? Ond alla i ddim... mae cymaint...

– Af i hebddot ti, Steffan, os bydd rhaid.

– Llio... Gwranda nawr... Rwy'n gwybod bod hyn braidd yn annisgwyl, ond... ond...

– Ie...?

– Wel, rwy'n credu... falle... Mae'n flin gen i am hyn, Llio, ond... Wy'n ofan bod rhaid i fi... .

– Ie... Poera fe mas wnei di Steffan!

–...bod rhaid i fi fynd i... bod rhaid i fi fynd i Leipzig ddydd Mercher... neu ddydd Iau fan bella...

Bu saib hir. Atebodd Llio mewn llais digynnwrf.

– O'r gorau, Steffan. Cer di i Leipzig... Af inne i weld y tŷ.

– Ac un peth arall... Mae'n rhaid i fi ffindio rhywun sy'n byw yn SA20...

Ond erbyn hyn doedd Steffan yn siarad â neb ond ef ei hun.

18 Pecyn Arall

DIDOLI, NID CYFUNO, oedd priod ddawn Steffan. O'r
herwydd, gwelodd fywyd fel rhyw bos hynod gymhleth ond
gorffenedig: pos yr oedd angen ei ddadansoddi a'i ddatrys ond
nid ei ymestyn. O na, nid ei ymestyn, nid ar unrhyw gyfrif:
roedd y pos yn hen ddigon dyrys yn barod. Yn hyn o beth,
roedd e'n wahanol iawn i Lio. Iddi hi, clai oedd bywyd, a
deisyfai clai fysedd y crochenydd i'w fowldio a'i wneud yn
gyflawn. Iddi hi, pethau i'w creu'n gyson oedd cyflawnder
a dedwyddwch. I Steffan, pethau i'w darganfod wedi hir
chwilio oeddent, os oeddent yn bod o gwbl, fel gwobr ar
ddiwedd helfa drysor.

I Steffan, ar hyn o bryd, math o groesffordd oedd y pos
hwnnw. Yr oedd, mewn gwirionedd, wedi cyrraedd sawl
croesffordd yr un pryd, os oes modd dychmygu'r fath beth. Yr
oedd ei galon yn pendilio yn y tir neb rhwng fan hyn, ei unig
gartref, a rhyw fan draw nad oedd eto'n bod. A gan ei gymar
yr oedd yr unig fap i'r diriogaeth honno. Yr oedd ei feddwl,
wedyn, yng nghrog rywle rhwng y Weinyddiaeth a... wel,
a Leipzig, mae'n debyg, ond bod y lle hwnnw, hefyd, yn fan
draw nad oedd eto'n bod. Yn yr achos hwn, nid oedd ganddo
fap na chwmpawd na chymar, hyd yn oed, i'w roi ar ben y
ffordd, dim ond cliwiau cryptig un – ac un yn unig – o'r 172
ymgeisydd am gymhorthdal.

Ac wrth gyrraedd adref am saith o'r gloch y nos Lun
honno, byddai Steffan wedi bod yn falch o'r cyfle i rannu â'i
gymar ei holl bryderon ynglŷn â'r dewis mawr a'i hwynebai,
am bwysigrwydd pwyso a mesur manteision ac anfanteision y

bywyd trefol o'i gymharu â'r bywyd gwledig, eu rhestru hyd yn oed a rhoi gwerth rhifyddol i bob un. Byddai wedi hoffi cnoi cil gyda hi dros wydraid o win, efallai, ynglŷn â throeon yr yrfa.

Ond ni chafodd y cyfle hwnnw. Nis cafodd oherwydd, i Llio, nid oedd ond un ffordd ymlaen a gorau po gyntaf y câi ei dilyn. A oedd angen rhoi bysedd traed ym mhob môr cyn mentro nofio mewn un? Wel, nac oedd, siawns. Malu cachu dynion a ofnai fywyd oedd yr holl sôn yma am groesffyrdd a dewisiadau anodd. Na, doedd Llio ddim gartref. Ond roedd wedi gadael neges ar fwrdd y gegin.

Steffan

Wedi mynd at Mam a Dad heno er mwyn cael diwrnod cyfan i'r gwaith fory.

xxx Llio

Hyd yn oed wedyn byddai Steffan, o weld eisiau ei gariad, o fod yn edifar am arllwys dŵr oer ar ei breuddwydion, o fod ychydig yn ofidus, petai'n dweud y gwir, ynghylch eu dyfodol gyda'i gilydd... Hyd yn oed wedyn, byddai siŵr o fod wedi codi'r ffôn ac, ar ôl y sgwrs ddefodol â mam Llio, byddai wedi dymuno pob lwc iddi a gofyn iddi frysio yn ôl gyda'r newyddion i gyd, a dweud cymaint yr oedd yn ei charu a chymaint yr oedd yn edrych ymlaen... ac yn y blaen, ac yn y blaen.

Ond o dan y nodyn yr oedd pecyn. Ac roedd yr ysgrifen arno'n lled gyfarwydd. Na, yn rhy gyfarwydd, er mai dim ond unwaith yr oedd wedi'i weld o'r blaen. Oedodd am funud.

Gobeithiai'i feddwl y câi ynddo'r cyfarwyddiadau manwl y dyheai amdanynt er mwyn dianc o'i ddryswch. Ofnai ei galon na cheid ynddo ond mwy o ddryswch. Agorodd y pecyn a thynnu allan lythyr a theipysgrif. Darllenodd.

Annwyl Swyddog

Maddeuwch i mi am ddod i'ch tŷ heddiw. Roeddwn yn digwydd bod yn y cyffiniau ac roedd yn fwy cyfleus na

---

Am ddod i'ch tŷ chi heddiw? Swmpodd Steffan y pecyn. Roedd yn rhy fawr i fynd trwy'r drws. Ac felly… Ystyriai Steffan nifer o bosibiliadau, ond roedd pob un yn arwain at yr un casgliad. Roedd hi wedi bod yn ei dŷ. Roedd hi wedi bod yn ei dŷ! Ac o fod yn y tŷ… Roedd hi wedi cwrdd â Llio, mae'n rhaid. Ceisiodd ddychmygu'r cyfarfyddiad. O'i ddychmygu, o'i droi'n gyfres o fân ddigwyddiadau, efallai y câi ofnadwyedd y syniad ei ddofi rywfaint. Cnoc ar y drws. Syndod Llio. Esboniadau Rhif 86. Llio'n derbyn y pecyn gyda diolch, a gwên. Ie, gwên… ac wedyn? A oedd 'na 'wedyn'? Ac wedyn, o bosib, o ran cywreinrwydd, efallai, ei gwahodd i ddod i'r gegin, os nad oedd ar frys, a chael cwpanaid. A siarad wedyn, dros goffi neu de, am hyn a'r llall, amdano fe, efallai, am waith Llio. Byddai Llio'n siŵr o fod wedi dangos rhai o'i cherfluniau iddi, y rhai pren, fwy na thebyg… 'Gymrwch chi gwpanaid bach arall, Kate…?' A Rhif 86 yn eu hedmygu nhw'n fawr, fel y bydd pobl, gan amlaf, ac yn canmol ac yn dweud rhywbeth am ei gwaith hi ei hun, efallai, ei gwaith sgrifennu. Ei gwaith sgrifennu… Edrychodd Steffan eto ar y llythyr.

mynd i'r Weinyddiaeth. Dyma ddarn o'm gwaith
creadigol. Fel y gwelwch chi,drama ydyw. Fe'i
hysgrifennais sawl blwyddyn yn ôl. Tybiwn na fyddai'n

Gwaith sgrifennu! Roedd Steffan yn dala sampl o waith
sgrifennu Rhif 86! Nid lluniau, nid cliwiau, nid triciau clyfar
ond gwaith creadigol! Nid rhith mohoni. Nid twyllwraig.

Darllenodd weddill y llythyr.

dderbyniol gennych am fod y cyfrwng yn wahanol i'r
gwaith y dymunaf ei wneud gyda'ch cymhorthdal; ond
dyma hi, gan eich bod chi'n pwyso. Nis cyhoeddwyd.
Nis darlledwyd. A dweud y gwir, nis gwelwyd gan neb
arall ond gennych chi. Ei theitl yw LLAIS BACH.

– Heb feddwl ymhellach am ffonio, a heb feddwl ychwaith,
hyd yn oed, am sut yr oedd Rhif 86 wedi cael gafael ar ei
gyfeiriad, na pham yr oedd hi wedi dewis cyflwyno'r gwaith
heddiw yn hytrach nag wythnos yn ôl, fe daflodd Steffan ei
hun i ymchwydd y geiriau, yn llawn hyder y tro hwn y câi
ddychwel i'r lan yn ddoethach dyn.

19 Y Ddrama

D<small>ECHREUODD</small> S<small>TEFFAN</small> yn y dechrau.

LLAIS BACH

Cymeriadau:

GWYN	gŵr Jean, tua 45 oed
JEAN	gwraig Gwyn, tua 40 oed
GARETH	eu mab, 14 oed
NOSTRODAMUS	byji

GOL 1 Breuddwyd/ Gareth yn codi

Golygfa freuddwydiol. Gwelir Gwyn, y tad, yn gweiddi (heb sain) ac yn gwneud ystumiau bygythiol (yn syth i'r camera) i gyfeiriad ei fab, Gareth. Mae dryll yn dod i'r golwg, yn llaw Gareth. Mae'n tynnu'r glicied. Mwg. Gwelir twll crwn, taclus, heb waed, yn nhalcen Gwyn a syndod ar ei wyneb. Serch hynny, ymhen ychydig eiliadau mae'n ail-gychwyn ar y gweiddi a'r ceryddu, yn fwy gwyllt nag o'r blaen.

Trwy hyn i gyd mae Jean yn paratoi te yn y cefndir. Ar ôl y saethu, mae hi'n nesáu at y camera a chynnig teisen i'w

mab, â gwên ar ei hwyneb. Mae Gareth yn pwyntio'r dryll ati
ond mae'r dryll eisoes wedi troi yn gloch law. Mae Jean yn ei
chymryd oddi arno, gan ddal i wenu, a'i hysgwyd. Mae'i sŵn
yn troi yn sŵn cloc larwm. Mae Gareth yn dihuno'n sydyn, yn
ei wely, gan hanner codi.

GARETH   (llais drosodd) Un…dau…tri…

JEAN   Gareth! Dere lawr y funud yma neu chei
                di ddim brecwast.

Mae Gareth yn troi ei wyneb at y drws yn ddisgwylgar.

GARETH   *(llais drosodd)* Un…dau…tri…

Saib byr… yna daw ei dad at y drws.

GWYN   Glywaist di dy fam, y pwdryn? Neu wyt
                ti'n meddwl aros fan'na trwy'r dydd? E?
                Ti'n gwrando, y mwydyn bach budur?

Trodd Steffan dair tudalen.

                ac am ba hyd 'yn ni'n sefyll… myn uffarn
                i. Byt dy blydi brecwast 'nei di yn lle pipo
                arno'i fel…fel…*(yn edrych ar Jean am*
                weledigaeth)… fel…

JEAN   *(yn undonog, yn gyflym ac yn oeraidd)*…fel
                tylluan y nos ym mola'r goedwig yn
                llygadu'i ysglyfaeth o bell ni ŵyr y truan
                ddim amdano, ddim oll, y plu meddal yn
                lladd pob sŵn, dim gobaith, ond dim ofn

chwaith, dim, heblaw falle ryw wich fach
sydyn ar y diwedd, gwich fach fach, y
creadur yn gwingo falle, y mymryn lleia, a
dyna i gyd...

GWYN   *(yn edrych yn amheus ar Jean, yna'n parhau
â'i fonolog, gan bwyntio ar y byji)* Dyna
greadur sy'n gwybod lle mae fe'n sefyll yn
y byd ma, a phopeth wedi'i drefnu'n deidi
iddo fe... *(y tri yn syllu ar y byji)... (wrth
Gareth)*...gallet ti ddysgu lot wrth y deryn
'na... Trefen... Ma fe'n gwybod beth yw...

*Clywir sŵn ambiwlans y tu allan. Mae Gwyn yn mynd at y
ffenest, fel Jean cynt. Daw yn ôl dan siglo'i ben. Mae Gareth
wedi dechrau bwyta ei frecwast o'r diwedd.*

GWYN   Trefen... Wnewch chi *ddim byd* heb
drefen... *dim b*... A ti! Paid ti â byta pan
wy'n siarad â ti!...Ti'n meddwl mynd yr un
ffordd â dy frawd, wyt ti...? Y...? Smo ti'n
deall beth *yw* trefen, wyt ti? Trefen... *(gan
daro'r bwrdd)* Y drefen sy'n dod... *(taro'r
bwrdd eto)* sy'n dod... *(y llais yn codi)* sy'n
dod... *(mae'n edrych yn ymbilgar ar Jean)*...

JEAN   *(yn llafarganu, yn gyflym ac yn undonog)*...
sy'n dod o'n tarddiad a'n gwreiddyn ni oll
fel yr oedd yn y dechreuad ac a fydd byth
bythoedd hyd yn dragywydd ac efallai wedi
hynny pwy a ŵyr...

GWYN   Wyt ti'n deall beth wy'n trial gweud
wrthot ti?

Aeth Steffan at ganol y deipysgrif.

GOL 9 Y Lolfa (yr hwyr)

(Mae'r lolfa yn eithriadol o daclus ond yn llawn petheuach. Clywir sŵn aneglur gêm bêl-droed ar y teledu, a Gwyn yn ei gwylio, y teclyn rheoli yn ei law; hefyd synau o'r gegin – y rhain yn raddol gynyddu yn ystod yr olygfa. Mae Gareth yn eistedd o flaen cyfrifiadur, ffonau clust ar ei ben, yn edrych tua'r nenfwd ac yn barod i deipio. Mae'r byji tu ôl iddo.)

GWYN   Paid â sefyll fan'na, y twpsyn gwirion – mae ca' cyfan 'da ti a ti'n dewis sefyll fan'na... 'Na...'Na fe, 'na welliant... Gad nhw ddod atot ti gynta 'chan... Gad nhw ddod mas...'Na fe...'Na fe...*(yn cynhyrfu)*... Nawr... Nawr... Rho'r bêl drwyddo nawr... *(Clywir sŵn torri llestri o'r gegin)* Iesu gwyn... Be ti'n 'neud mas fan'na, fenyw?

JEAN   Teisen. *(Saib. Mae Gareth yn ciledrych ar ei dad, sy'n ystyried yr ateb am eiliad ac yna'n pwyntio'r teclyn at y teledu er mwyn weindio'r fideo'n ôl. Rhaid dangos mai'r un darn y mae'n ei wylio drosodd a throsodd)*

GWYN   Iesu gwyn o'r nef, pwy ochor wyt ti'n chwarae, bachan... Cer lan y ca' wnei di, ti'n dda i ddim byd yn sefyll fan'na fel

	lemwn… 'Na ti, 'na ti… Reit, dala hi nawr, dala hi, 'na fe… Reit *(yn cynhyrfu)*… Mas â hi!!…Mas â hi wedes i!!… Mawredd!! *(Sŵn tebyg i ddryll yn tanio o'r gegin. Saib. Mae Gwyn a Gareth yn edrych i gyfeiriad y gegin.)*… Ydy'r deisen 'na'n iawn, Jean?
JEAN	Ody, glei… Pam wyt ti'n gofyn…? *(Saib. Mae Gwyn yn troi yn ôl at y teledu.)*
GWYN	Meddwl o'n i… Meddwl o'n i… *(Mae'n cydio yn y teclyn ac yn weindio'r fideo'n ôl unwaith eto)*… falle bod y ffwrn…*(Mae'n ail-ddechrau'r fideo o'r un man)*… bod y ffwrn… *(Mae'n ymgolli yn y gêm eto)*
GARETH	*(yn siarad yn ei feddwl; llais drosodd)* Y Wahadden… gan Gareth Thomas… Dosbarth Pedwar C… Yn ein gardd ni mae gennym… wahaddod. Mae gwahaddod yn siarad iaith y pridd a'r mwydod, dail ac esgyrn bach. Y mae geiriau yn yr iaith honno, yr un peth â'n geiriau ni…Ond dydyn nhw ddim yn swnio'n debyg i'n geiriau ni… Maen nhw'n swnio fel… fel… fel rhyw grensian bach 'da'r dannedd… pan mae'r wahadden yn llowcio mwydyn… rhyw duchan tawel pan mae'n tyrchu'r ddaear… rhyw snifflan-snwfflan, sgrwnshan-sgrenshan o iaith

Trodd Steffan bum tudalen.

110

GWYN   *(y papur newydd o'i flaen, yn darllen*
eitemau. Jean yn porthi'n achlysurol)
'Daeargryn yn… Balukistan… Lladdwyd
200 o bobl'… 200… Wyt ti'n credu
'ny?… Ar ei ben, 200? Ers pryd bu
daeargrynfeydd mor gymen…? Beth oedd
yn bod â 199? Ond na, rhif anfoddhaol yw
199… Rhaid cael 200 ar ei ben, dim llai,
dim mwy…

GARETH   *(yn siarad yn ei feddwl; llais drosodd)*
Unwaith, yng nghartref y gwahaddod…

GWYN   Dyma un gwell. 'Damwain yn Nigeria.
Lladdwyd 23 o bobl gan gynnwys pump
o blant… a thair dafad oedd yn teithio
gyda nhw… Anafwyd 43…' Ydy hynny'n
cynnwys y defaid, tybed?

GARETH   *(llais drosodd)* Unwaith, yng nghartref
y gwahaddod, cartref lle mae Mr
Gwahadden yn chwydu geiriau yn yr un
modd ag y bydd e'n chwydu coesau pryfed
a dail a mwydod ac esgyrn bach… a Mrs
Gwahadden yn ei ddilyn i bobman ac yn
glanhau ar ei ôl e…

GWYN   *(wrth Gareth)* Pam na wedi di rywbeth?

JEAN   Gwed rywbeth wrth dy dad, Gareth bach.

(Saib. Gwyn a Jean yn edrych ar Gareth.)

GWYN   Dyw e ddim yn gweud dim byd achos
does 'da fe ddim byd i weud… Nac oes?
(Mae'n troi yn ôl at ei bapur. Yn darllen yn
araf o bedantig)… mm… Mae'n dwym
iawn ym Mallorca… ond disgwylir glaw

	yfory...
GARETH	*(llais drosodd)* Unwaith, yng nghartref y gwahaddod, a oedd fel pin mewn papur, ganed gwahadden fach. Roedd y wahadden hon yn fud, druan fach... Ond o dipyn i beth, dros y blynyddoedd...
GWYN	Yn Washington, prifddinas Unol Daleithiau America, mae'n bwrw eira... Mae'n bum gradd dan y rhewbwynt...
GARETH	*(llais drosodd)...* dros y blynyddoedd, er na ddysgodd siarad ei hunan...
GWYN	Yn... Ulan Bhaatar... Does dim sôn am eira... ond *(yn anghrediniol)* mae'n ddeugain gradd dan y rhewbwynt!...

Trodd Steffan at dudalennau olaf y deipysgrif.

GOL 12 Y Lolfa (yr hwyr)

(Fflachiadau achlysurol i'w gweld trwy'r ffenestri, ynghyd â sŵn ffrwydradau. Does neb yn cymryd sylw. Mae Gwyn a Jean yn ymarfer dawnsio llinell, i gyfeiliant cerddoriaeth briodol, gan wisgo'r dillad priodol.)

GARETH	*(llais drosodd, gan deipio'r geiriau ar ei gyfrifiadur. Mae'r byji wrth ei ymyl, yn trydar.)* o dipyn i beth, dros y blynyddoedd, dysgodd siarad â thafodau gwahaddod eraill...

GWYN   Dere 'mlân Jean fach... Cadw lan... Un...
dau... tri... pedwar... Un... dau... tri...
pedwar... 'Na ti... Ond rhaid i ti gadw
dy gefen yn syth, fel hyn, drycha... A
rownd... A dechre 'to... Ymlaen, yn ôl,
chwith, de... 'Na ti... Ond cadw dy gefen
yn syth... A dal dy freichiau fel hyn... *(yn
colli amynedd) Na*, fel hyn...

GARETH   *(llais drosodd)*...ac roedd gan y wahadden
fach bethau pwysig i'w dweud ar ôl bod
mor dawedog ar hyd ei hoes, ond iddi
ddod o hyd i'r geiriau i'w dweud nhw.
*(Sŵn ffrwydrad mwy nag o'r blaen, a'r
llestri'n siglo ar y dresel; neb yn cymryd
sylw.)* Ar y dechrau, dim ond geiriau bach
y gallai eu plannu ym mhennau ei theulu o
wahaddod... Ambell i... *dwmplen!*

GWYN   Dwmplen! dau, tri, pedwar...

GARETH   *(llais drosodd)*... Ambell i rech...

GWYN   Un, dau, tri, *(rhech)*...

GARETH   *(llais drosodd)*... Rhyw wich fach...

JEAN   *(yn gwichian)*

GARETH   *(llais drosodd)*... Ambell i waedd...

GWYN   We – he!!

GARETH   *(llais drosodd)* Ond gydag amser, daeth yn
fwy hyderus... yn fwy mentrus. Dysgodd
nid yn unig i reoli gwefusau a thafodau ei
gyd-wahaddod. Dysgodd feddiannu eu
meddyliau hefyd...

GWYN   Cadw i'r rheolau sydd eisiau...a byddi di'n

113

iawn... *(Gareth yn parhau i deipio)*...Cadw
i'r *(Mae'n gwneud ystumiau fel petai'n ceisio
peidio ag ynganu'r geiriau ac yn ymladd
yn erbyn ei dafod ei hun, ond gan barhau i
ddawnsio)*... 'Cadw... cadw... fi...' *(Mae'n
gorfod ildio ac yn dechrau canu'r emyn, i gyd-
fynd â'r dawnsio â'r gerddoriaeth)*... 'Cadw
fi'n ddiogel, beunydd ar fy nhaith, arwain fi
mewn chwarae, arwain fi mewn gwaith...'
(ffrwydrad arall)

JEAN   *(yn canu ar ei phen ei hun)* 'Boed fy ngwaith
yn onest, rho i'm galon bur, nertha fi i
ddewis rhwng y gau a'r gwir...'

GWYN a JEAN *(yn canu gyda'i gilydd)* 'Diolch iti,
Arglwydd, yw fy llawen gân; canaf nes
i'm gyrraedd bröydd Gwynfa lân...'
(ffrwydradau trwy gydol y pennill)

GWYN   *(yn bryderus am eiliad; peidia'r dawnsio)*...
Glywaist ti... Glywest ti *(yna'r ddau yn
siarad ar garlam, fel act digrifwyr gwael ar
lwyfan, gan adrodd y jôc wrth Gareth, sy'n
teipio'n fwyfwy egniol)*... Glywaist ti'r un
am y dyn aeth i weld Cymru'n chwarae
Lloegr...?

JEAN   ... Ac mae'n gweld bod sêt wag rhyngto fe
a'r dyn nesaf.

GWYN   Wel na beth od, mae e'n gweud wrth y
dyn arall... Ro'dd da hwn diced i gêm
fwya'r flwyddyn a dyw e ddim wedi troi
lan.

JEAN   Ac mae'r dyn arall yn egluro taw sêt ei

	wraig odd hon, a'u bod nhw'n arfer mynd i bob gêm 'da'i gilydd… Ond bod hi wedi marw, 'twel… A dyma'r tro cynta iddo fe ddod ar ei ben ei hunan.
GWYN	Wel, na beth ofnadw, mae'r dyn cynta'n gweud. Allech chi ddim ffindo rhyw ffrind neu ryw berthynas i ddod 'da chi?
JEAN	Na, mae'n gweud, a golwg drist ar ei wyneb. Mae pawb wy'n nabod wedi mynd i'r angladd… Ha! Ha! Ha! Ha!

(Chwerthin afreolus gan Gwyn a Jean; Gareth yn stopio teipio)

Ni ddarllenodd Steffan yr olygfa olaf. Wedi rhoi'r deipysgrif yn ôl yn yr amlen, aeth at ei gyfrifiadur.

20 Neges Fach Gwrtais

Annwyl Ms O'Brien

Diolch am gyflwyno'r ddrama, *Llais Bach*, i ategu eich cais. Yn anffodus, nid yw'n amlwg sut y mae'r gwaith hwn yn berthnasol i'ch prosiect arfaethedig, sef nofel gofiannol, nac ychwaith i'r defnyddiau eraill a gyflwynwyd gennych eisoes. Er enghraifft, ni ddeallaf ym mha ffordd ac i ba raddau y bydd y nofel yn gofiannol. Cofiant pwy yn union a fwriedir? Ynte, ai ffigwr dychmygol ydyw'r 'gwrthrych? A fyddech carediced ag anfon gair o eglurhad, os gwelwch yn dda?

Yn gywir

Ni chysgodd Steffan yn dda'r noson honno.

21 Dau Orchwyl Pwysig
a Llawer o Godau Post

BYDDAI CYMYDOG LLYGATGRAFF wedi sylwi, y bore dydd
Mercher hwn, bod rhywbeth wedi bwrw Steffan oddi ar
ei echel oherwydd, yn lle dilyn trywydd 2, fel y dylai fod
wedi 'i wneud yn ôl ei amserlen ar gyfer y trydydd o'r
cylchoedd wyth diwrnod, fe ddilynai'n union yr un llwybr â
ddilynasai'r diwrnod cynt. Gwnaeth hynny heb ddangos yr
un anesmwythyd, na'r un newid yn ei osgo na'i gerddediad.
Yr oedd fel petai'n ceisio ail-afael yn y dydd Mawrth a aeth ar
gyfeiliorn a'i dywys yn ôl i rigolau normalrwydd a threfn.

Pan gyrhaeddodd y swyddfa ni wastraffodd amser trwy
fân siarad nac ymofyn coffi na darllen y negeseuon e-bost
niferus a oedd wedi cronni yn ystod y dyddiau diwethaf. Petai
wedi gwneud, dichon y byddai wedi arbed cryn drafferth a
phoendod meddwl. Ond ni wnaeth. Yr oedd ganddo ddau
orchwyl pwysig i'w cyflawni a châi popeth arall aros ei dro.

Ystyriodd orchymyn y Pennaeth i ddyfarnu grant
i rywun a oedd yn byw yn SA 20. Yr oedd y dyddiad
cau ar gyfer ceisiadau wedi hen fynd heibio ac roedd y
system gyfrifiadurol, er diogelu atebolrwydd, yn rhwystro
swyddogion rhag ychwanegu enwau ar ôl y dyddiad hwnnw.
Fel arall, byddai Steffan wedi pwyso ar un o'i gydnabod yn
y byd llenyddol – un a oedd yn fwy crwydrol na'r rhelyw,
efallai – i arddel cyfeiriad yn y fro ddifreintiedig dan sylw. Dan
amgylchiadau tebyg yn y gorffennol, llwyddodd i ddarganfod
awduron a chanddynt ryw fodryb neu fam-gu neu gyfnither

yn yr ardal benodedig, weithiau'n ddiarwybod i'r awduron eu hunain. Unwaith, trwy ddamwain, wrth adleoli un o'r ymgeiswyr hyn, cafodd hyd i blentyn a genhedlwyd gan yr awdur ryw ddeng mlynedd ynghynt ac a fu'n dihoeni mewn dirfawr dlodi byth oddi ar hynny. Mawr fu llawenydd y fam, a'i diolch i Steffan, am ddod o hyd i dad y truan a'i alluogi i roi cynhaliaeth deilwng iddynt. Torrodd ton o foddhâd a balchder dros galon friw Steffan wrth gofio'r achlysur. Rhyfedd, hefyd, na chawsai'r un gair o ddiolch gan y tad.

Heddiw roedd ei dasg yn wahanol. Rywsut, byddai'n rhaid iddo adleoli awdur *ar ôl* iddo ymgeisio a chael ei gofrestru. Ei adleoli, felly, wedi iddo roi ei gyfeiriad, a heb newid y cyfeiriad hwnnw. Edrychodd trwy dabl cyfeiriadau'r ymgeiswyr. Yr oedd SY23 a CF11 a LL55 yn frith trwyddynt, a cheid clystyrau go fawr yn SA44 ac NP7 hefyd. Roedd amryw ohonynt wedi hepgor côd post ar eu ffurflenni cais, ond erbyn hyn yr oedd y cyfrifiadur, o'i wybodaeth gynhwysfawr am drefi a maestrefi a phentrefi a phlwyfi Cymru, wedi canfod côd i bob un. Neu, yn hytrach, bron pob un. O edrych eilwaith ar y sgrîn a sgrolio i lawr yn ofalus ac yn bwyllog, sylwodd Steffan fod yna un… na, dau… fod yna ddau gais heb gôd iddynt, dim ond y rhifolyn, '0'. Dynodai hyn, i Steffan, naill ai nad oedd dim codau'n bod yn yr achosion hyn – peth annhebygol iawn – neu fod y cyfrifiadur wedi methu dod o hyd iddynt. Casâi Steffan foelni'r '0' yna. Gwrthodai rannu ei chyfrinachau. Doedd dim rhyfedd bod brenhinoedd yr Oesodd Canol yn carcharu pobl am ei defnyddio, gan ofni pob math o dwyll ac anhrefn – a bod y Weinyddiaeth, saith canrif yn ddiweddarach, yn gweithredu gwaharddiad tebyg. Rywsut neu'i gilydd, byddai'n rhaid i Steffan gael gwared arnynt, cyn cyflwyno ei adroddiad terfynol ddydd Llun. Cael gwared arnynt, a rhoi rhywbeth ystyrlon yn eu lle.

Cais 29 oedd un. A! Yr ymgeisydd o blaned Romula. Doedd hynny'n syndod yn y byd. Ond, o graffu ar y sgrîn, gwelodd Steffan fod nodyn wedi cael ei ychwanegu o dan y cofnod hwn gan un o'r Swyddogion Dilysu.

Rhoddwyd Rhif 29 yn y Cywiriadur Enwau Lleoedd a barnwyd mai un ai Rhiwlen, Maesyfed (1349) neu Aber-miwl, Trefaldwyn (1694), a olygir wrth Romula. Dewiser un o'r rhain yn ôl gofynion y Strategaeth.

Yr oedd 29, felly, nid yn unig yn bod, ond yn byw yn ein byd ni: yn *gorfod* byw yn ein byd ni er mwyn cyflawni amcanion y Weinyddiaeth. A oedd yr iaith Rihannsu yn bod hefyd, felly? Does bosib y gellid cael un heb y llall. Ond sut y byddai'n gwerthuso'r gwaith pan ddôi i mewn? Trwy ofyn am gyfieithiad...? Mm... Digon i'r diwrnod, meddyliai Steffan. Gan nad oedd gwahaniaeth rhwng y ddau le ym mlaenoriaethau'r Strategaeth, plannodd y Romulad yn Rhiwlen, gan fod yr enw'n bertach na'r llall a rhoes 'LD2' yn lle'r '0' annynol, anghyfannedd yna.

Cafwyd yr ail '0' o dan gyfeiriad Kate O'Brien. A dim syndod yno, chwaith. Ond lle nad oedd y cyfrifiadur wedi canfod yr un Romula ddilys, ddilychwin i gartrefu Rhif 29, gormodedd posibiliadau oedd y maen tramgwydd yn achos Rhif 86. Hyd y gwyddai Steffan, nid oedd ond un Cwm Clydach i'w gael, sef y dyffryn dwfn, cul, coediog a thywyll yn ymyl Bryn-mawr: hynny yw, y Bryn-mawr mawr yn yr hen Sir Frycheiniog, nid y Bryn-mawr bach ger Llanaelhaearn nac ychwaith y Bryniau Mawrion llai byth yn siroedd Mynwy, Meirionnydd a Dinbych. Nid oedd wedi rhagweld y gallai fod unrhyw amwysedd ynglŷn â manylyn mor syml, mor bendant. Ond na, mynnai'r cyfrifiadur fod cymaint ag wyth Cwm Clydach i'w cael. Yn ogystal â Chwm Clydach

Bryn-mawr (NP7), gallai fod yn Gwm Clydach Uchaf (SA6), yn Gwm Clydach Isaf (SA6 hefyd), neu hyd yn oed yn Rhyndwyglydach (SA6 o hyd). Gallai fod yn Gwm Clydach ar bwys Tonypandy (CF40), yn Gwm Clydach Ynysybwl (CF 37) neu yn… neu yn Gwm Clydach… Ha!… yn Gwm Clydach SA20! Ac yr oedd, nid un, ond DAU Gwm Clydach yno! A gâi fentro dodi Rhif 86 yn un o'r rheiny? Ni fyddai'r cyfrifiadur ddim callach. Ac wedi'r cyfan, doedd dim byd i brofi *nad* oedd Rhif 86 yn byw yn SA20. O! Y fath lawenydd! Y fath ryddhad!

Wrth gwrs, petai Rhif 86 yn llwyddiannus – na, *pan* fyddai'n llwyddiannus – byddai'n rhaid bod yn garcus. Beth petai hi'n siarad â'r wasg, er enghraifft, a gadael y gath o'r cwd? Mm. Yr oedd y Llwybr Papur yn hen beth iawn yn ei le, meddyliai Steffan, ond ar adegau fel hyn, nid oedd yn glir sut y gallai neb ond creaduriaid papur ei droedio heb fynd i gorsydd dwfn. Yn y byd go iawn, ni ddaliai llwybr papur bwysau llygoden, heb sôn am y behemoth o gelwydd yr oedd ar fin chwythu anadl einioes i mewn iddo. Oni bai, wrth gwrs… Oni bai mai yn un o'r ddau Gwm Clydach hynny yr *oedd* hi'n byw mewn gwirionedd. Dyna'r gobaith.

Ond sut mae mesur gobaith? Beth yw ei hyd a'i led? Dyna'r cwestiwn a boenai Steffan nesaf. Ai 33.333% oedd y tebygolrwydd mai yn y Cwm Clydach hwn y trigai Kate O'Brien? Os felly, roedd yn obaith ychydig yn denau i un fel Steffan, a fynnai sicrwydd a sadrwydd ym mhob agwedd ar fywyd. Ynte, a effeithiwyd ar y cyfartaledd hwnnw gan boblogaeth y gwahanol Gymoedd Clydach? Treuliai Steffan weddill y bore yn amcangyfrif, ar sail Cyfrifiad 2001, boblogaeth y cymoedd perthnasol. Erbyn amser cinio daeth i'r casgliad mai oddeutu 1.7% oedd y gwir debygolrwydd. Ni wnaeth y darganfyddiad hwn ddim i godi 'i galon. Serch hynny, gan nad oedd ond un dewis ganddo, cartrefodd Rif 86 yn SA20 a throi ei sylw at y daith fawr drannoeth.

Nid oedd lle ar y *flight* o Lundain i Leipzig ddydd Iau na dydd Gwener. Rhyw ŵyl lyfrau wedi llenwi'r awyrennau i gyd am wythnos gyfan, mae'n debyg. Ond er mawr syndod i Steffan, yr oedd modd teithio pob cam o'r ffordd i Leipzig ar y trên, a hynny am bris tipyn yn llai na hedfan: ffactor pwysig, o gofio bod Steffan yn gorfod talu am y daith o'i boced ei hun. Ni feiddiai hawlio treuliau gan y Weinyddiaeth. Er y gallai lenwi deunaw o'r un cwestiwn ar hugain ar y ffurflen dreuliau – y rhai a ymwnâi â hyd y daith, amserau ymadael a dychwelyd, prisoedd cymharol llogi car, hwylio, mynd ar y bws, ac ati – ofnai na allai, a'i law ar ei galon, roi atebion boddhaol i'r cwestiynau ynglŷn ag amcan y daith, canlyniadau disgwyliedig y daith, a pherthynas y daith i'r Strategaeth. Gwell talu nag ymrafael â mwy o gwestiynau nad oedd ganddo atebion iddynt. Gwell talu na difaru.

22 Steffan yn mynd i Leipzig
   a chwrdd ag Anton ar y ffordd

I MI, mae dala trên am bum munud ar hugain wedi saith yn golygu codi am hanner awr wedi pump. Rhwng ymolchi, gwisgo, siafo, brecwesta, glanhau dannedd, chwilio am allweddi, colli sbectols, sylwi eu bod nhw yn fy llaw trwy'r amser, colli'r allweddi eto – 'chi'n gwybod y drefn – ac yna cael tacsi i'r orsaf hanner awr cyn i'r trên ymadael, er mai taith bum munud yw hi, jyst rhag ofn – prin bod dwy awr yn ddigon.

Ac mae codi am hanner awr wedi pump yn golygu, i bob pwrpas, noson ddi-gwsg, waeth pryd yr af i i'r gwely, waeth faint o glociau larwm a ddodaf ar bwys y gwely yn dweud 'Paid â becso, Steffan bach, cer di i gycu-bei nawr ac mi goda i'r fath dwrw ymhen wyth awr, bydd y stryd gyfan yn meddwl bod eu tai nhw ar dân'.

Dyma fi, felly, yn gwneud fy ngorau glas i gysgu a hithau ond yn ddeg o'r gloch a hanner fy meddwl yn dweud, drosodd a throsodd, 'Cofiwch chi fod 'da chi drên i 'ddala am hanner awr wedi saith... Cofiwch!... Cofiwch!... A'r hanner arall yn cyfri faint o funudau fydd eu hangen i lanhau dannedd, cael cachad – tybed a fydda' i'n gallu cachu'r amser yna o'r bore, a'm perfeddion yn dal i gysgu yn eu gwely bach clyd? Ac erbyn i mi neilltuo dwy funud i'r dasg hon a saith munud i'r dasg arall, a rhyw nifer o funudau i bob un o'r naw tasg ar hugain – naw ar hugain! – sydd gen i i'w cyflawni cyn mentro rhoi troed ar y trên – gan ganiatáu deg munud ar gyfer y weithred dyngedfennol honno – rwy wedi colli cyfri ac yn gorfod mynd yn ôl i'r dechrau. Stori gyfarwydd i chi i gyd, rwy'n siŵr. Cwympo i gysgu, o'r diwedd, am bum munud wedi pump. Erbyn hynny, mae dwy ran o'm meddwl wedi colli pob diddordeb a fu ganddyn nhw mewn tasgau a threnau a chwsg a diffyg cwsg, ac yn dechrau sibrwd yn dawel yn fy nghlust, 'Cysga am byth, cysga am byth, sgam byth, sgam byth, gambyth, *gamby, amby, tic, toc, tic, toc, tic, toc…* '

clic… BRRRRRRR!!!!

Mae pedair rhan i'r daith. Y daith ar y trên, felly. Nid y daith o'r tŷ – nac o'r gwely, o ran hynny. Byddai'n rhaid ychwanegu rhyw ugain o deithiau bychain, neu ragor, efallai, petawn i'n cynnwys cerdded draw i mofyn coffi o'r pot coffi, a rhyw betheuach dibwys ond anhepgor felly… Cannoedd efallai. Pedair rhan, o gadw at yr amserlen. 7.25 o Gaerdydd. 10.27 o Lundain. 14.52 o Frwsel. Ac 18.10 o Köln, gan gyrraedd Leipzig am un funud i ganol nos. A hyd y rhannau'n cynyddu, fesul cam, fel mae'n digwydd.

Wrth gymryd fy sedd, rwy'n gwbl effro. Rwy'n gweld popeth, rwy'n clywed popeth fel petaen nhw'n wreichion a darawyd o einion. Mae popeth yn llachar, yn groch, yn ormesol o fyw. Y papur newydd yn haid o wylanod, eu hadenydd yn gyllyll trwy'r awyr, yn siswrn yn y glust. Fy nwylo fy hun, yn gwningod bach anystywallt, yn rhedeg fan hyn a fan draw a byth yn gwrando ar fy llais, petai gen i lais, ond mae'r llais yn cysgu yn y dowlod ac yn breuddwydio bod rhywun wedi rhoi'r sgubor ar dân ac yntau'n ffaelu galw am help, ffaelu symud…

Dihunaf yn Reading ac yno, am y bwrdd â mi, yn darllen y papur, mae'r llyfrgellydd a roes gymaint o help i mi ynglŷn â'r *Western Mail*… Ac ynglŷn â'r ymholydd arall 'na… Pwy *oedd* yr ymholydd arall…? Mae cofio amdani yn codi pwl cas o losg cylla... fel 'se llysywen ddanheddog yn cnoi fy ymysgaroedd.

– Shw mae... y...

Ceisiaf eto.

– Helô 'na!... Cyd-ddigwydd...

Mae'n codi ei lygaid o'r papur... Ond na, dim fe yw e. Dim y llyfrgellydd... Dim byd tebyg iddo fe, a dweud y gwir.

– Mae'n... Mae'n flin gen i...

Mae'r geiriau, fel defaid blwydd, yn anfoddog iawn i gael eu corlannu, ond o'r diwedd mae brawddeg gyntaf y bore yn brefu ei hymddiheuriad bloesg.

– Mae'n flin gen i... Meddwl taw rhywun arall oeddech chi...

– *Bitte?*

Y peth diwetha rwy'n moyn 'neud ar hyn o bryd yw tynnu sgwrs â dieithryn. Ond cymaint yw fy nghywilydd, sut alla i beidio?

– Gweithio yn Llundain ych chi, felly...? meddwn i.

– *Nein, nein...* medd yntau.

– A! Mynd i weld eich mam-gu...

– Na, na! daw'r ateb, a rhyw dinc fach ddilornus ddigon annymunol ambwyti fe hefyd... Na... Mynd adrrref ydw i...

Ac yn ei 'weud e jyst fel 'na... A-d-rrrr-e-f... Yn y-ng-a-n-u p-o-b c-y-t-s-a-i-n fel 'se fe'n trial torri cneuen â'i wefusau.

– A... !

A dyma fi'n dechrau pwlffacan â 'ngeiriau, rhag ofn i mi wneud ffradach o bethau eto.

– Ydych... ydych chi... ym... o fan hyn... bell... ydych chi'n... byw... ymhell... o fan hyn... mm?

– Leipzig.

– Leipzig…? Na, na, meddwn i, yn amyneddgar, gan feddwl bod y dyn wedi gwneud camgymeriad, ac yntau heb ymgyfarwyddo eto â holl deithi'r iaith, chwarae teg iddo. Na, na, meddwn i, fi sy'n mynd i Leipzig, Pwyntiaf fy mys at fy mrest. Fi… a dyma fi'n troi fy mys i gyfeiriad lled-gyfandirol… Leipzig. A chi…?

Codaf fy aeliau a throi cledrau fy nwylo tuag at y nenfwd mewn ystum ymholgar ond cwrtais. Cwrtais, ie, ond gyda rhywbeth tamaid bach yn nawddoglyd yn ei chwt hi hefyd, rhaid cyfaddef, dyna fu 'ngwendid i erioed, fel mae Llio o hyd yn fy atgoffa.

– Yn Leipzig yr wyf yn byw… medd yntau, yn araf, fel 'se fe'n siarad â phlentyn… *Leipzig ist meine Heimatstadt*… Leipzig yw fy nghartref…

– A!

– Ac Anton yw fy enw…

– Anton?

– Anton Carl Schwartzenburg.

– A! meddwn i, yn ceisio cofio pam roedd yr enw'n gyfarwydd. Mm… Cyd-ddigwyddiad.

Dychwelaf i'm cragen, ond mae'n amlwg bod hwn yn un da am dynnu gwichiaid o'u cilfachau.

– Methu hedfan oherwydd y *buchenliebe*… Pobl y ffair lyfrau wedi mynd â'r seddau i gyd…

– A, ie!… Y ffair lyfrau… Pwy fyddai'n meddwl?… Ie… Llyfrau…

– A fi a'm cyfeillion yn gorfod… *kriechen*… cropian?… Cropian ar draws Ewrop…

– Eich cyfeillion?

– *Ja*... Rydyn ni'n perthyn i Gangen Cymru o Gymdeithas Anton Bruckner...

– Anton...?

– Ie, cefais fy enwi ar ei ôl...

– A!

Ar hynny, mae'n troi at ddyn a menyw oedrannus, eiddil a chrebachlyd yr olwg, sy'n hepian yn y sedd y tu ôl iddo.

– Dyma Gwyneth a Cedric...

– A...

– A'r tu ôl iddyn nhw mae Gwenda, Helmud, Diarmuid...

– Wel, mae...

– Ac yn y pen draw... Stephanie, Sabine... Gruff, Tom, Mererid... Angharad... Penelope...

Rwy'n gwenu cyfarchiad ar bob un yn ei dro, ond ofnaf fod y wên yn dechrau colli peth o'i sglein cyn cyrraedd Penelope druan... ond byddai wedi pylu beth bynnag, am fod Penelope yn edrych yn hynod o debyg i Audrey... Bron na ddwedwn i... ond mae hi yn y cysgodion, braidd, a'i thrwyn mewn llyfr...

– Ac mae'r gweddill yn y cerbyd nesaf... gobeithio!

– Y gweddill?

– *Ja, ja*... Rydyn ni'n bymtheg i gyd... Dyma hufen y Gymdeithas, gallech chi ddweud... Ac rydyn ni'n mynd i Leipzig i ddathlu ei sefydlu... Ach! Mae eich rhifau mor gymhleth!... *Funfundsiebzig Jahre*... Pymtheng mlynedd... a thrigain...? *Ja*, pymtheng mlynedd a thrigain yn ôl... Yn y ddinas honno.

– A! Wel... hen gyd-ddigwyddiad rhyfedd, yntefe...?

Beth arall alla i 'ddweud?

Erbyn ail-ddechrau'r daith yn Waterloo rwyf wedi dadebru rywfaint. Edrychaf ar yr amserlen. Clamp o lyfr yw hon, nid y daflen fach lipa gewch chi yn y rhan fwyaf o lefydd. Bues i'n pendroni'n hir a ddylwn i ddod ag e neu beidio, ond mae'n gas gen i rwygo tudalennau mas o lyfr, ac mae'r un mor gas gen i fynd ar fy nhrafels yn ddiymgeledd, fel petai. Felly, dyma fe. Ac rwy'n ddiolchgar amdano. Mae e yma, yn gloddfa o wybodaeth, rhag ofn. Mae 'na gysur yn hynny. Yn y ffaith ei fod yn rhestru pob trên, pob gwasanaeth, o Ddulyn i Vladivostock. Yn y ffaith bod y daith rhwng Moskva (Yaroslavski) a Vladivostok yn cymryd wyth niwrnod. Bod y trên yn pendilio yn ôl ac ymlaen mewn cylchoedd wyth niwrnod... Heb sôn am longau i Seydisfjordur... A Sassnitz-Mukran... A bysus y Romantische Strasse.

– Rydych chi'n hoff o'u darllen nhw, felly?

– Yn hoff...? A! Yr amserlen! Na, eisiau gweld pryd 'ryn ni'n debyg o gyrraedd Brwsel... 'Na i gyd...

– Mi ydw innau'n hoff iawn o'u darllen nhw hefyd.

A dyma fe'n tynnu'r clorwth coch allan o'i *briefcase* a throi yn syth at Dabl 20.

– Mae'n debyg na sylwoch chi ar hyn...

– Mm...?

– Ond mae ein taith i Frwsel yn cymryd dwy awr a thri deg tri o funudau... Ar ei ben...

– Ddim yn ffôl, ydy e, gyda'r twnnel a phob...

–... Sef, cant pum deg tri o funudau...

– Wel, o'i ddweud e fel 'na...

– *Schön, sehr schön*... Rhif hynod o dlws!

– Tlws?

Mae'n tynnu beiro o'i boced.

– Fel hyn… $153 = 1^3 + 5^3 + 3^3$… nodwedd anghyffredin iawn…

– Mm… Ie…

– A'r rhan nesaf o'r daith… o Frwsel i Köln… Cant… saith deg… pump o funudau… Ddim cystal… Ond eto'n ddiddorol. Edrychwch… $175 = 1^1 + 7^2 + 5^3$… Hyfryd o gymesur, ynte?

– Ydy, g'lei…

– Ond mae'n mynd ar chwâl wedyn… Tri chant pum deg un munud o Köln i Leipzig… Na, wna hwnna ddim o'r tro… Tri chant pum deg tri dyle fe fod…

Ond tri chant saith deg un o funudau yw hyd y daith, mewn gwirionedd, wrth i ni dynnu i mewn i Hauptbahnhof Leipzig, ugain munud yn hwyr, am 00.19 fore dydd Gwener. Erbyn hynny mae Anton yn ymorol am ei braidd ac nid yw'n clywed fy ffarwel.

Disgynnaf o'r trên. Y mae'r orsaf yn fawr. Yn frawychus o fawr, a dweud y gwir, ar ôl treulio dwy awr ar bymtheg yng nghroth y trên. Ar y llaw dde, mae'r platform yn ymestyn hyd at y gorwel, bron. Ac ar y llaw chwith hefyd. A'r cledrau i gyd, yn ôl pob golwg, yn terfynu fan hyn. Nad oes neb eisiau mynd y tu hwnt i Leipzig? Ond does fawr neb yn mynd nac yn dod yn yr anferthedd hwn heno, heblaw fi a phymtheg o ffans Bruckner sy'n araf ymlusgo tuag at yr *Ausgang*. Gadawaf iddynt fynd o'r golwg cyn dechrau fy ffordd fy hun tuag at y diddymdra mawr.

23 Gwesty Glan Môr

NID YW'R SEASIDE HOTEL yn ymyl y môr, wrth reswm. Y
mae'r môr agosaf siŵr o fod tua dau gan milltir i'r gogledd.
Pam yr enw, felly? Ni wn, ac ni fynnaf wybod. Rwy'n
benderfynol o ddisgyblu fy hun yn ystod yr ymweliad
hwn, gan gyfyngu fy sylw at faterion perthnasol yn unig,
sef materion sy'n ymwneud â chais Rhif 86… A Leipzig…
A Bach… A'r Thomaskirche… A'r organ… A'r syrcas, o
bosib… A 1946… A hippo o'r enw Anton? Na, yn bendant,
na. Ddim yr hipo. Dyna le rwy'n tynnu'r ffin.

Na, dyw e ddim ar lan y môr. Ond *mae* e o fewn canllath
i'r orsaf, a dyna pam y dewisais i'r lle. Os ych chi'n amau
hynny, mynnwch fap o Leipzig. 'Welwch chi'r Willy-Brandt-
Platz? Mae e reit ar bwys yr Hauptbahnhof. A Richard-
Wagner-Strasse? Yn rhedeg ar hyd un ochr y sgwâr? Dyna le
mae e. Rhif 7. Wel, chwiliwch amdano ar y we, os ych chi'n
dal i amau. *www.seaside-hotels.de* Neu ffoniwch nhw ar Leipzig
98520. (Ffacs: 985-2750.) Mae gen i stafell yng nghefn y
gwesty, stafell 281, a honno ar y pedwerydd llawr, ta beth am
y rhif.

Mae canol Leipzig, o ran ei siâp, rywbeth yn debyg i
Gymru. Ac o feddwl amdani fel 'na, mae'r orsaf rywle ar
bwys Wrecsam. Ar Ynys Môn y ffindiwch chi'r Gerddi
Swolegol, rhyw faestref foethus yng Nghaerdydd yw'r Neues
Gewandhaus, ac un o gapeli mawreddog Aberystwyth, yn
briodol ddigon, yw'r Thomaskirche. Mae'r gwesty, yn ôl y
map hwn, heb fod ymhell o Flaenau Ffestiniog. Ac mae iddo
ddwy nodwedd arbennig sy'n gwneud iawn, i ryw raddau,

am y diffyg golygfa o'r môr. Y cyntaf yw'r *façade* drawiadol. Dwi ddim yn gwybod a ddaeth y *façade* hon trwy'r rhyfel yn union fel y mae, yn gwbl ddianaf; ynte, a gafodd ei chodi o'r newydd. Ond gallech chi ofyn yr un cwestiwn am bob 'hen' adeilad yn y ddinas, am wn i. A faint callach fyddech chi o wybod? Y mae'n *façade* nobl iawn, ta p'un.

Ac ail brif nodwedd y gwesty yw'r stafell lle rwy'n brecwesta ar hyn o bryd. Byddai lawn cystal gen i beidio â sôn am y stafell hon, ond i gofnodi'r ffaith bod y cyfan wedi cael ei wneud ar batrwm cerbyd trên, cerbyd o'r hen Orient Express, a bod yn fanwl, a'r byrddau a'r goleuadau i gyd yn cydymffurfio â'r ddelwedd. Cerbyd go urddasol, rhaid cyfaddef, yn ysgogi atgofion oes fwy sedêt, mwy moethus. Nid atgofion fel y cyfryw, wrth gwrs, ond atgofion o ffilmiau a lluniau ac ati. Ond cerbyd trên yw cerbyd trên, a gwelais ddigon o'r rheiny ddoe, diolch yn fawr. Pobl yr Ŵyl Lyfrau yw'r gwesteion eraill, mae'n debyg, a barnu yn ôl y *buche*, y *books*, y *livres*, y *libros*, a'r *libri* sy'n britho pob sgwrs. Mae'n dda gen i ddweud na allaf glywed yr un *llyfr* yn unman.

24 Thomaskirche

WEDI CERDDED rhyw hanner canllath ar hyd Richard–
Wagner-Strasse, cymeraf y tro cyntaf ar y chwith, Nikolai-
Strasse. Rwyf wedi marcio'n ddu, ar fap o'r ddinas, y llwybr
y mae'n rhaid i mi ei ddilyn. Nid yw map, wrth reswm, yn
dangos y ceir, y lorïau, y tramiau, y bysus, na'r bobl na'r llwch
na sŵn y gwaith adeiladu a'i ddrilio a'i daro a'i lifio byddarol,
a dim ond awdur llyfr taith fyddai'n trafferthu i'w disgrifio.
Nid yw diflas bethau o'r fath yn rhan o'm pwrpas i ac, er eu
bod nhw'n pwyso'n drwm iawn ar fy mhum synnwyr, gwnaf
fy ngorau i'w hanwybyddu. Ar ôl munud o gerdded, trof
i'r dde, i Bruhl, yna i'r chwith, ar hyd Katharinestrasse, tuag
at y farchnad. Ac yno, am y tro cyntaf, gwelaf fynegbost yn
cadarnhau mai hon, yn wir, yw'r ffordd i'r Thomaskirche, yn
ogystal â'r cyfleusterau cyhoeddus a'r ganolfan gynadledda.
Ac o droi am y tro olaf, gwelaf, rhwng y blociau moel, y tŵr
gwyn, wythochrog.

Rwy'n petruso nawr. Mae'r map yn mynd â mi at ddrws
yr eglwys. Ond nid yw'n mynd ymhellach. Y gwir amdani yw
nad oes gen i ddim cynlluniau pellach. Un math o gynllun fu
gen i o'r cychwyn, sef dala'r trên, newid ar yr amser priodol
yn y lle priodol, gwneud fy ffordd i'r gwesty priodol ac yna
dilyn y strydoedd priodol er mwyn cyrraedd drws yr eglwys
hon. Pob cam wedi'i baratoi a'i ddilyn, un ar ôl y llall. Ac yn
arwain at... y cam nesaf. Cymryd y cam nesaf, dyna i gyd sydd
eisiau ei wneud. Ond erbyn hyn, mae'r camau'n lleihau, o un
i un. Cyn bo hir, byddan nhw wedi diflannu'n gyfan gwbl.

Yn lle mentro, rwy'n gogor-droi. Cerddaf draw at gerflun o'r dyn ei hun. Cerflun digon celfydd yw e, hefyd, os braidd yn farwaidd. Gwell gen i'r hen dŷ'r ochr draw i'r sgwâr, lle mae o leiaf ryw atsain o fywyd a fu. *Neue Bachgesellschaf...* Amgueddfa, am wn i. Ac o dipyn i beth, wrth grwydro'n ddigyfeiriad, mae sŵn y ddinas yn cilio. Mae gwynder yr eglwys, yr ychydig goed o bob tu iddi, eu glendid a'u destlusrwydd – mae'r cyfan yn troi yn ynys fach lle gallaf anghofio'r llwydni sy'n fy amgylchynu. Anadlaf yn ddwfn. Am unwaith mae'r cam nesaf yn ddirgelwch i mi. A does gen i'r un daten o ots. Rwy'n barod i agor y drws ac wynebu fy nhynged, ac os nad...

– *Der Vergrauung...*

– Beth?

– A! *Gutten Morgen, mein freund...*

– Anton...

– Dweud oeddwn i... wrth fy ffrindiau...

A gwelaf Audrey eto, ond mae hi'n edrych draw i gyfeiriad y cerflun, fel na allaf weld ei hwyneb yn llawn.

– Dweud roeddwn i fel y cafodd Leipzig ei throi'n llwyd i gyd. Oherwydd y cloddio am lignit, wyddoch chi. Yn claddu popeth dan amdo o lwch... A dyna pam rydyn ni'n sôn am y *Vergrauung...* Am bopeth... Pob adeilad, pob ffenest, pob coeden, pob lliw yn graddol droi'n llwyd... Ac yna'n malurio... Yn gwywo... Y *Vergrauung* ydy'r enw ar hynny...

Rwy ar fin dweud wrth y dyn gymaint o syndod i mi yw ei weld e, a'i giwed ufudd, yn ymweld â'r eglwys hon, a nhwthau'n bobl Bruckner i fod, a rhyw ffwlbri fel'na oherwydd, yn y bôn, rwy'n teimlo braidd yn flin ei fod e wedi tarfu ar fy myfyrdod. Ond mae e eisoes yn hysio ei braidd tuag at y drws.

– Croeso i chi ymuno â ni… Rydyn ni un yn brin, beth bynnag…

– Un yn brin…?

– Mae pymtheg yn well na phedwar ar ddeg.

– Ydy?

Rwyf bellach yn un o'r praidd. Awn i mewn i'r eglwys. Y mae'n llawer llai nag yr oeddwn i'n disgwyl.

– Mae'n llai nag yr oeddech chi'n disgwyl…? medd Anton, fel petai wedi darllen fy meddwl.

– Ydy…

– Ac eto, beth sydd yn fach a beth sydd yn fawr? Ydy saith deg metr o hyd yn fach…? Neu bum metr ar hugain o led…? Neu ddeunaw metr o uchder…? Corff yr eglwys, rwy'n 'feddwl, wrth gwrs… mae twr yr eglwys yn llawer uwch… chwe deg wyth metr os cofiaf yn iawn… Ond mae popeth yn gymharol, on'd yw e? Y mawr a'r bach. Yr hardd a'r hyll.

– Ydy…

– Ac eto, nid popeth 'chwaith… Sylwoch chi ar y to cyn dod i mewn? Naddo? Mae'r to, wyddoch chi, ar oledd anghyffredin o serth, hyd yn oed am do o'i gyfnod. Wyddoch chi ba mor serth yw e? Rhowch gynnig…

–…

– Dewch ymlaen… Rhowch gynnig…

– Tri deg…?

– Tri deg? Tri deg! Ha! *Nein, nein, mein Freund.* Y mae'n…

Ac ar hynny, mae'n rhoi ei wefusau yn dynn wrth fy nghlust, rhag imi golli'r un sill.

– Chwe… deg… tair… gradd!

Mae Anton yn aros am ryw arwydd o syfrdandod, mae'n debyg, ond rhaid i mi gyfaddef nad ydw i'n ddigon hyddysg

mewn pensaernïaeth eglwysig i werthfawrogi arwyddocâd y
ffigurau hyn. Ymbalfalaf am ateb.

– Wel... rhyfedd... bod gan eglwys... mor fach... mm... do
mor serth...

Anton sy'n edrych yn syn y tro hwn.

– Ond, Steffan, 'does gwahaniaeth yn y byd beth yw *maint* yr
eglwys, na'r to o ran hynny. Y *mae'r* to yn serth iawn. Ffaith!
A dyna ddiwedd arni. Mae 'na derfyn ar ba mor serth y gall to
fod... Ni *allai* fod yn fwy nag wyth deg naw pwynt naw naw
naw naw naw... Ond does dim terfyn ar ei *maint*...

– Heblaw, wrth gwrs... heblaw am...

Ond cyn i'r sillaf gyntaf hercian dros fy ngwefus, mae
Anton yn troi a chyfarch ei gyfeillion.

– Ac i'r eglwys hon, i'r eglwys fach hon, sydd eto'n fawr,
am iddi sefyll yn gyfan pan ddrylliwyd y byd o'i chwmpas yn
gyrbibion man... I'r eglwys hynafol hon, hynafol ac eto'n
newydd, yn newydd am ei bod hi'n gwisgo harddwch ein
hundod newydd fel cenedl... I'r eglwys hon y gorymdeithiai'r
miloedd bob wythnos, yn dorf ganhwyllog braf, i ddatgan
mai dim ond trwy chwyldro y caent ail-afael yn eu hanes,
mai trwy newid eu byd o'r bôn i'r brig, y caent ail-afael yn
eu hunaniaeth ddigyfnewid. Yn y fan hon y cychwynnodd y
Wende...

Mae Anton erbyn hyn wedi ein hebrwng i ganol yr
eglwys. Ac yno, o flaen yr allor, gwelwn sarcophagus gwyn,
ac arno bedwar tusw o flodau coch a gwyn a phlac pres yn
dwyn yr enw **Johann Sebastian Bach**.

– A dyna'r rheswm pam y daethon ni yma, gyfeillion. Canys
yma y claddwyd athrylith mwyaf ein cenedl. Yn y gweddillion
hyn yr oedd gwreiddyn, yr oedd bywyn y gobaith newydd...

'Fues i erioed yn hoff iawn o deithiau tywysedig. Profiad

annifyr, i mi, yw cael fy mugeilio o fan i fan gyda'm cyd-greaduraid, ac mae'r annifyrrwch yn tanseilio unrhyw bleser y byddwn yn ei gael fel arall o fod mewn llecyn diddorol a difyr. Gwell gennyf dorri fy nghwys fy hun. Ond, ar y llaw arall, dyma ddyn sydd, mae'n amlwg, er gwaethaf ei nodweddion annymunol niferus – ei ffordd chwyddedig, hunanbwysig o siarad, ei duedd bedantig i orfanylu, ei frawddegau hirwyntog, aml gymalog – ie, er gwaethaf y rhain i gyd, ac eraill, dyma ddyn sy'n meddu ar gryn wybodaeth am ddau, o leiaf, o'm cliwiau. Byddwn yn ffôl i golli'r cyfle i fanteisio ar yr wybodaeth honno. Ond ar hyn o bryd, gadawaf iddo barhau â'i berorasiwn dysgedig, gan feddwl, yn y man, pan ddaw'r cyfle, y caf gydio yn ei fraich a mynd ag e naill ochr... Draw ar bwys y pwlpud, efallai... Mynd ag e naill ochr a'i holi... Ei holi am... Am yr hyn sy'n clymu'r cwbl...

Ond mae Anton yn siarad o hyd, gan bwyntio at y peth hwn a'r peth arall, at y bedyddfaen lle bedyddiwyd plant bach Bach, y rhai a fuont fyw yn ddigon hir i gael eu bedyddio, at y ffenestri o'r bedwaredd ganrif ar bymtheg, y pwlpud a'r drws, heb dynnu anadl bron, a heb gymryd sylw o'r ymwelydd arall yn ein mysg, y gŵr main, moel, mewn siwt lwyd sydd, erbyn i mi ei weld, eisoes yn datgloi drws yng nghefn yr eglwys. Mae Anton yn ceisio egluro, os wyf yn deall yn iawn, fod yr eglwys a welwn ni heddiw yn hŷn nag yr oedd hi yn amser Bach, peth digon amlwg i bawb, dybiwn i, ond na, yr hyn mae'n ei feddwl yw hyn, sef bod yr eglwys yn fwy *gwreiddiol* nag yr oedd hi yn y cyfnod hwnnw, ei bod hi wedi tynnu ei dillad benthyg, fel petai, ac erbyn hyn yn ymddangos yn ei noethni llathraidd, gogoneddus cyntaf. Ond na, nid hynny'n hollol ychwaith, am fod rhai pethau 'gwreiddiol', mewn gwirionedd, wedi cael eu hychwanegu yma a thraw er mwyn atgyfnerthu'r gwreiddioldeb hwn, ei wneud yn fwy dilys, megis. A hefyd, mae'n ymddangos, am fod yna ryw gorff mwy gwreiddiol byth yn llechu o'r golwg rywle, rhyw gorff, ysywaeth, nad oes digon ohono ar ôl i gael ei ddiosg. A'r tu hwnt i'r corff hwnnw wedyn, ryw ysgerbwd gorweddog o eglwys arall eto… Ac yn y blaen ac yn y blaen.

A thrwy'r cwbl, clywaf y dyn main, moel yn cloi'r drws y tu ôl iddo, ac yna sŵn traed ar risiau cerrig – yn dringo, ddywedwn i, nid yn disgyn – yna ryw *glec-gric-glac*, rhyw *sgwffl-sgraffl*, rhyw *twoc-twoc-twoc*… Ac yna, lle cynt y bu sŵn, neu dawelwch, yn awr y mae'r eglwys yn llawn o rywbeth nad yw'n sŵn nac yn dawelwch, ond yn rhyw gyfuniad gwyrthiol o'r ddau. Y mae cynghanedd yr organ yn taflu ei rhwyd drosom, yn gwthio ei bysedd melfed i bob twll a rhic a hollt, yn gyrru ei chryndod hyd berfeddion pob un a phob peth.

– Nid yw'r organ hon…

Nid oes pall ar huodledd Anton. Fe'u clywaf, bob yn ail, yr organ ac Anton, yn cyd-weu eu brawddegau maith a chymhleth.

– Nid yw'r organ hon, wrth reswm, mor hen â'r eglwys. Mae rhai yn honni bod organ o ryw fath i'w chael yma mor gynnar â 1384, ond wyddon ni ddim byd am yr offeryn hwnnw, ond am y ffaith ♪♪♪♪♪ Wilhelm Sauer ydy hon, organ gafodd ei hadeiladau mor ddiweddar â 18 ♪♪♪♪ ei hadeiladu, ac eto, nid ei chwblhau, gan fod Wilhelm, yn nodweddiadol o'r rhai fu'n dilyn ei alwedigaeth trwy hanes, bob amser yn gwneud rhyw welliannau ♪♪♪♪♪ ar ôl darganfod bod ei gi wedi bwyta ei swper! Rhyfedd o fyd, ynte? Ac weithiau, fe welwch nad yw ambell organ byth, byth yn cael ei chwblhau. Ystyriwch yr hen organ a gafodd ei symud oddi yma yn 1885 drawsbren fan hyn, rhyw biben fan draw a fu'n rhan o'r 98, a honno yn ei thro yn cynnwys defnyddiau o ryw hen, hen organ o ryw hen, hen fynachlog nes ei bod hi'n amhosibl dweud, dyma organ hwn-a-hwn, a wnaed yn y lle a'r lle, yn y flwyddyn a'r flwyddyn…

Mae Anton yn myfyrio am ychydig eiliadau, a minnau'n dechrau pendwmpian.

– Ta waeth, fel y dywedais i, byddai Wilhelm o hyd yn gwneud rhyw welliannau i'r organ… rhai ohonynt, a rhai yn unig, o'i ddewis ef ei hun, ond y rhan fwyaf ohonynt ar ôl i ryw Bastor trwm ei glyw achwyn nad oedd digon o sŵn ganddi… *I beth oedd y fath ganu grwndi'n dda…?* Neu ar ôl i ryw organydd ffroenuchel grybwyll, wrth ymadael â'r eglwys, fod gan organ Thomaskirche Köln lawer mwy o stopiau na hon, *oedd wir.* Neu ar ôl i ryw Ac felly, er bod chwe deg pump o stopiau yn ddigon i bawb arall, yn 1907 fe ychwanegodd dri ar hugain atynt. A phetai hynny ddim yn ddigon, flynyddoedd wedi hynny, fe gafodd sawl un eu newid eto er mwyn gwneud yr organ yn fwy dilys hynafol. Ha! Ac yn fwy [nodau cerddorol] nes ei bod hi'n amhosibl dweud p'un ai nodau o 1746 a glywn ni, ynte nodau o 1860, neu 1605... Wrth gwrs, ar un ystyr, fel mae'n digwydd, nodau 1746 *ydy*'r rhain a glywch chi nawr, does dim dwywaith...

Gwelaf Anton yn codi ei olygon at yr organ.

—... Neu, dyna pryd y cafodd e'r wŷs gan y Brenin i sgrifennu'r darn, beth bynnag...

Y brenin? Ac ym mwrllwch fy nghysgadrwydd, synhwyraf fod rhywbeth o bwys yn cael ei ddweud. Ceisiaf hoelio fy sylw o'r newydd.

—... A'r llaw dde, fel y clywch chi [nodau cerddorol] yn chwarae'r alaw...

Gwrandawaf. Ond, i mi, suo-gân yw'r dôn, ac mae'n cydio yn fy llaw ac yn mynd â mi yn ôl i wlad y cymylau gwlanog.

—... Ac ar yr un pryd... *ar yr un pryd!* ..mae'r llaw chwith yn chwarae'r un alaw... [nodau cerddorol] ond dydych chi ddim yn ei hadnabod hi, nac ydych? [nodau cerddorol] Dydych chi ddim yn ei hadnabod hi... [nodau cerddorol] oherwydd [nodau cerddorol] mae hi'n chwarae'r alaw [nodau cerddorol] o chwith [nodau cerddorol] Fedrwch chi ei chlywed hi nawr?

Ac yn fy mreuddwyd, gwelaf ddwy law'r organydd…

−… Fel bod yr alaw'n symud yn ôl ac ymlaen, yn ôl ac ymlaen…

Gwelaf ddwy law'r organydd yn symud yn ôl ac ymlaen, yn ôl ac ymlaen…

−… Yn ôl ac ymlaen, fel dwy granc…

Y ddwy law, fel dwy granc…

−… Yn ôl ac ymlaen, am byth

− Na, gyfeillion, ni wyddom beth yw tarddiad yr un nodyn. Y cwbl y gallwn ni 'ddweud yw bod gan yr organ, bellach, wyth deg wyth o stopiau, bod pump ar hugain o'r rhain ar y seinglawr cyntaf, yn cynnwys Bordon un droedfedd ar bymtheg, Doppelfloete wyth troedfedd a Rohefloete pedair troedfedd, ac yn y blaen, bod un stop ar hugain ar yr ail seinglawr, yn cynnwys Salicional wyth troedfedd arbennig o bersain, Piccolo dwy droedfedd hynod o felys, Principal ac ar y trydydd seinglawr…

− Mae tri?

Dihunaf o'm breuddwyd. Mae fy nghwestiwn yn atseinio trwy'r eglwys.

− *Bitte*?

– Mae'n flin gen i… Dwedoch chi fod tri… be'ch chi'n galws…

– Seinglawr…? Do. Oes. *Ja*. Tri seinglawr.

– Ers faint?

– Sut?

– Wel roeddech chi'n dweud fel mae pethau wedi cael eu hychwanegu a'u tynnu i ffwrdd… Oedd 'na adeg pan nad oedd dim ond dau…?

– Nac oedd, yn bendant…

– Wel, oes 'na organ arall…

– Oes…

–… A sawl..

– Pedwar… Mae pedwar seinglawr gan honno…

– Ond yn y llun…

Rwy'n chwysu, fel petai rhyw dwymyn arna i. Ond ofn yw e. Ofn a dryswch. Ofn a dryswch ac anobaith. Ofn a… Yna'r embaras. Rwy'n dod yn ymwybodol o'r embaras. O'r diffyg cyfatebiaeth rhwng achos ac effaith. Y diffyg tystiolaeth am y fath gyflwr difeiriol truenus. Ac mae hynny'n gwneud i mi chwysu'n fwy. Mae rhywun yn gwthio gwlân cotwm i'm clustiau. Mae rhywun yn taflu dail te tu ôl i'm llygaid, mwy a mwy ohonyn nhw, nes bod popeth yn troi'n fagiau te, yn fagiau te pygddu. Mae'r coesau'n gwegian… Audrey! Audrey…? Mae Anton a'r fenyw sy'n edrych yn debyg i Audrey'n cydio yn fy nwy fraich ac yn… Ond mae 'mhen yn troi cymaint, rwy'n ffaelu gweld dim byd yn iawn…

★ ★ ★

Wedi eistedd ar y fainc am ychydig funudau, rhaid i mi gyfaddef fy mod i'n bur ddi-amynedd gydag Anton. Mae

gormod o ffwdanu a tendans wedi codi 'ngwrychyn i erioed.
Prysuraf i ddweud mai dim ond rhyw bwl o'r bendro sydd
wedi dod drosto'i… Heb gael digon i'w fwyta, angen mynd
'nôl i'r gwesty, diffyg coffi, gormod o wres, a rhyw ddwli fel
'na, nad ydyn nhw'n llyncu'r tamaid lleiaf ohono fe. Rhaid
bod yn ofalus. Feiddia i ddim dangos dim gwendid rhag ofn
bod rhain yn cario claps yn ôl i'r Weinyddiaeth, neu…

– Dod yma i glywed yr organ wnaethoch chi, te?

– Nage, meddaf i. Ddim yn hollol.

– Nage? Wel, pam daethoch chi yma te?

A dyna fynd at graidd y mater.

– Ga'i ymddiried ynoch chi, Anton?

– *Ja, ja*. Cewch, siwr. Ond beth ydych chi'n bwriadu ei
ymddiried ynof i?

Mae hyn yn troi'n fwyfwy tebyg i gêm nadredd ac
ysgolion. Oedaf eilad cyn taflu'r dîs.

– Cwestiwn.

– Os felly, nid y chi sy'n ymddiried ynof i. Fi sy'n gorfod
ymddiried ynoch chi.

– Sut hynny?

– Gan mai fi, mae'n debyg, sydd i fod i gynnig yr ateb. Ond
mae hynny'n dibynnu ar y cwestiwn, on'd yw hi…? Mm…?
Wel…? Fy nghyfaill! Beth
yw'r cwestiwn!?

– A! Y cwestiwn! Y cwestiwn yw hyn…

A dywedaf y cyfan wrtho. Y cyfan am Rif 86 ac am ba
mor bwysig yw cael gwybod pwy yw gwrthrych y cofiant
arfaethedig, a thipyn o hanes yr Eneiniog hefyd, hyd yn oed,

er mwyn iddo gael deall y cyd-destun, er mwyn iddo ddeall gwraidd fy ngofidiau, rhai ohonynt beth bynnag, yn ogystal â chrynodeb o ofynion llym y Weinyddiaeth a rhyw fanion eraill a ddaw i mi, wrth imi geisio ffordd trwy ddrysfa'r diwrnodau diwethaf. Nid peth hawdd yw tynnu'r edafedd at ei gilydd. Nid peth hawdd o gwbl. Ac yna, wrth gwrs, ceisiaf ddisgrifio'r dystiolaeth yr wyf eisoes wedi'i chasglu.

– Wel, mae gen i ddyddiad. 1946… Ac mae gen i enw. Thomas… Mae gen i frenin hefyd. Mae gen i… Mae gen i lun o organ… Ond nid yr organ hon, gwaetha'r modd… A hen ddyn yn ei chwarae… Mae gen i ddarn o gerddoriaeth gan Bach… A… A gorchymyn i chwilio…

– I chwilio? I chwilio am beth?

– Ie. Yn union. Dyna'r cwestiwn. Dyna'r cwestiwn rwy am ofyn i chi. Am beth ydw i'n chwilio?

– Am beth ydych chi'n chwilio? Ond mae hynny'n hawdd, gyfaill. Doedd dim angen i chi gynhyrfu gymaint. Rydyn ni'n dau yn chwilio am yr un peth. Ac rydyn ni'n dau wedi dod o hyd iddo fe.

– Do fe?

Daw gwên i'w wyneb.

– A does dim un cofiant yn gyflawn hebddo.

– Mm…?

– *Gebeine, mein freund. Bachs Gebeine.* Esgyrn Bach.

25 Y Trydydd Pecyn

EISTEDDAF ETO yn yr Orient Express. Syllaf trwy'r ffenestri ar y golygfeydd dau ddimensiwn. Ar *Wien*. Ar *Venedig*. Ar *Konstantinopel*. Cymaint i'w weld mewn un awr ginio. Daw'r *concierge* ataf ag amlen yn ei law.

– *Herr Steffan?*

– Ie.

– *Zimmer Zweihundert einundachtzig?*

– Y… *Ja*

A dyna i gyd sydd ar yr amlen. Dim cyfeiriad. Dim sôn am y gwesty hyd yn oed. Ac mae'r ysgrifen yn gyfarwydd.

Agoraf yr amlen. Clywaf aroglau sinamon. Mae ynddi… bedair, pump, chwe thudalen A4. Mae'r gyntaf yn dechrau ar ganol brawddeg. Teipysgrif anghyflawn, felly? Ac yn fwriadol anghyflawn, o bosib? Neu dystiolaeth bod rhywun heblaw'r sawl a'i hanfonodd wedi ymyrryd â'r cynnwys?

– *Kaffe mit Milch, bitte…*

Ym moethusrwydd y cerbyd, taniaf sigaret a dechrau darllen.

was doch heut Nacht ein Sturm gewesen. A! Wyt ti'n cofio'r gân honno? Y tro hwn, ysywaeth, parodd y storm trwy'r bore hefyd. Roedd yn arllwys y glaw. O flaen y gybolfa o bren pwdr, llaid ac esgyrn, hawdd oedd anobeithio.

Ond, er gwell neu er gwaeth, roedd y Pastor Tranzschel
yn benderfynol o rygnu ymlaen. *Nach dem Mitagessen!*
Cawn ni well lwc a gwell tywydd ar ôl cinio! meddai'n
harti, a ninnau wedi bod wrthi ers wyth y bore. Roedd y
gweithwyr yn anniddig, rhai yn dadlau bod a wnelo Duw
â'n diffyg llwyddiant, yn dweud mai peth rhyfygus, os
nad cableddus oedd tynnu rhywun o'i orffwysfan olaf. A
minnau'n dweud, heb lawer o frwdfrydedd, mai gwneud
gwell gorffwysfan iddo oedd ein bwriad. Hwythau'n
gweiddi yn ôl bod hynny'n iawn i'r etholedig un, ond pwy
faliai fotwm corn am y trueiniaid lu oedd yn cael eu malu
a'u stwnsho a'u hyrddio fan hyn a fan draw cyn i ni hyd
yn oed gael cip ar y gelain lwcus? Roedd y twll roedden ni
wedi'i balu yn llenwi â dŵr, a minnau'n blino ar y cyfan, ar
gofnodi pob bys, pob clun, pob dant.

Ars moriend! Selig sterben! Ha! Efallai mai nhw oedd
yn iawn. O farw'n ddedwydd, onid anghwrteisi, a dweud
y lleiaf, oedd cael dy ddatgymalu, dy chwarteru, dy
ddarnio'n yfflon, a'th stwffio blith draphlith i ryw gymanfa
fawr o esgyrn? Ac eto, dywedai rheswm wrthyf na
pherthynai unrhyw urddas na dedwyddwch i'r bedd, i'r
Stinckenden Todten-Aas, Koth und Unflath chwedl yr hen
August Pfeiffer. Yn y pydredd hwnnw, gyda'i gynrhon a'i
fwydod, lle roedd y dedwyddwch yn wir?

Ond pa mor bell mae rheswm yn mynd â ni? Bu
farw Herr Bach ar yr unfed dydd ar ddeg ar hugain o fis
Gorffennaf yn y flwyddyn mil saith cant pum deg. Dyna
un ffaith bendant. Gwyddom hefyd iddo gael ei gladdu
ym mynwent Johanneskirche, a dywed yr haneswyr na
osodwyd na charreg na chroes na dim i nodi'r fan. Ai
dyna derfyn ein gwybodaeth? Dyna, mae'n debyg, yr
unig ffeithiau diymwad sydd gennym ynglŷn â'i farw a

lleoliad ei gladdu. Ond gwyddai pawb ffordd yma, 'waeth beth ddywed yr haneswyr, fod côr y Thomaskirche, ers blynyddoedd lawer, yn arfer canu ym mynwent ei chwaer eglwys ar ddyddiad marw'r hen Gantor, ac yn canu bob tro yn yr un lle, sef chwe cham oddi wrth ddrws deheuol yr eglwys. Mi fûm i'n bresennol ar yr achlysur hwnnw fy hun lawer gwaith. Ar ben hynny, bu Taid yn ddisgybl yn y Thomasschule a chefais yr hanes i gyd ganddo fe droeon pan oeddwn i'n ifanc. Ai tystiolaeth yw hynny? Ynte chwedlau ein gwerin hygoelus?

Ac yna, Sabine, gwneuthum y darganfyddiad pwysig. Ie, myfi a'i gwnaeth, er yn gwbl ddamweiniol. Yn unol ag arferion y dydd, wrth gwrs, a phawb yn fwy parod i blygu'r glin, Gustav Wustmann, Cyfarwyddwr Archifau Leipzig, gafodd y clod, ond fi, a neb arall, a dynnodd ei sylw at y peth. Wrth astudio cyfrifon y ddinas ar gyfer 1750, darllenais i Herr Bach gael ei gladdu mewn arch dderw. Ac o'r 1400 o bobl a fu farw yn y ddinas y flwyddyn honno, deuddeg yn unig a gladdwyd mewn eirch derw. Llefnyn ifanc oeddwn i, wrth gwrs, ac wedi cynhyrfu'n fawr wrth glywed bod y Pastor Tranzschel, ar sail yr wybodaeth hon, wedi penderfynu cloddio am y gweddillion. Nid fy lle i oedd holi pam. Er mwyn eu dyrchafu yng ngolwg ei gydwladwyr, meddai yntau. A minnau heb ddeall ar y pryd beth, yn union, roedd hynny'n ei feddwl.

Ond, fel y dywedais eisoes, rhwng y glaw a'r llaid a'r ffieidddra, bu ond y dim i ni roi'r gorau iddi ar ôl cwta ddwy awr. Nid yw natur y pridd, nag arferion claddu, yn parchu trefn y blynyddoedd na gofynion dad-gladdu. Nid oedd yno ffin rhwng claddedigaethau 1750 a 1751, na rhwng 1679 a 1820 o ran hynny. O dan y ddaear, gwelir y

145

cenedlaethau, y tad a'r ferch, yr ewythr a'i nith, bonedd a gwreng, yn uno mewn aflan briodas.

Ond tua hanner awr wedi deg, cyrhaeddodd Wihelm His. Wyt ti'n ei gofio, Sabine? Gwnaeth dipyn o enw iddo'i hun oherwydd ei ddull o draethu am benglogau – yn eu darllen yn union fel petai'n darllen llyfr, meddai yntau. Cyhoeddodd lyfr wedyn yn rhoi i ni'r holl hanes. Y mae'r llyfr hwnnw o'm blaen wrth i mi ysgrifennu'r geiriau hyn. *Johann Sebastian Bach: Forschungen über dessen Grabstätte, Gebeine und Antlitz.* Adroddiad hynod fanwl, ond anghyflawn, fel y cawn weld yn y man. O na buaswn yn ddigon dewr i ddweud rhywbeth ar y pryd!

Wel, roedd yr hen Bastor wedi gwysio His i fod yn bresennol er mwyn archwilio'r gweddillion, pe deuai rhai addawol i'r wyneb. Y fath anhrefn a'i hwynebai! Y fath domen o siwtrws dynol! Cafodd fraw, a gofyn i Transzchel sut yr oedd disgwyl iddo ganfod penglog Herr Bach, heddwch i'w lwch, lle nad oedd braidd un benglog gyfan i'w gweld, dim ond darnau, a'r rheiny wedi'u gwasgaru dros bob man, rhyw dyllau llygaid yn rhythu fan hyn, rhyw geg yn gwenu'n fantach fan draw. Ond mynnai'r Pastor nad oedd dim i boeni yn ei gylch. Doedd yr un arch dderw wedi ei hamlygu ei hun. Eirch pinwydd oedd y cyfan o'r rhai y cafwyd hyd iddynt tan hynny. A gwendid yr arch a wnaed o binwydd, meddai'n awdurdodol, oedd ei bod hi'n pydru bron cyn gyflymed â'i chynnwys. Doedd hynny ddim yn wir am yr arch dderw.

Am ryw reswm, cododd presenoldeb Herr His ein calonnau ni oll, fel petai ei wybodaeth ddihafal ym maes penglogau yn ddigon i ddenu'r un a chwenychem i'r fei. Ac yn wir, ymhen ychydig funudau fe gafodd y pysgotwr craff ei frathiad cyntaf. Ernst, mewn gwirionedd, un o'r

146

pedwar a fu'n cloddio yno, a gafodd hyd iddi. Fe'i haliodd
o'r pydew a'i rhoi gerbron His. Doedd dim angen trosol
i dynnu'r clawr – yr oedd eisoes wedi dechrau dod yn
rhydd ar ôl yr holl golbio. Buan y gwelwyd mai rhywun
ifanc oedd y tu mewn – merch, mae'n debyg – ac, wedi
i mi gofnodi'r manylion angenrheidiol, fe'i taflwyd ar y
domen. Ond yr oedd gwynt y prae yn ffroenau pawb
erbyn hyn a doedd dim troi 'nôl i fod. Aeth y ceibwyr ati
fel lladd nadredd nes y bu'n rhaid i Herr His ofyn iddynt
ymbwyllo rhag difrodi tystiolaeth bwysig. Cyngor doeth,
ond rhy hwyr, mae arna i ofn, fel y cewch weld.

Yr oedd clychau'r eglwys wedi taro deuddeg ac roedd
y Pastor yn gweiddi y caem well lwc ar ôl cinio, fel y
dywedais eisoes, pan drawodd rhaw Sigmund yn erbyn
arch arall. Arch dderw oedd honno hefyd. Arch gyfan.
Erbyn hyn, wrth gwrs, yr oedd y gweithwyr yn erfyn eu
cinio a bu'n rhaid aros awr gron cyn i ni gael cyfle i'w
thynnu i'r lan a'i hagor. Yr oedd y gwaith o'i hagor yn
fwy llafurus nag a ddisgwyliem oherwydd y clasbiau pres
a ddefnyddiwyd i glymu'r clawr yn dynn wrth yr arch.
Ond wedi ei hagor, ni fu'n rhaid aros yn hir cyn i Herr His
ddedfrydu bod ysgerbwd anghyffredin iawn yn gorwedd
ynddi ac. Gafaelodd yn y benglog, ac o'i harchwilio'n
fanwl, datganodd fod iddi nodweddion urddasol a grymus
odiaeth a ddynodai fod gan ei pherchennog natur a
doniau ymhell uwchlaw'r cyffredin. Tynnodd ein sylw
at bont y trwyn, gan ochneidio ei edmygedd, *sehr kühn!*
sehr kühn! ac yna at helaethrwydd y *fenestra rotundra,*
at y bochau *stark entwickelt,* at olwg *kräftig* y cyfan. Ni
fedrai ymborthi digon ar harddwch yr ysgerbwd. Dyma,
meddai, nodweddion digamsyniol athrylith. Pwy ond
Herr Bach allai feddu arnynt?

Ni chefais fy synnu ar y pryd gan ei sicrwydd na chan y penderfyniad i beidio â chloddio ymhellach, a hynny er mai dim ond rhyw ddeg o benglogau yr oedd wedi'u harchwilio, rhwng ei gyrraedd hwyr a'i ginio cynnar, cymaint oedd ein parch at y gwyddonnydd yn ein plith, ac at ei bendantrwydd. A dyna ddiwedd ar fy ymwneud innau â'r mater hyd y dydd heddiw. Aeth His rhagddi i gasglu rhagor o dystiolaeth ac i astudio'n fwyfwy manwl gyfrinachau'r benglog. Yn wir, gyda chymorth nodwydd, tri deg saith o gyrff marw, a cherflunydd o'r enw Karl Steffner, aeth mor bell â chreu model o'r pen cyfan, y cnawd, y llygaid, y gwallt a'r cwbl. Pa wahaniaeth os nad oedd yn edrych yn debyg i'r portreadau? Onid oedd realiti'n rhagori ar gelfyddyd?

Meine Liebe Sabine, aeth hanner canrif heibio ers y digwyddiadau hyn. Aeth hanner canrif heibio ers i mi dy weld. Mae hon bellach yn ddinas o esgyrn. Fyddet ti ddim yn nabod y lle. Fyddet ti ddim yn dymuno nabod y lle. Ni allaf ddweud fy mod i'n fy nabod fy hun mwyach. Rwyf wedi blino ar y cyfan. Fel petawn i'n colli teimlad trwy fy nghorff, yn araf bach, bysedd fy nhraed yn gyntaf, un ar y tro, mae'r parlys yn traflyncu pob un, ac yna'r traed a'r coesau a'r breichiau. Maen nhw i gyd yno, yn y drych, ond dydyn nhw ddim yn perthyn i mi ragor. Maen nhw mor farw â hoelen. Maen nhw mor farw ag esgyrn Bach. Ac i feddwl ein bod ni'n dau, Bach a minnau, wedi dod trwy'r rhyfel yn ddi-anaf. Wyddost ti hynny, Sabine? Wyddost ti fod Johanniskirche wedi cael ei bomio'n rhacs yn y cyrch cyntaf? Ond chyffyrddodd e mo'r bedd. Daeth y bedd, daeth yr esgyrn, a'r rhwysg i gyd – daethant hwy trwy'r alanas yn fyw!

Sabine, darllenais ddoe yn *Die Zeit* fod yr esgyrn i

gael eu symud eto. Dyma, mewn gwirionedd, pam rwy'n
sgrifennu atat heddiw. Hanner y rheswm, beth bynnag.
Fe wyddost y llall. *Begegnung.* Nid y gân. Ond ti a fi.
Unwaith. Wrth bwy arall fedrwn i ddweud y pethau olaf
hyn? Blinais ar fyw ymhlith y meirw. Bûm yn byw yn eu
plith cyhyd nes fy mod i'n farw-fyw fy hun. Y meirw sy'n
addoli'r meirw! Sabine, claddwyd miloedd yma dan yr
adfeilion. Ond mae'r esgyrn hyn, yr esgyrn gwirion hyn,
eto'n fyw! Weithiau, rwy'n credu bod y meirw'n gryfach
na'r byw.

Pryd dechreuodd hyn i gyd, Sabine? Beth oedd
gwraidd y drwg? Ai pan dderbyniodd Herr Bach y
gwahoddiad i ymweld â'r Brenin? Oedden nhw, Bach
a'r Brenin, yn cyd-gynllwynio ein tynged ni heddiw, heb
iddyn nhw wybod? Ynte, ai'r drwg oedd gwybod, ganrif
a hanner wedi hynny, mai mil pedwar cant saith deg naw
pwynt pum centimetr sgwâr oedd arwynebedd mewnol
penglog Herr Bach, sef yr union gyfartaledd ar gyfer
Almaenwyr o hil ddihalog – a meddwl bod hynny o bwys?

Sabine, drwgdybiaf bawb y mae angen esgyrn arnynt.
Trueni na fyddwn i wedi eu drwgdybio ynghynt ac yn fwy
ffyrnig. Ond ofnaf hefyd na fydd yr hyn sydd gennyf i'w
ddweud wrthyt yn pylu dim ar eu chwant am farwolaeth.
Fe fynnant eu heilunod, doed a ddêl. Ac nid wy'n
amau dim na fydd y mwyaf ohonynt wedi'i ddyrchafu
eto a'i osod, fel llo pasgedig, o flaen allor ei annwyl
Thomaskirche.

Sabine: nid esgyrn Bach oedd yn yr arch honno.
Doedd yr esgyrn hynny ddim yn ddigon hen. Roeddwn i'n
amau hynny fy hun, oherwydd cyflwr yr arch. Ond roedd
y clasbiau pres yn profi hynny tu hwnt i bob amheuaeth.
Gwyddwn, o astudio'r cyfrifon, nad oedd y fath glaspiau

wedi cael eu harchebu tan 1811, mwy na thrigain mlynedd ar ôl marw Herr Bach. Doedd eu 'tystiolaeth' yn werth dim.

Ni soniwyd ar y pryd am y drydedd arch dderw a dynwyd o'r ddaear y bore hwnnw. Ni wn i p'un ai ysgerbwd Bach oedd ynddi ai peidio. Yn anffodus, yr oedd y pren yn bwdr, a chwalwyd y benglog yn yfflon gan y ceibio cyn i neb gael cyfle i weld beth oedd yno. Gwrthododd pawb arall – His, Transzchel, Wustmann – roi'r un munud o'u hamser i ysytried y darnau. Bron na fedrwn i eu clywed yn meddwl, Na, na, frodyr, nid Bach yw hwn, siawns, ni fyddai'r Almaenwr mwyaf a fu yn pydru, yn dadfeilio, yn malu mor rhwydd â hwn.

Ac eto, Sabine, roedd bysedd y llaw dde yn gyfan!!

Byddaf yn y pridd fy hunan, mae'n debyg, cyn iddynt roi eu cynlluniau ar waith. Yn y ddinas hon, heddiw, ni allaf, ni feiddiaf herio'r awydd i atgyfodi esgyrn. Ond i ti, Sabine, efallai y daw'r cyfle, ryw ddydd, i ddweud, yn dawel ac yn rhesymol, sut yr oedd pethau, a byddi di'n dangos yr esgyrn hyn iddynt, yr hen esgyrn melyn, bregus a amgaeaf gyda hyn o lythyr, nid i'w haddoli, ond i brofi eu bod nhw'n farw, ac yna byddi di, neu efallai dy blant… A briodaist ti erioed, Sabine? Dwi ddim yn gwybod hynny, hyd yn oed… Neu eu plant hwythau, gallant eu claddu nhw am byth, mewn lle dirgel, diarffordd, lle na fydd neb yn eu darganfod, heb na charreg na chroes, er cof amdanom ni'n dau, am ein *Begegnung. Der Bursche träumt noch von den Kussen, die ihm das susse Kind getauscht, er steht, von Anmuth hin gerissen, derweil sie um die Ecke rauscht.*

– *Begegnung... Begegnung*
– *Wie, bitte?*
– Coffi arall... os gwelwch yn dda.

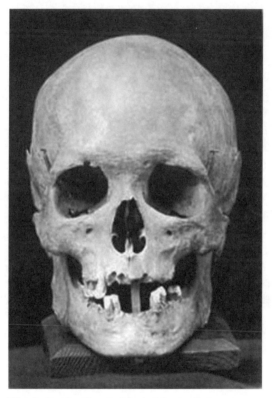

Ffig. 9   Penglog Honedig Johann Sebastian Bach

26 Begegnung

MAE'N HANNER AWR WEDI PUMP. Rwy wedi penderfynu
ymghylch tri pheth. Ymhen awr a chwarter bydda i'n codi,
bydda i'n talu'r bil, a bydda i'n mynd i'r orsaf er mwyn
dala'r trên dwy funud i wyth. Dyna'r peth cyntaf. Yn ail,
dwi ddim am ildio i gynllun mileinig, dirgel, dan-din rhyw
greadur truenus sydd heb amlygu dim o'r doniau llenyddol, y
weledigaeth ffres a gwreiddiol, yr arddull fyrlymus a chyhyrog,
yr ieithwedd liwgar a chyfoethog, y dioddef dirdynol, a'r
pethau eraill y byddwn yn troi atynt er mwyn esgusodi
ymddygiad yr Artist anystywallt. Ac yn drydydd, dwi ddim
yn dymuno troi yn gymeriad mewn rhyw nofel ôl-fodernaidd
eilradd.

Rwy'n gwybod beth sydd gen ti mewn golwg, Rhif 86.
Rwy'n gwybod nad caethiwed cyffredin fydd fy nhynged i, os
cei di dy ffordd. Wrth gwrs, os *wyt* ti'n fy nilyn i'n gorfforol,
sydd yn bur debygol, o ystyried dy becyn diwethaf, mae
hynny'n warthus. Onid trosedd yw stelcio? Ond nid hynny,
nid y canlyn corfforol, sy'n bwysig i ti, nage? Petaet ti am fy
wynebu, fy herwgipio hyd yn oed, gallet fod wedi gwneud
hynny eisoes. Pwysicach o lawer i ti, Rhif 86, yw fy nal, ac
yna fy nghadw'n gaeth, nid â rhaffau na chadwyni, ond yng
nghell afiach dy ddychymyg. Man a man i ti gyfaddef hynny'r
funud hon. Rwyt ti am i mi fod yn gymeriad yn dy gofiant
bondigrybwyll, rwyt ti eisiau chwarae mig â'm meddwl. Un
o'r rheiny wyt ti.

A phaid ti â meddwl am eiliad 'mod i'n dechrau ffwndro,
yn dechrau simsanu. Rwy wedi darllen digon o'u nofelau

bach pitw nhw i wybod yn iawn beth yw dy gêm di. On'd ydyn nhw'n dod ata i yn eu hugeiniau bob blwyddyn i fynnu nawdd er mwyn diflasu'r byd â'u trallod benthyg, eu hunigedd benthyg, eu habswrdiaeth benthyg? Do, fe ges i 'nhemtio ar un adeg. Daeth rhyw wefr fach, am sbel, o ddychmygu fy mod i'n gwacsymera yng nghwmni Melog, neu'n gogordroi â rhai o'm hen ffefrynau, Molloy, a K, a Meursault. Ond erbyn hyn, galla i weld sut mae'r plot yn ymddatrys, i le mae'n arwain, neu, yn hytrach, galla i weld nad yw'n arwain i unman, dim ond yn plygu 'nôl, i gnoi ei gynffon ei hun, ac rwy wedi colli diddordeb ynddo fe, do, ac mae'n flin gen i, ond alla i ddim bod yn bresennol, diolch yn fawr iawn i ti, ar achlysur y diweddglo amwys, awgrymog.

Gogordroi? Na. Symud ymlaen. Peidio ag edrych yn ôl. Gadael i'r meirw gladdu'r meirw, dyna sydd ei angen. Nid troi yn fy unfan. Gogordroi. *Begegnung*. Does dim rhaid i mi fod yn gaeth i batrwm o eiddo rhywun arall. Yn enwedig patrwm nad oes dim llinell derfyn iddo. Yn wahanol i amserlen, er enghraifft. Nawr te! Dyma beth *yw* patrwm derbyniol. Mewn amserlen, gelli di weld dechrau a chanol a diwedd. Gelli di weld dechrau a chanol a diwedd pob amrywiad, hefyd, pob is-batrwm, y cysylltiadau, y rhwydweithiau, a'r cwbl yn glytwaith o linellau syth, a phob llinell wedi'i rhannu'n gamau rhesymol, at ddefnydd hwylus y teithiwr. Ni fydd llinell felly byth yn bwyta ei chynffon ei hun. O na, daw i'w therfyn, rywbryd, rywle, yn debyg i'r rheilffordd yn Leipzig, er enghraifft, ac yno, bydd y teithiwr diddig yn chwilio am y cam nesaf, am fod pob diwedd hefyd yn ddechrau. Ni all llinell ogordroi. *Begegnung*.

Symud ymlaen. Ac mae hynny'n golygu peidio â chael fy nghamarwain, peidio â chael fy nargyfeirio gan bethau dibwys. *Begegnung*. Dyna enghraifft i chi. Bu ond y dim i mi ildio i swyn ei gytseiniau di-ystyr. Ond fe ymataliais. Fe galliais. Os abwyd ffug yw hwn, dwi ddim gwaeth o'i anwybyddu. Ac os oes rhyw ystyr yn llechu yno rywle, fe ddaw i'r wyneb gyda threigl amser. Wedi'r cyfan, does dim disgwyl i mi gwblhau'r llyfr ar ran yr ymgeisydd, siawns! Na, dim ond llenwi'r ffurflen gofrestru. Disgrifio amcan y cais. Nodi gwrthrych y nofel gofiannol. Cloriannu ei addewid. Rhoi rhesymau dros ei gymeradwyo. Ei gysoni ag uwch-amcanion y Weinyddiaeth.

Begegnung. Na, af i ddim ar ôl y sgwarnog gyntaf i siglo'i chynffon y tro hwn. Myfyriaf. Ystyriaf yr opsiynau cyn tynnu'r edafedd at ei gilydd. Mae gen i bedwar diwrnod cyfan. Pedwar diwrnod i sgrifennu rhyw ddau gant o eiriau... tri chant ar y mwyaf... a thicio rhyw ugain o flychau.

Mae'r trên yn cychwyn ar amser, amheuthun o beth yn wir. *Begegnung.* Gallai fod yn air dibwys, wrth gwrs, heb arwyddocâd, os nad heb ystyr. Rhaid cofio hynny. A dibwys neu beidio, arwyddocaol neu beidio, y mae'n debyg o olygu pethau gwahanol i wahanol bobl. I fi, ar hyn o bryd, dan ddylanwad ffwlbri diweddaraf Rhif 86, ond heb eiriadur boddhaol, does gan y gair ddim ystyr. Nac oes. Ac eto y mae'n drwm o ystyron. Y mae'n drwm o ystyron oherwydd ei gysylltiadau: â hanes Sabine; â'r archifydd di-enw; â'u perthynas fyrhoedlog. Mm. Fydda i ddim gwaeth, siwns, o holi. Cyhyd â 'mod i'n cadw meddwl agored. Cyhyd â 'mod i ddim yn credu'r peth cyntaf sy'n dod o enau'r creadur cyntaf a welaf, dim ond achos mai'r peth cyntaf yw e. Na'r ail, na'r ugeinfed, o reidrwydd. Nid llinell yw pob cyfres. Nid cam yw pob arhosfan.

– Esgusodwch fi...

Mae'r fenyw wallt golau ganol oed sy'n eistedd gyferbyn â mi yn codi ei llygaid o'i llyfr ac yn edrych dros ei sbectol.

–... *Bitte?*

–... y... *Begegnung...*

– *Begegnung*?

– Maddeuwch i mi ond dyw'r gair ddim yn y tipyn llyfr sydd gen i... Tybed a wyddoch chi ei ystyr?

– *Entschuldigung. Ich verstehe nicht. Meine Tochter soll Ihnen helfen.*

A gyda hynny o eiriau, geiriau nad ydynt yn golygu dim i mi, mae hi'n troi at y ferch sy'n eistedd wrth ei hymyl. Rhyw ddeuddeg mlwydd oed yw hon, efallai, ac yn amlwg yn ferch i'r llall, am fod gan y ddwy yr un steil gwallt a'r un olwg ddifrifol. Ac mae'r ferch, hefyd, wedi ymgolli'n llwyr mewn llyfr. Rhydd ochenaid fach wrth i'w mam geisio denu ei sylw, ac yna edrych arnaf – yn ddisgwylgar, efallai, ond hefyd, os nad wy'n cangymryd, braidd yn amheus.

– A! Wel... maddeuwch i mi... Digwydd dod ar draws y gair yma... *Begegnung*... Fallai 'mod i'n ei ddweud e'n anghywir... Byddai'n well, fallai, petawn i'n ei sgrifennu i lawr... Ie, digwydd dod ar ei draws mewn... mewn llythyr, a minnau heb allu deall Almaeneg... Tybed... Ydych chi'n gwybod... Mm?

Mae'r ferch yn cymryd anadl ddofn ac yn sythu ei hysgwyddau a'i chefn fel petai ar fin cystadlu mewn gornest lefaru.

– *Begegnung* ydy'r gair Almaeneg am...

Yn y saib sy'n dilyn, fe'm trewir yn sydyn gan y posibilrwydd arswydus mai gair anweddus o ryw fath, gair brwnt neu air awgrymog yw *begegnung*, ac mai dyna sy'n gyfrifol am eu dwyster, am eu gwefusau tynn, am y saib hir...

–... cyfarfod!

– Cyfarfod?

Wel, teitl braidd yn ddi-fflach i gân, dybiwn i. Ond os felly, diolch amdano. Di-fflach yw di-ffwdan. Gallaf anghofio am y cwbl wedyn. Ac eto, mae sawl math o gyfarfod…

– Cyfarfod, felly. Wel, diolch i ti am hynny. Ond… Os ca i ofyn un cwestiwn bach arall… Os nad oes gwahaniaeth gen ti… Ydy e, tybed, yn meddwl rhyw fath arbennig o gyfarfod? Ydy e'n golygu… cyfarfod mawr mawr, a llawer iawn o bobl… Neu fel arall, ydy e'n golygu… rhyw *rendezvous* bach bach rhwng dau…?

Mae'r ferch yn troi at ei mam, ond mae honno, erbyn hyn, wedi ail-afael yn ei llyfr.

– Gofynnaf i'm brawd.

Bachgen eithriadol o swrth yw ei brawd. Eistedda ar ei ben ei hun yr ochr arall i'r gerbyd, yn syllu trwy'r ffenest, ac mae'n anfoddog iawn, ddywedwn i, i gymryd rhan mewn rhyw gêm 'dyfalwch beth yw ystyr hwn'. Ond mae'r ferch, chwarae teg iddi, yn egluro fy nghais, tra 'mod innau'n gwenu'n llydan er mwyn cyfleu diniweidrwydd y cais hwnnw a'm gwerthfawrogiad mawr o'u hymdrechion i'm helpu.

– *Die Begegnung? Ach! Der blode Idiot. Er will zum Fussballspiel – Deutschland gegen England – aber er sitzt im falschen zug – so ein Arschloch!*

– Mae fy mrawd yn dweud croeso mawr i'n gwlad, a gobeithio y cewch chi hwyl yma, ac mai gêm bêl-droed yw'r *begegnung* rydych chi'n sôn amdano. A'ch bod chi ar y trên anghywir.

– A! Diolch yn fawr iawn i chi'ch dau. Gornest, felly. Cyfarfod rhwng gwrthwynebwyr. Purion!

Ond pam, wedyn, y byddai'r archifydd yn hel atgofion rhamantus, melys am gêm bêl-droed? A oedd caneuon pêl-droed i'w cael mor bell yn ôl â 1896? Oni bai bod y gân yn sôn am fath arall o ornest, wrth gwrs.

Mae'r ferch, druan fach, yn edrych ychydig yn bryderus

erbyn hyn, gan feddwl, siŵr o fod, fy mod i'n gefnogwr pêl-droed sydd wedi colli'i ffordd, ac sydd, o'r herwydd, ar fin colli ei limpyn hefyd a dechrau malu ffenestri a phennau pobl. Neu, efallai, am ei bod hi'n methu deall pam fy mod i mor ddifater ynglŷn â'r peth. Prysuraf i'w chysuro.

– Na. Nid gêm bêl-droed oedd gen i… Rhyw feddwl oeddwn i mai enw cân oedd *Begegnung*. Hen gân, hefyd. Cân sy'n mynd yn ôl gan mlynedd a mwy. Ond dyw e ddim yn bwysig…

– Gofynnaf i Dad-cu.

Yn ddiarwybod i mi, mae hen ddyn wedi bod yn cysgu yn y sedd y tu ôl i'r ferch. Cymeraf mai cysgu oedd e, gan fod angen i'r ferch roi pwt i'w fraich i dynnu ei sylw, ond mae'n bosibl, wrth gwrs, ei fod ar ddi-hun trwy gydol yr amser ond yn drwm ei glyw. Ta p'un, mae'r ferch, sydd erbyn hyn, rwy'n credu, yn dechrau ymfalchïo yn ei galwedigaeth newydd fel lladmerydd ac ymchwilydd a chyflafareddydd y gerbyd, yn bwrw ati i gyfleu byrdwn fy neges i'w thad-cu.

– *Ein Altes Lied! Beg-eg-nung! Ein Altes Lied!* mae hi'n 'ddweud, ac yna'n ei weiddi, drosodd a throsodd, gan roi pwyslais ar bob sill, a'r llais yn codi ac yn arafu, codi ac arafu, nes bod pob teithiwr yn y gerbyd yn rhythu arnom mewn syfrdandod, a minnau'n teimlo'n bur aniddig o'r herwydd.

– *E I N A L T E S L I E D ! B E – G E G – N U N G ! E I N A L T E S L I E D !*

Parha'r gweiddi yn ddidrugaredd. O'i ran ef, nid yw'r hen ddyn yn edrych fel petai wedi'i gynhyrfu'n ormodol gan y cythrwfl. Mae'n cwpanu ei glust, ac yn plygu ryw fymryn bach tua'r ochr, a dyna i gyd a welaf: braich mewn llawes lwyd, llaw grynedig, a chorun moel.

– *E I N A L T E S L I E D ! B E – G E G – N U N G ! E I N A L T E S L I E D !*

Saib. Distawrwydd. Mae pob clust a phob llygad wedi'i
hoelio ar yr hen ddyn. Daw pob sgwrs i ben. Rhoddir pob
llyfr i lawr. Y mae hyd yn oed y brawd hŷn yn deffro o'i
syrthni. Estynaf fy ngwddf gymaint ag y gallaf er mwyn
clywed ymateb yr hen ddyn, os daw. Ond ni ddaw. *Cric,
di-cric, Cric, di-cric, Cric, di-cric, Cric, di-cric, Cric, di-cric, Cric,
di-cric.* Sŵn yr olwynion ar y cledrau yw'r unig sŵn a glywir.
Cric, di-cric, Cric, di-cric, Cric, di-cric, Cric, di-cric.

Ni ddaw am funud gyfan.

Ni ddaw am funud arall.

Ond pan ddaw, fe ddaw o bell. O bell, bell, trwy afonydd
a gwrychoedd, dros weunydd a mynyddoedd, i lawr simneiau,
rownd corneli strydoedd cefn myglyd lle mae'n bwrw llwch
yn barhaus, fe ddaw fel lleisiau o bell, bell ar donfeddi tramor
ers llawer dydd, yn mynd a dod, yn mynd a dod ar adenydd y
gwynt, yn tuchan a thewi, yn peswch a thewi, yn ochneidio a
thewi, yn chwerthin, yn chwerthin! a thewi.

– Mae Tad-cu'n dweud…

Mae'r ferch yn ei pharatoi ei hun i draddodi eto.

–… yn dweud nad yw'n gwybod dim am y gân. Ond mae e'n
cofio'r ffilm.

– Y ffilm? meddaf i, o ran cwrteisi yn fwy na chwilfrydedd.

– Mae Tad-cu yn dweud… iddo gofio… iddo gofio… *ein
Liebesgeschichte*… Dwi ddim yn gwybod y gair… Mae'n cofio
mynd â Mam-gu i weld y ffilm, am ei bod hi'n stori… Sut
mae ei dweud hi? *Liebesgeschichte*…

– *Liebe*… A! Stori serch wyt ti'n feddwl? Cariadon… Dyn a
menyw…

– *Ja, ja*, ac mae e'n dweud mai'r noson honno y gofynnodd
iddi ei briodi… A dyna pam mae e'n cofio'r ffilm…

– Diwedd hapus i'r stori, te!

–… A dyna pam, mae Tad-cu yn dweud, dyna pam aethon nhw i weld y ffilm dair gwaith, a dysgu llawer o'r geiriau, er eu bod nhw'n gorfod darllen yr… *die Untertitel*… i gael deall…

– *Unter*…?

–… Ac mae Tad-cu yn dweud sut roedd Mam-gu yn esgus bod… yn esgus bod yn… Selma Jensen…? A sut roedd Tad-cu'n esgus bod yn… Trau… von… Neualt…? Mae'n flin gen i… Dwi ddim yn siarad Saesneg yn dda iawn…

Wel, dyma beth yw hwyl, yn wir, ac rwy'n dechrau gorfoleddu oherwydd amherthnasedd a diniweidrwydd y cyfan, ac mae rhyw lonyddwch, rhyw fodlondeb mawr yn arllwys ei enaint drosof. Mae fel petawn i'n nesáu at ddiwedd twnel ac yn gweld, aleliwa! nad cul a thywyll a dyrys yw bywyd wedi'r cwbl, ond helaeth, a heulog ac yn hyfryd o amlwg.

Nesáu. Ond nid cyrraedd. Dim eto. Dim cweit. *Begegnung*.

Mae'n gân.

Mae'n gêm bêl-droed.

Mae'n gyfarfod.

Mae'n ffilm serch.

Pedwar tyst, dyna i gyd. Dim llawer i ddyn mor brofiadol â minnau mewn arferion monitro a chloriannu a dadansoddi, siawns. Ac eto, pedair tystiolaeth go wahanol. Gwahanol – ac efallai'n gwbl anghymarus â'i gilydd. Doedd dim dal. Ac ar ben hynny…

Ac ar ben hynny dyweder 'mod i'n dilyn pob un o'r pedwar trywydd hyn. Ac wrth wneud hynny, dyweder 'mod i'n darganfod bod pob un ohonynt, yn ei dro, yn arwain at bedwar trywydd arall. Ble byddwn i wedyn? Ble byddwn i o gymryd, dyweder, dim ond pedwar cam pellach?

Un cam. O bedwar i un ar bymtheg.

Dau gam. O un ar bymtheg i… chwe deg pedwar.

Tri cham. O chwe deg pedwar i… ddau gant pum deg chwech.

Pedwar cam. O ddau gant pum deg chwech i… mil, dau ddeg pedwar.

Mil, dau ddeg pedwar o bosibiliadau gwahanol, a minnau wedi symud dim ond pedwar cam ymlaen. A phob un o'r posibiliadau'n hawlio'r un sylw manwl, jyst rhag ofn mai yno, ym mocs saith cant tri deg a naw, neu yno, ym mocs wyth cant naw deg a phedwar, y mae'r allwedd i'r pos cyfan.

Ond *does* dim pos, oes e?

27 Camau yn y Gegin

CYRHAEDDODD STEFFAN ADREF am naw o'r gloch y nos
Sadwrn honno. Roedd hi'n briwlan. Roedd seiren ambiwlans
yn *wa-wa-wawan* trwy ganol y ddinas. Roedd y carfannau
cyntaf o ferched breichnoeth, bogeiliog, sgrechlyd, ac o
fechgyn sarrug, bygythiol eisoes yn meddiannu'r strydoedd.
Doedd neb yn y tŷ i'w groesawu. Ond doedd hynny'n menu
dim ar Steffan. Yr oedd, yn ei dyb ef ei hun, wedi darganfod
ffordd, nid i ddatrys y broblem, ond yn hytrach i dynnu ei
cholyn. Ac fe orlifodd ei foddhad i wacter y byd o'i gwmpas.

Gwnaeth Steffan gwpanaid o de ac eistedd wrth fwrdd
y gegin. Gydag ochenaid o ryddhad, dechreuodd flasu'r
fuddugoliaeth y tybiai ei fod newydd ei hennill, a gwnaeth
hynny yn yr unig ffordd a wyddai, sef trwy ei fesur. Fel
roedd Steffan eisoes wedi darganfod ar y trên, byddai datrys
y broblem ddiweddaraf wedi golygu cymryd mil, dau ddeg
pedwar cam. Tipyn o dasg, yn wir. Ond, wrth gnoi'r syniad
yn hamddenol am ychydig funudau, sylweddolodd y gallasai'i
gyfyng-gyngor fod yn llawer gwaeth na hynny. Yn llawer,
llawer gwaeth. Y fath ddihangfa! Cymerodd ddarn o bapur
a thynnu llun o'r pedwar cam damcaniaethol ar lwybr un,
a llwybr un yn unig, sef y gosodiad mai *cân* oedd prif ystyr
Begegnung.

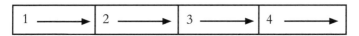

*Ffig. 10   Y Pedwar Cam Damcaniaethol ar Lwybr Un, sef y gosodiad
mai cân oedd prif ystyr Begegnung*

Rhaid pwysleisio, yn y fan yma, nad oedd dim
gwahaniaeth gan Steffan ynglŷn â natur y pedwar cam hynny:
hynny yw, y natur y gellid ei phriodoli iddynt yn y byd
go iawn. Er enghraifft, o fwrw mai cân oedd *Begegnung*, fe
allai rhywun â gwir ddiddordeb yn y ffaith honno ymweld
â'r llyfrgell, fel y gwnaeth Steffan yntau ar ddechrau
ei ymchwiliadau, a chanfod mai un o gyfansoddiadau
Schubert oedd hon. Neu Schumann neu – pwy arall,
tybed? – dywedwn ni Hugo Wolf. Ie, a chanfod mai un o
gyfansoddiadau Hugo Wolf oedd *Begegnung*. A dyna gam
un ichi. O'r man cychwyn hwnnw, gellid dilyn y llwybr
cofiannol ymlaen i ryw garwriaeth ddirgel, efallai. Carwriaeth
rhwng Wolf a menyw o'r enw Melanie. Melanie Köchert.
A dod o hyd i lun ohoni, hanes ei thylwyth, a'r cwbl. Ac
yna, wedi rhagor o dwrio, dod at ryw fardd neu'i gilydd.
Dywedwn ni mai Eduard Friedrich Mörike yw ei enw. A
chael gan Mörike y gerdd sy'n crynhoi, i Wolf, holl angerdd
a gwewyr y garwriaeth. A darllen llythyr wedyn lle mae
Hugo'n disgrifio sut yr aeth ati i gyfansoddi cân ar y gerdd
honno. *Begegnung*. Ond sut gallwn ni fod mor sicr mai
ymdrin â'r berthynas honno y mae'r gân? Y mae'r llythyr
braidd yn amwys. A dyna'r llwybr yn cymryd tro sydyn, ac
yn ein harwain… I ble, dwedwch? Wel, petaen ni'n anlwcus,
gallai arwain, yn ôl ffasiwn yr oes ymhlith cyfansoddwyr
rhamantaidd, trasig, at syffilis, gwallgofrwydd a marwolaeth
annhymig yn seilam Fienna. A dyna gyrraedd y pedwerydd
cam, a diwedd y llwybr, a diwedd y cofiant hefyd, o
gymhwyso'r cyfan at amcanion ymgeisydd Rhif 86. Yn gwbl
ddamcaniaethol, wrth gwrs.

Ond doedd gan Steffan ddim diddordeb mewn camau o'r
fath. Nid ar hyn o bryd, beth bynnag. Am y tro, roedd wedi
codi uwchlaw'r byd a'i drybestod. Yr haniaeth, y patrwm
oedd popeth.

Ac yna, tynnodd lun o'r pedwar llwybr arall a allai, yn ôl gweledigaeth Steffan, fod yn ymestyn o bob un o'r camau hyn.

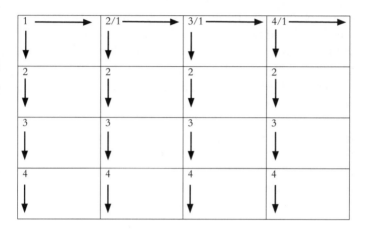

Ffig. 11   Pedwar llwybr a allai fod yn ymestyn o bob un o'r pedwar cam ar Lwybr Un

Ni fynnai Steffan weld yr un sylwedd yn y camau hyn, chwaith. Pe dymunai, diau y gallai fod wedi cymryd Cam Pedwar ar y Llwybr Cyntaf (4/1) – gwallgofrwydd Hugo Wolf yn seilam Fienna yn yr enghraifft ddamcaniaethol uchod – fel man cychwyn newydd ac, o ddefnyddio ei drwyn llenyddol, tyrchu am ryw amheuthun bach annisgwyl. Cerdd arall, ddywedwn ni. Cerdd yn dwyn y teitl *'When Hugo Wolf Went Mad'* – byddai hynny'n ddigon o abwyd iddo, byddech chi'n tybio. Cerdd gan rywun annhebygol ond gogleisiol. Rhywun fel Charles Bukowski, efallai: y gwallgofddyn o Los Angeles. A chael, o orffen ei daith yn y fan honno, mai Bukowski oedd gwir destun cofiant Rhif 86, nid Bach, na

Wolf, na Melanie Köchert, nag Eduard Friedrich Morike chwaith. Ie, o orffen y daith yno. Ond wnâi e ddim, na wnâi? Ymhen hir a hwyr byddai'n darganfod bod rhyw fand pŷnc, rhyw fand Cymraeg hyd yn oed – fe'u galwn yn *Rheinallt H. Rowlands* – wedi rhyddhau cryno-ddisg dan y teitl *Bukowski* – beth arall? – a honno'n deyrnged i'r bardd gwallgof… A dyna'r hen lwybr cyfrwys wedi gyrru ei gwlltwr danheddog trwy ei ardd gefn ei hun!

Ond na, doedd gan Steffan ddim diddordeb mewn llwybrau o'r fath: llwybrau y gallai pobl o gig a gwaed eu troedio. Yr hyn a boenai Steffan erbyn hyn oedd y ffaith nad oedd ei bapur yn ddigon mawr, na'i ddyfeisgarwch fel arlunydd yn ddigon disglair, i gynnwys yr un llwybr ar bymtheg arall a ymestynai o'r camau hyn, na'r chwe deg pedwar a darddai o'r rheiny… Ac yn y blaen.

Edrychodd Steffan ar ei lun bach syml. Fe'i trôdd ar ei ochr, ac edrych eto. Fe'i trôdd ar ei ben. Arllwysodd gwpanaid arall o de. O dipyn i beth, a heb feddwl unwaith am y lluoedd o feirdd a chyfansoddwyr gwallgof a allai fod yn cuddio y tu ôl i'r blychau cymen, fe ddôi yn ymwybodol o erchylltra gwirioneddol y llun a'i oblygiadau. Y fath ddihangfa, yn wir! Oherwydd, yn ogystal â'r llwybr syth, 1, 2, 3, 4, a darddai o gam 1, efallai, o gyrraedd cam 2 ar y llwybr hwnnw, meddyliai Steffan, y dôi llwybr arall i'r golwg. Llwybr troellog, o bosibl. Llwybr a grwydrai rhwng – ie, *rhwng!* – y llwybrau cyfarwydd, llwybr nad oedd yn dilyn y drefn rifyddol gyfleus yr oedd Steffan wedi'i dyfeisio iddo, ond yn hytrach a gamai yn ôl ac ymlaen heb unrhyw resymeg amlwg. Llwybr di-lwybr. Llwybr nad oedd ond enw arall ar Anhrefn. Ceisiai Steffan olrhain y fath lwybr trwy ei lun.

Ond yna, o ddilyn y llwybr newydd, dieithr hwnnw, pwy allai ddweud na fyddai Steffan yn canfod mai hwnnw, yn wir, *oedd* y llwybr syth a'r lleill i gyd ond yn llwybrau defaid, heb

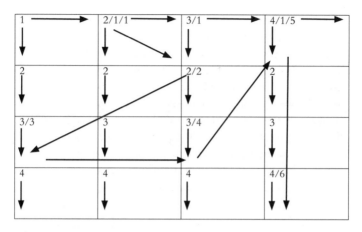

Ffig. 12 Y llwybr troellog a olrheinodd Steffan trwy'r pedwar llwybr a allai fod yn ymestyn o bob un o'r pedwar cam ar Lwybr Un

arwain i unman? Gan bwy mae'r hawl i ddeddfu beth sydd yn syth a beth sydd yn gam? Onid perspectif yw'r cyfan?

Ond ar ben hynny eto, meddyliai Steffan, byddai'n rhaid cydnabod posibilrwydd pellach, sef bod yna lwybrau eraill i'w canfod, a'u dilyn hefyd, *rhwng y grŵp hwn o bedwar llwybr a'r tri grŵp arall.* Rhaid caniatáu, na, rhaid *mynnu* bod sylw teilwng yn cael ei roi i bob un o'r llwybrau dichonadwy *rhwng mil dau ddeg pedwar o gamau.* A faint o lwybrau fyddai hynny? Ni wyddai Steffan, wrth gwrs. A oedd mwy na miliwn? Yn gymaint â biliwn? Yn gymaint â gronynnau'r tywod ar y traeth? Doedd taten o ots. Doedd taten o ots am nad oedd modd i'r un dyn byw ddilyn cymaint o lwybrau. Doedd gobaith i holl ddynion y byd gyda'i gilydd eu dilyn, neu efallai'r holl ddynion, byw a marw, a fu erioed. A dyna ogoniant y peth i Steffan. Am y tro cyntaf erioed, roedd wedi bwrw iau ei hen ddyletswydd. Roedd wedi ymwrthod â'r hen orfodaeth i archwilio pob agwedd ar bob mater cyn dod i benderfyniad.

Ac onid gronynnau un traeth yn unig oedd y rhain? meddyliai Steffan. Siawns nad oedd mwy o draethau lle'r oedd y tonnau wrthi'n ddyfal ddiddiwedd yn malu creigiau'r oesoedd yn llwch mân. Canys pwy a benderfynodd mai pedwar cam yn unig a geid ar bob llwybr? Gallai fod pedwar cant! Neu bedair mil! A beth fyddai'r cyfanswm wedyn? Na, meddyliai Steffan: *petai* 'na bos a oedd yn werth ei ddatrys, *allen* ni byth ei ddatrys e, am fod y llwybrau a'r camau yn ymestyn o'n blaenau am byth. Ac os felly, os oedd y dewis yn wirioneddol ddiddiwedd, a wnâi unrhyw wahaniaeth pa un a ddewisai? Na, doedd dim diben poeni am y peth.

Yn ysgafn ei droed a'i galon, camodd Steffan at y ffôn i weld a oedd unrhyw negeseuon wedi cael eu gadael iddo yn ystod tridiau ei absenoldeb.

– Steffan… Fi sy' ma…

Llais Llio.

– Jyst yn ffonio i ddweud 'mod i'n aros yma am ddiwrnod neu ddau arall… Mae Mam yn helpu twtio ychydig, a Dadi'n chwynnu'r ardd a rhyw betheuach fel 'na… Allwn ni ddim gwneud llawer mwy nes 'yn bod ni'n arwyddo'r contract… Der draw ddydd Sul os gelli di, i gael gweld y lle… Gest ti amser da yn yr Almaen? Paid gwneud gormod o waith… 'Wy'n dy garu di… O! Cofia ddod â'r map a'r cyfeirnod rhag i ti fynd ar goll! Wela i di ddydd Sul, te… O!… Ac fe ges i wared â'r hen enw Saesneg 'na, a dodi'r hen enw nôl… Enw od… Sa i'n 'i ddeall e, a gweud y gwir… Sori, Mam yn galw… Gorffod mynd… Cymer ofal.

Ffoniodd Steffan yn ôl yn ddiymdroi. Peiriant a'i hatebodd.

– Llio… Diolch am y neges… Ces i amser… buddiol iawn… yn yr Almaen… Cei di'r hanes i gyd fory… Do i draw i'r tŷ newydd marce tri… Ddaliais i ddim o'r enw, gyda llaw…

Rho wybod os cei di gyfle... Yn dy garu dithau hefyd... Ta-
ta...

Do, bu'r daith yn fuddiol iawn. Yn fethiant ysgubol,
hefyd, ar un ystyr, ond yn fuddiol yn y pen draw oherwydd
fe ddysgodd i Steffan nad oedd angen chwilio ymhellach am
atebion i gwestiynau blinderus Rhif 86. Un cam bach arall a
byddai'n gallu cwblhau'r ffurflen gofrestru. Byddai ei waith
ar ben. Yna, hwyrach, gallai ddechrau mwynhau, fel yr oedd
Llio eisoes yn ei wneud, y dyfodol cyflawnach a dedwyddach
a'i disgwyliai. Un cam arall. Rhaid nodi testun ar gyfer y
nofel gofiannol arfaethedig. Ni allai osgoi hynny. Dyna a
fynnwyd gan Lwybr Papur y Weinyddiaeth. Nid *y* testun,
wrth reswm. Nid y testun terfynol, absoliwt. Na, oherwydd,
os oedd hwnnw'n bod, yr oedd y tu hwnt i grafangau pitw ei
law feidrol ef. Na, nid hwnnw. Ac eto, yr oedd angen testun
a feddai ar *ryw* hygrededd. Testun a oedd, o leiaf, yn gyson
â'r deunydd arall a gyflwynwyd. Testun a wnâi synnwyr o
Begegnung. Wel, 'doedd ond pedwar posibilrwydd. A chan nad
oedd gwahaniaeth pa un a ddewisai, cystal iddo ddewis ar hap.
Torrodd ddarn o bapur yn bedair ac ysgrifennu ar y darnau fel
hyn.

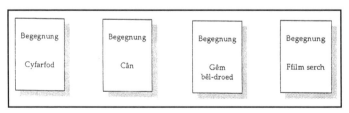

Ffig. 13   Pedwar darn o bapur

Yna, rhoes y pedwar darn o bapur yn rhes ar ganol bwrdd
y gegin a mynd o'r ystafell.

Pan ddaeth yn ôl ymhen rhyw ddeg munud, roedd
Steffan yn cario pot jam â chlwtyn drosto. Dododd y pot ar

ei ochr ar ben y bwrdd, ac aros. Clywodd sŵn chwerthin
o'r tŷ drws nesaf. Clywodd sŵn ambiwlans arall yn y pellter.
Ond hoeliodd ei sylw ar y pot. Tipiai'r eiliadau heibio. Un,
dau, tri .. ..Chwerthin… Dau ddeg saith, dau ddeg wyth…
Gwaedd… Tri deg pedwar, tri deg… Corryn. Yn araf ac
yn betrus, camodd corryn o'r pot jam. Stopiodd. Gwibiodd
i'r chwith, stopio eto. Yna gwibio ymlaen. I'r chwith. I'r
dde. Ac yna dros ymyl y bwrdd, i hafan gysgodol braf. Dyna
gorynnod i chi, meddyliai Steffan.

Aeth Steffan o'r ystafell eto. Pan ddychwelodd, aeth trwy'r
un ddefod eto, ond gan roi ambell bwt bach i'r pot.

– Dere mlaen, mochyn bach, dere mlaen.

Ond ni thyciai. Doedd y belen fach lwyd ddim am symud.
Dyna foch coed i chi, meddyliai Steffan.

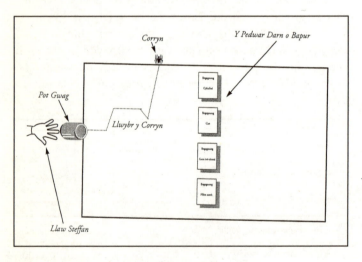

Ffig. 14 Taith y corryn ar draws bwrdd y gegin

Yna, aeth Steffan allan trwy'r drws ffrynt. Ar ôl chwarter awr, dychwelodd â chath dan ei fraich. Nid oedd i'w gweld yn gath ddiddig iawn, a dichon fod y crafiadau coch ar ddwylo Steffan yn tystio i hynny. Ond, ym meddwl Steffan, byddai ychydig o aniddigrwydd, o bosibl, yn ychwanegu at haprwydd ei hymateb pan ddôi'r galw. Rhoes y gath ar y llawr a chau'r drws, er mwyn rhwystro unrhyw ymgais i ddianc. Mewiodd ei hanfodlonrwydd. Yna, dododd Steffan y darnau papur ar y llawr, ryw ddeg troedfedd oddi wrth y creadur, fel hyn:

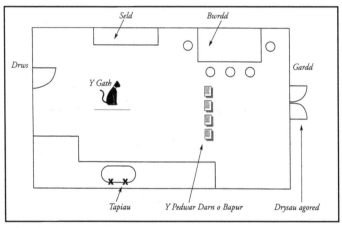

Ffig. 15   Y gegin, y gath a'r darnau papur

Wisht! Wisht! Wisht! ebychodd Steffan. Edrychodd y gath yn syn arno a mewian eto. *Wisht!*

Doedd Steffan ddim yn ddyn creulon. I'r gwrthwyneb; yr oedd yn hoff o gathod ac yn edmygu eu cymeriad annibynnol. Cadwai gathod pan oedd yn grwt, nid yn haid, wrth reswm, ond un ar y tro. Pan gâi un ei lladd ar yr hewl – a digwyddai hynny'n ddi-fael bob tri neu bedwar mis – fe brynai'i fam un newydd iddo, hyd at yr unfed gath ar ddeg, yr hon a laddodd y byji. Ta waeth, yn wahanol i Steffan, yr oedd y dyn drws nesaf, ffotograffydd wedi ymddeol ac yn tynnu am ei gant ac yn gwisgo het wellt, yn caru blodau ac yn casáu cathod, am eu bod nhw'n baeddu ym mhob man, ac yn enwedig ymhlith ei fegonias. Ei ddull ef o gadw cathod draw oedd tywallt neu chwistrellu dŵr drostynt. Wedi gwneud hynny ryw bump neu chwech o weithiau, byddai codi potel neu biben yn ddigon i'w dychryn ac fe ruthrai'r creadur oddi yno fel, wel, fel cath i gythraul. Fe ddôi yn ei hôl yn ddigon buan, bid siŵr, ond doedd hynny ddim yma nac acw i Steffan ar y funud. Y nod oedd cael y gath i symud. Doedd diawl o wahaniaeth beth a wnâi wedyn.

Rhoes soseraid o laeth ar y llawr er mwyn tawelu ei gynorthwy-ydd anfoddog. Arhosodd nes bod y gath yn llymeitian nerth ei thafod. Yna, cerddodd i ben arall y gegin ac agor y drws cefn led y pen. Daeth yn ôl a llenwi cwpan o'r tap. Heb oedi, taflodd y cyfan dros ei chynffon. Nid arhosodd y gath i ystyried achos ei gwlybaniaeth annisgwyl. Gyda sgrech annaearol, saethodd ar draws y gegin a diflannu i'r gwyll y tu allan. Ac wrth wneud hynny – diolch i'r nefoedd! dywedodd Steffan dan ei anadl – wrth wneud hynny, fe sathrodd ar un o'r pedwar darn o bapur. Neu, yn hytrach, fe gyffyrddodd y gwynt – y gwynt a grewyd gan y gath wrth gythru heibio ryw lathen i'r dde – fe gyffyrddodd y gwynt hwnnw â'r papur fel y bydd awel ysgafn yn codi deilen, gan ei symud dim ond y mymryn lleiaf, ond ei symud serch hynny.

Aeth Steffan i weld.

Begegnung… Cân…

Cân! Na! meddai Steffan yn uchel. Doedd e ddim am ddilyn y llwybr helbulus hwnnw eto. Gwastraffodd ddigon o amser yn barod yng nghwmni Anton a Bach a'i esgyrn bondigrybwyll. Ofnai Steffan, hefyd, mai hwn oedd y llwybr yr oedd Rhif 86, gyda'i holl obsesiynau cerddorol, yn awyddus iddo ei ddilyn. A phwy a ŵyr nad oedd yn iawn, o feddwl am helyntion y cyfansoddwyr a'r beirdd gwallgof y daethom ar eu traws wrth roi cynnig ar ddilyn y llwybr hwnnw ein hunain? Dechreuodd amau, hyd yn oed, bod 86, trwy ei dyfeisgarwch mileinig, wedi rheibio'r gath. Ta waeth am hynny, os nad oedd ots pa gam a ddewisai, pam na allai ddewis un a roddai ychydig o ddifyrwch iddo ar ôl ei helyntion diweddar. Y gêm bêl-droed efallai? Yr oedd amherthnasedd trawiadol y dewis hwn yn apelio'n fawr. Ar yr un pryd, nid oedd yn siŵr sut y gallai gyfuno ymchwil i ryw gêm bêl-droed yn yr Almaen â gweddill y dystiolaeth a gasglwyd. A pha gêm bêl-droed? Doedd dim llawer o reswm, chwaith, dros feddwl mai cofiant pêl-droediwr oedd mewn golwg gan Rif 86. Na, roedd amser yn rhy brin i gychwyn ar lwybr newydd mor anaddawol. Cyfarfod, felly? Rhy ben-agored. A gormod o berygl y byddai'n cwrdd â rhyw hurtyn arall. Ffilm serch o'r Almaen? Braidd yn anodd cael gafael ar wybodaeth…

Bu bron i Steffan fynd allan i mofyn cath arall pan gofiodd beth ddywedodd y ferch ar y trên. Bod ei thad-cu wedi gofyn i'w gariad ei briodi ar ôl gweld y ffilm. A hynny yn fuan wedi diwedd y Rhyfel. Yr oedd yno, felly, ryw arwydd o fywyd newydd, o ddedwyddwch i ddod, a da hynny. A fuon nhw'n ddedwydd wedyn? A oedden nhw'n ddedwydd o hyd? Doedd e ddim wedi meddwl gofyn. Ond os mai hwn oedd y cam olaf ar y llwybr, pennod olaf y stori, roedd hi'n argoeli'n well na'r lleill. Y ffilm serch amdani, felly: diweddglo hapus, cymhorthdal hael i Rif 86, tŷ newydd iddo ef a Llio, a dim rhagor o ogordroi mewn pwll o ddiflastod.

28 Y Siop Fideo

BU STEFFAN YN SEFYLL am hanner awr y tu allan i'r siop fideo cyn i rywun ddod i'w hagor. A hithau'n fore Sul, yr oedd mochyndra'r noson gynt, y chŵyd, y tsips, y caniau gwag, y cyri, y stympiau ffags, y platiau budr, y pisio, yn stremp ddrewllyd hyd y pafin, a Steffan fel ynys fach o syberwch yn y canol.

– Meddwl taw am ddeg ro'ch chi fod agor...

– Ie, i fod... 'Sneb fel arfer yn dod mor fore â hyn... Rhaid 'bo chi'n despret!

Edrychodd Steffan ar y llanast a'i amgylchynai, ac yna cofiodd nad oedd wedi cael cyfle i siafo'r bore hwnnw, na chribo'i wallt, na chael hyd i ddillad glân... Na chau ei gopis.

– Dim yn y ffordd rych *chi'n* 'feddwl, na!

Wedi tynnu ei zip, eglurodd wrth y dyn ifanc mai ffilm ddu a gwyn hen a pharchus oedd gwrthrych ei gwest, a byddai'n falch o gael gwybod am unrhyw ffilm â'r gair *meeting* yn y teitl, neu *tryst* neu unrhyw air tebyg...

– *Meet Whiplash Willie?*

– Ydych chi'n tynnu 'nghoes?

– Na'dw... Mae'n ffilm ddu a gwyn... Am dwyll...

– Am dwyll? Mm...

Edrychodd Steffan yn amheus arno.

– Na. Mae hon, y ffilm dwi'n chwilio amdani, mae hi'n... Wel, stori serch yw hi... Stori serch weddus hefyd...

– Amau dim... Amau dim

–... Ffilm fyddai wedi cael ei gwneud tua diwedd y Rhyfel, neu cyn hynny falle...

– Ffilm serch? Du a gwyn? Parchus? Adeg y Rhyfel? No problemo! Dewch miwn i'r siop a chawn ni weld beth gallwn ni 'wneud i chi.

Yr oedd yn y bachgen hwn ryw gyfuniad o'r gwybodus a'r pedantig a oedd yn atgoffa Steffan o Anton, gyda'i hoffter o rifau a mesuriadau. Ond dyddiadau aeth â bryd y gweinydd fideo hwn, er mai rhif a mesur yw dyddiad hefyd, bid siŵr. Syllai ar y llawr tra dadlwythai beth o'i archif ymenyddol.

– Wel... Gad i ni ddechre 'da *Meet*... Mae 'na ryw faint o serch yn *Meet Me at Dawn*... 1946... Ac yn *Meet Me Tonight* hefyd... 1952... Ond mae honno 'bach yn rhy ddiweddar i chi, g'lei...? Wedyn, mae *Meet Mr Penny*... 1938... A *Meet Nero Wolfe*... Ond mae lot o saethu a clatsio yn honno... O's ots bod y serch yn dod â 'bach o drais gyda hi...?

– Wel...

– Neu *Meet Me in St Louis*... 1944... gyda Judie Garland a Margaret O'Brien...

– O'Brien...?... Mm... Oes rhywun ag enw Almaeneg ynddi?

– Ddim a bo'fi'n cofio, nac oes, ond Fred Finklehoffe sgrifennodd y sgript, mae hwnna'n swnio'n eitha' Jyrman i fi... Ond os taw Jyrmans ych chi moyn, dylech chi weld *Meet the Baron*... 1933... Ffilm ambwyti rhyw sincyn bac simpil yn esgus bod yn Baron Munchausen... Mae lot o Jyrmans yn honna... Ond nage stori serch yw hi, a gweud y gwir... Nawr te... Sefwch funud...

Edrychodd Steffan ar ei wats. Roedd hi eisoes yn chwarter i un ar ddeg. Dwy awr a hanner cyn y byddai'n rhaid cychwyn am y gorllewin.

– Wedyn, mae gyda chi *Meet John Doe*… 'Na gwd ffilm i chi… Gary Cooper a Barbara Stanwyck… Maen nhw'n ffindo'r tramp ma, 'chwel, ac maen nhw'n hala fe miwn i bolitics…

– Na, na, sa i'n credu… Ac, wel, a bod yn onest… Rwy ar dipyn o hast…

– Wel gwedwch be chi moyn, te, ac fe weda'i wrtho chi'n strêt odi'ddi gyda ni… Wy yn gwneud 'y ngore i chi, 'chmod..

– Ie, ie, rwy'n gwybod 'ny…

–… Trial rhoi bach o ddewis i chi, 'na i gyd ro'n i'n 'neud…

– Ie, ie, rwy'n gwybod hynny, a diolch o galon am eich cymorth… Plîs, plîs, bwrwch ymlaen… Ych chi'n help mawr..

– Iawn, wel… Ble o'n i…? 'Wedes i *Meet John Doe*?

– *Meet John*? Do.

– Wel, wedyn mae *Meet Dr Christian*… *Meet the People*… *Meet the Missus*… Weden i bod honno dipyn yn nes ati, falle… Ond taw cwmpo mas maen nhw yn honna… A *Meet the Stewarts*… Cwmpo mas sy yn honna hefyd… A dyna'r Meets i gyd wy'n credu. Odych chi moyn i fi ddechrau ar y *Rendezvous*?

– Oes lot?

– Dim cymaint, nac oes… Dim ond un, 'gweud y gwir… *Rendezvous With Annie*… 1946… ond Americanwyr sy' yn honna… Mae mwy o *Reunions* i ga'l…

– *Reunions*?

– Ie, fel *Reunion in France* 1942… gyda John Wayne a Joan Crawford… a *Reunion in Vienna*… 1933 wy'n credu…

– Fienna, wedoch chi?

– Ie.

– Oes Almaenwyr yn honno?

– Wel, nawr te, mae John Barrymore… a Una Merkel… Ydy Una Merkel yn Jyrman?

– Sa i'n gwybod… Nawr te, gwrandwch… Dwi ddim eisiau eich brysio chi… Ond ydych chi wedi clywed am actor o'r enw…

Tynnodd Steffan ddarn o bapur o'i boced.

–… Actor o'r enw Trau Von Neuhalt, neu rywbeth tebyg…? Ac un arall… Selma Jansen?

Daeth gwên i wyneb y bachgen.

– A chi sy'n tynnu 'nghôs i nawr, yntefe?

– Mm?

Cerddodd y bachgen draw at adran y ffilmiau clasurol.

– Dyma chi. Pam na wedoch chi 'nghynt?

– Ond…

– Dwy bunt, plîs.

29 Brief Encounter

EDRYCHAF AR Y BOCS. *Brief Encounter*. Rwy'n siŵr 'mod i
wedi gweld y teitl rywle o'r blaen, ond yn fy myw, alla i ddim
cofio ble. Gyda Trevor Howard… a Celia Johnson… Trau
von Neuhalt a Selma Jensen? Mae rhyw debygrwydd, sbo.
Oes digon? A beth yw 'digon'? Oes modd olrhain yr enwau
yn ôl i'w tarddiad rywsut, fel y mae'r ieithgwn yn olrhain
y Gymraeg yn ôl i'r Frythoneg a'r Lladin? Oes rhyw ddull
gwyddonol o ganfod sut y byddai enwau'r actorion yn cael
eu llygru a'u llurgunio a'u hystumio wrth dreiglo drwy gof a
chlust a thafod dau Almaenwr, un yn fyddar bost a'r llall yn
rhy ifanc i wybod y gwahaniaeth? Ac yna drwy fy nghlustiau i
fy hun, wrth gwrs, rhaid peidio ag anghofio'r rheiny: clustiau
dibrofiad a diniwed ym materion hen sêr ffilm ac acenion
Almeinig. Ac o ddefnyddio'r dull gwyddonol hwnnw, wedyn,
a fedrwn i brofi y tu hwnt i bob amheuaeth mai'r *Begegnung*
hwn, ac nid rhyw gyfarfyddiad arall, oedd y gwrthrych a
deilyngai fy sylw? Na fedrwn, wrth gwrs. Beth petai'r llipryn
plorynnog yn y siop fideo yn cyfeiliorni? Neu'n fy nhwyllo'n
fwriadol, yn dial arna i am fynd ar ei nerfau, am darfu ar
heddwch ei fore Sul, am golli fy amynedd gydag e? A beth,
wedyn, petai'r enwau yn rhai dilys? Yn rhai adnabyddus, hyd
yn oed. *Dych chi ddim wedi dilyn campau carwriaethol rhyfeddol
Selma Jensen?* Naddo, syr. *Dych chi ddim wedi clywed am yr
enwog Trau von Neuhalt, hwnnw sydd â gên fel cŷn newydd ei hogi?*
Naddo, madam. *Rhag eich cywilydd!*

Ond os wyf wedi dysgu un wers yn ystod yr wythnosau diwethaf yma, dyma hi: y mae 'na amser i bwyso a mesur yr opsiynau, ac y mae 'na amser i weithredu. Ac mae hwn yn amser i weithredu, os buodd un erioed. Gweithredu. Yn bendifaddau. Ond nid gweithredu byrbwyll, chwaith. Mwya'r hast, mwya'r rhwystr, ys dwedan nhw. Na, nid yw brys yn caniatáu blerwch... Y ffilm hon, mae'n debyg, fydd y darn olaf o dystiolaeth y byddaf yn ei gyflwyno. Nid y darn olaf yn y jig-so, wrth gwrs, er imi gael fy nhemtio i ddefnyddio'r hen ystrydeb honno. Na, gwn yn iawn nad oes dim jig-so yn bod. Ond, o ran ffurfioldeb gweinyddol, hon, synnwn i ddim, fydd yr eitem olaf i'w rhoi yn ffeil Rhif 86, ac oherwydd hynny mae'n haeddu'r un trylwyredd â phob un arall o'm hymchwiliadau, os nad mwy. Ac os na chaf i ddim byd o werth o'r ffilm – ac mae hynny'n bur debygol, rhwng popeth – byddaf wedi cofnodi'r dim-byd-o-werth hwnnw a'i roi ar gof a chadw i bawb a fynno ei weld. Byddaf wedi diogelu safonau uchel y Weinyddiaeth o ran bod yn Atebol, yn Dryloyw, ac yn Hygyrch. Byddaf wedi cadw trefn. Byddaf wedi cadw Y Drefn. Trueni, hefyd, na fu'n bosib imi fod yn fwy Lleol, nac yn fwy Strategol. Ond, er tegwch i mi fy hun, dwi ddim yn credu mai fi yn unig sydd ar fai am hynny.

Trof y laptop ymlaen a gosod i lawr y manylion ffeithiol agoriadol, gan gofio darparu ar gyfer cofnodi amser, rhag ofn y bydd angen cyfeirio yn ôl at ambell olygfa, er mwyn cadarnhau cywirdeb dyfyniad, enw cymeriad, ac yn y blaen.

Cynllun 5/B Cymorthdaliadau i Awduron BLWYDDYN
2004/5

SWYDDFA C/I ADRAN HC/2/C

SWYDDOG SM

Ymgeisydd 86 Kate O'Brien

Testun:

BRIEF ENCOUNTER (cyf. David Lean, seiliedig ar
ddrama lwyfan gan Noel Coward)

Amser   Sylwadau

Gwnaf yn siŵr fod y *Real Time Tape Counter* ar 0. Gwasgaf y
botwm ar y teclyn. Dechreuaf deipio.

0.00.01 Hen drên stêm yn rhuthro trwy orsaf.

Gwasgaf *pause*. Pa orsaf yw hon, tybed? Wela' i ddim enw'n
unman. Oes ots?

0.00.20 Chwiban y trên yn torri trwy farrau agoriadol
Rach. 2

Y mae'r trên ar amser yn ôl y wên ar wyneb y gorsaf-feistr.

Y gerddoriaeth yn dweud rhywbeth gwahanol, wrth gwrs. Rhywbeth llawer mwy…

> 0.03.23 Bwffe yn yr orsaf. Cwpl canol oed wrth y
> bwrdd. Tawel. Trist? Yn amlwg mewn cariad. Cariad
> anobeithiol? Clebran hurt dau wth y bar.

… ie, llawer mwy angerddol. Ac eto, mae popeth mor llwm… mor llwyd… Dyna sut y gwelwn ni bob ffilm du-a-gwyn heddiw, mae'n debyg. A'r cymylau mwg yn codi'n barhaus o'r trenau. A'r llwch yn setlo. *Vergrauung.*

> 0.04.06 Ffrind, menyw barablus yn tarfu ar y cwpl a'u
> distawrwydd trist/ melys. Dolly. Alec a Laura yw'r
> cariadon. Yntau'n ymadael. Llaw ar ei hysgwydd.hi.
>
> 0.07.10 Uchelseinydd: 'The train for Ketchworth is
> now arriving at platform 3.'

I Ketchworth mae Alec yn mynd, mae'n debyg. Ydy Ketchworth yn bod, tybed? Cloc… Amser yn thema ganolog, 'falle… Stopiaf y ffilm eto… Dyna arwydd yn dangos enw'r orsaf.

> 0.10.35 Milford Junction yw enw'r orsaf, a man
> cyfarfod y ddau gariad.

Milford. Mae sawl Milford i gael, siŵr o fod… Oes gwahaniaeth pa un? Gall fod yn Milford dychmygol. Gall fod yn fwy nag un lleoliad wedi'u cyplysu er mwyn ymddangos fel un… Gall fod…

Dolly a Laura hwythau yn codi a mynd am y trên.

Trên gwahanol, siŵr o fod. Maen nhw'n pasio… na, sefyll… o flaen ffenest. Ysgrifen ar…

0.11.58 Dolly a Laura'n sefyll o flaen swyddfa (yn yr orsaf?)

Rwy'n methu â gweld yr ysgrifen yn iawn. Mae'r ffenest yn goleuo a thywyllu eu hwynebau bob yn ail. Mae'r effaith yn drawiadol. A bwriadol, siŵr o fod. Goleuni a chysgod.

Stopiaf eto. A weindio 'nôl. A gweld

AY & SON

NEERS

& VALUERS

Dyna i gyd. Symudaf y fideo ymlaen un ffrâm ar y tro. Mae Dolly'n cuddio'r llythrennau ar y chwith. Yna Laura'n cuddio'r llythrennau ar y dde.

WAY & SON

CTIONEERS

AGENTS & V

Maen nhw'n gwahanu ryw ychydig. Petaen nhw ond yn symud modfedd, dwy fodfedd ar y mwyaf, ond fyddai David Lean byth wedi caniatáu hynny, rhag chwalu cydbwysedd y llun.

BBWAY & SON

CTIONEERS

AGENTS & V

Maen nhw'n closio at ei gilydd eto. Dau gymeriad arall yn cerdded heibio. Methu gweld dim. Llaw Dolly ar law Laura. Dolly'n plygu lawr.

Y & S

NE

&

Yn chwerthin... Yn gwahanu eto...

BBWAY & SO

CTIONEE

Petai Dolly ddim wedi gwisgo'r got drwchus â'r coler ffwr. Dau arall yn cerdded o flaen y ffenest. Gweld dim. Mae Laura'n mynd. A Dolly'n ei dilyn? Na. Codi ei llaw. Symud mymryn bach... Ie, da chi, shifftwch, wnewch chi!

E...ai B yw hwnna?

ESTA

Mae Dolly'n pallu symud. Mae hi'n meddiannu canol y ffrâm. Yn dilyn cyfarwyddiadau Lean eto, mae'n debyg. Eu dilyn i'r dim. Ond rhaid iddi symud rywbryd, siawns? Ydy, mae hi'n symud... Ond na, nid hi sy'n symud... Y llun sy'n newid... Mae'r llun yn dechrau ymdoddi i lun arall...

0.14.01 Laura'n mynd adref.

Stopiaf y fideo a gwylio'r un darn eto. Gwelaf yr un peth.
Wrth reswm. Dim llai. Dim mwy. Rhan o enw ar ffenest
swyddfa. Neu'r cyfan, o bosib, ond darnau gwahanol bob tro.
Amhosib dweud. Oes ots? Ydy e'n rhan o'r stori? Go brin.
Manylyn digyswllt. Cyd-ddigwyddiad amherthnasol.

> 0.15.54 Gartref. Gŵr Laura'n gwneud pos croeseiriau.
> 'Romance... ..fits in with delirium and Balukistan'.

Balukistan?

> Laura'n cilio i fyd ei hatgofion.

A! Mae'r diwedd ar y dechrau, felly. Ryn ni'i gwybod o'r
cychwyn taw gwahanu fydd hanes y cariadon yn yr orsaf.

> 0.16.28 Y trên yn mynd a dod, yn torri ar draws y
> cyfan
>
> 0.17.02 Laura'n mynd i'r llyfrgell. Yn gofyn am y llyfr
> diweddaraf gan

Mae'n ddeuddeg o'r gloch. Pwysaf *fast forward* a gweld y
cymeriadau'n troi'n bypedau, nes bod...

0.27.45 Alec a Laura'n mynd i'r sinema. Organydd yn chwarae.

Amseru gwael fan'na. Dwy ti ddim yn berffaith wedi'r cyfan, Mr Lean! Weindiaf yn ôl... Da gen i weld nad yw'r organ na'r organyddes na'r hyn mae hi'n ei chwarae o'r pwys lleiaf. Gwasgaf *fast forward*.

0.34.13 Alec yn feddyg. Pneumoconiosis

Pyllau glo. Acenion... Cockney, am wn i. Ble yn y byd maen nhw...?

    Fast forward...

0.51.27 Rhwyfo ar lyn. Yntau'n cwympo i'r dŵr.

Maen nhw mewn ffilm, wrth gwrs. Dyna le maen nhw.

0.56.59 Ar y trên eto.

Fast forward...

0.65.32 Y diwedd yn dod, fel y gwelwyd ar y dechrau. Laura'n ystyried lladd ei hun trwy neidio o flaen trên.

> **0.75.43 Yna hithau gyda'i gŵr (a'i bib) a'u plant. Yntau i Affrica.**

A Laura'n ymddangos yn falch o gael dychwelyd i normalrwydd. Gollyngdod. *The End*.

The End. The End! Ac er nad oeddwn i'n obeithiol iawn ar y dechrau, rhaid cyfaddef – ac nid am y tro cyntaf, chwaith – fy mod i braidd yn siomedig na chefais rywbeth mwy sylweddol i gnoi cil arno. Ie, yn siomedig iawn, a dweud y gwir plaen, a minnau wedi bod yn cwrso'r hen *Begegnung* bob cam o Leipzig. Sôn am sgwarnog! Na, mae hyn yn waeth. Gwell hela sgwarnog na dy gwt dy hunan. A gyda'r siomiant hwn y mae cwmwl bach o bryder yn bwrw ei gysgod trosof unwaith yn rhagor, oherwydd rwy'n dechrau meddwl, nid am yr hyn a welais – a'r ychydig oedd o werth ynddo – ond yn hytrach am yr hyn a gollais. Am yr holl funudau, yr holl olygfeydd, yr holl ing ac angerdd, i gyd wedi'u troi yn sioe bypedau wirion gan y *fast forward*. A sioe fud, hefyd. Faint gollais i, tybed? Cymaint â hanner y ffilm? Ystyriwch y miloedd o eiriau. Ac ystyriwch fod yr allwedd i'r cyfan ynghudd mewn rhyw gyfarchiad bach ymddangosiadol ddibwys, mewn rhyw sylwadau ffwrdd-â-hi yn y caffi, neu'r bwffe, neu'r llyfrgell, mewn rhyw olygfa gwbl ymylol i rediad y stori. A minnau wedi'i cholli.

The End. Ac ar ben hynny, ai dyna *yw*'r diwedd, mewn gwirionedd? Petai *sequel* yn cael ei wneud, ai dyna fyddai'r diwedd o hyd? Ai byw'n hapus byth wedyn fyddai tynged Laura, gyda'i gwlanen o ŵr? Ac Alec, y meddyg: a fyddai fe'n hapus o gael gwireddu ei freuddwyd am wella afiechydon yn

Affrica? Neu a fyddai rhyw ysfa, rhyw bang o golled, yn cnoi ei ymysgaroedd o hyd? Ai dyma fyddai testun y *sequel*? Does dim diben gofyn. Ond does gen i ddim ffydd yn y diwedd hwn. Mae'n rhy dwt. Yn rhy gyfleus. Ac yn fwriadol felly, o bosib, er mwyn hala'r gwyliwr i chwilio am yr ystyr rywle arall. Ac i sbarduno Rhif 86 hithau...? Ha!

Mae'n hanner awr wedi deuddeg. Ni alla' i aros ddim mwy. Oes gen i ddigon o wybodaeth? Digon, felly, i gwblhau'r ffurflen werthuso? Digon i ddadlau, gyda rhyw faint o hygrededd, mai'r hyn sydd gan Rif 86 mewn golwg yw cofiant dychmygol sy'n dilyn hanes Laura, arwres *Brief Encounter*, wedi diwedd y ffilm? A digon, hefyd, i gyfiawnhau dyfarniad cadarnhaol?

30 Colton Junction

MAE GEN I BUM MUNUD, efallai, i fwrw golwg ar y we. Dim ond o ran chwilfrydedd.

Clic! *Ketchworth*

Yn union fel 'ro'n i'n disgwyl. Dim ond tri chanlyniad. A dau o'r rheini'n cyfeirio at *Brief Encounter*. Enw gwneud. Lle ffug. Wedi'i greu'n arbennig.

Clic! *Milford Junction*

5210 o ganlyniadau. Oherwydd y Milford, mae'n debyg. Enw cyffredin. Rhydyfelin. Eto, fel ro'n i'n disgwyl. Ac mae mwy nag un Milford Junction i'w gael hefyd, mae'n ymddangos. Clic! Un yn Surrey. Ac un... Clic!... yn Swydd Efrog...'Co ni... Milford Junction... Ar bwys Castleford. Taith un munud ar ddeg i fod yn fanwl, yn ôl y Diesel Preservation Society ar y pumed o Ebrill, 2003. Castleford 09.40 Milford Junction 09.51 Church Fenton 09.57 Colton Junction 10.03 York 10.11. Ac mae amryw byd o luniau o drenau'n mynd trwy'r lle. Anoracs! Milford Junction. Ai fan hyn y buon nhw'n ffilmio, felly?

Ond, hefyd, Colton Junction. Colton... Colton? Colton!

Na! Dyw hynny ddim yn iawn. Does gan Colton ddim byd i'w wneud â Milford Junction. Dim â'r Milford Junction iawn. Yr un yn y ffilm, rwy'n meddwl. Na. Does dim eisiau cofnod ar Colton. Fy nghartref newydd i yw Colton. Mae 'na Colton arall, dyna i gyd. Coal Town? Pentre glofaol. Dim byd mwy cyffredin. Y ffilm sy' dan sylw nawr. Y ffilm yn unig.

A Milford Junction.

Clic! *Milford Junction. Brief Encounter Locations.*

Ie. Dyna'r cloc. Tair munud i un...

Carnforth Station had many of the features needed by David Lean.

Carnforth.

Mae Milford Junction yn bod. Ond nid Milford Junction sydd yn y ffilm. Dim ond ei enw. Carnforth sydd yn y ffilm, ond heb ei gydnabod, heb ei enwi. Yn cymryd arni fod yn Milford Junction. Milford Junction ymhlith y pyllau glo. Milford Junction yn ymyl Colton. Coal Town.

Mae Carnforth yn bod... Wrth gwrs... Ro'n i'n gwybod hynny o'r blaen...

Clic!

Carnforth: Yn 263.25 milltir o Lundain. 62.75 milltir o
Gaerliwelydd. 28.5 milltir o Barrow in Furness. 27.25
milltir o Preston. Agorwyd yr orsaf yn 1846 a'i hail-
fedyddio yn Milford Junction ar gyfer y ffilm. Cafodd
523 eiliad o Brief Encounter (tua 10.63%) eu ffilmio yn
Carnforth.

Ac ar ben hyn i gyd… Carnforth yw fy nghartref i. A fu fy
nghartref i. A enwyd gan rywun arall. Beth sydd mewn enw?

Oes modd trafaelu o Carnforth i Colton?

Mae'n chwarter wedi un.

31 Taith Amhosibl

www.nationalrail.co.uk

Clic!

Carnforth — Colton

Clic!

Ffig. 16 Laura ac Alec yng Ngorsaf Milford Junction (Carnforth)

Ydy. Mae gorsaf Carnforth yn bod o hyd. Ond dyw'r
amserlen ddim yn adnabod Colton. Na Colton Junction. Mae
'na drên sy'n gadael Carnforth am bum munud ar hugain wedi

un… Yn mynd trwy Castleford… Trwy Church Fenton. Does dim sôn am Colton. Ond mae rhif ar gyfer ymholiadau.

Ffoniaf 0845 6040500.

– Toes 'na ddim Colton ar y rhwydwaith cenedlaethol, syr. Mae 'na Collington. Mae 'na Colchester. Mae Bae Colwyn. Hwyrach mai lein breifat ydy o… Hen injans stêm a ballu… Ond yr amserlen 'di'n busnas ni, syr, nid y gorsafoedd… Triwch Network Rail yn Euston.

Ffoniaf 08457 114141.

– Swyddfa'r Gogledd dach chi isio, ia…

Ffoniaf 0161 8381240.

– Mm… Ddim yn gwbod 'yn hunan, sori… Ato Phil Schreiber 'fi wastad yn mynd ambwyti pethach fel 'na…

Ffoniaf 0161 2288814.

– Sgrwnsh… Y?…Sgrwnsh… Phil Schreiber?… Sgrwnsh… Sgrwnsh… Sgrwnsh… Ma' fe mas yn cael ei gino, bach… Sgrwnsh… A phob un arall 'fyd… Sgrwnsh… Heblaw fi… Sgrwnsh… Ond nage fe fyddai'n gwbod, ta p'un.. Sgrwnsh… Ffonwch John Pengelly… Sgrwnsh… yn yr Adran Fasnachol… Bydd John yn gwbod…

Ffoniaf 0161 2284410.

– Bant ar ei wyliau mae e… Ond dyw e ddim byd i 'neud â ni… GNER sy'n gyfrifol am y gwasanaethau ffor' 'na…

Ffoniaf 08457 225125.

– Triwch Arriva.

Ffoniaf 0870 6023322.

– Colton Junction…? Ydw.

Saib.

– Oes rhywun 'na?

– Colton Junction wedoch chi?

– Ie. Ydych chi'n gyfarwydd â'r lle?

– Yn gyfarwydd â'r lle...? Dylen i fod. Bues i'n gyrru trêns trwy Colton Junction am ugain mlynedd... Wel, ydy, mae'r lle 'na o hyd... Na, dim fel'ny... Na, does neb yn cael mynd lawr fyn'na...A does neb wedi gwneud erioed, cyn belled â bo' fi'n gwbod... Mae'n un o'r junctions ffasta'n y byd... Trêns yn cyrraedd cant dou ddeg pum milltir yr awr... Mis Ebrill hyn, wedoch chi...? Jyst yr un dwarnod...? Sa i'n gwbod... .'Fallai bod y trên wedi stopyd i'r teithwyr gael pip trwy'r ffenest... .Ar y junction a'r seidins... Mae 'na bobl sy' miwn i bethe fel 'na...Na, 'sdim platfform 'co... Na seins... Na, dim enw... Lle i fynd trwyddo, 'na i gyd... Os gallwch chi alw hwnna'n lle!...Y pumed o Ebrill, wedoch chi?

– Ie. Diwrnod fy mhen-blwydd.

32 Cartref Newydd Steffan

CYCHWYNNODD STEFFAN ar ei daith am bum munud
wedi dau. Gadawodd neges ar ffôn symudol Llio i ddweud
y byddai ryw awr yn hwyr. Bu'n pendroni am ychydig
ynglŷn â pha ffordd i fynd. Gallai ddilyn y draffordd. Gallai
fynd dros y Bannau. Gallai ddyfeisio rhyw ffordd gyfrwys
rhyngddynt, ffordd fyrrach o bosibl, ond ffordd fwy troellog,
a ffordd a fyddai'n debyg o fod yn berwi gan dractors.
Dan amgylchiadau gwahanol, byddai wedi pwyso a mesur
manteision ac anfanteision pob un. Heddiw, dewisodd y
Bannau. am fod *B* am Bannau yn nes at ddechrau'r wyddor na
T am Traffordd a *T* am Tractors. Edrychodd eto ar fap 146 ac
ar y groes a ddynodai'r tŷ. 767 287. Chwaraeai â'r rhifau yn
ei feddwl. Cyfanswm… 37… neu 1054… Tynnu… 480…
Lluosi… 42… mm… Gobeithiai na fyddai'r injan yn tanio.
Ond fe wnaeth.

Cyrhaeddodd Steffan Drecastell am hanner awr wedi tri.
Trodd i'r chwith a dilyn y ffordd gul dros Fynydd Wysg a
heibio'r Gaer Rufeinig. Yn Cross Inn cymerodd y tro am
Fyddfai. Arafodd, er mwyn craffu ar enw pob fferm a bwthyn
a *villa* a byngalo rhag ofn y collai Colton… neu Colomendy,
neu Pant Collen, neu pa enw bynnag yr oedd Llio wedi'i
ddodi ar eu cartref. Rhag ofn nad oedd y cyfeiriad yn gywir.
Rhag ofn bod Mr Ebbway…

Nant-yr-adyn… Glyn-hir… Gelli Meirchiau… Tir-
paun… Gelli Mydog… Pwll-calch…

Croesodd bont a stopio. Edrychodd ar y map eto, a sylwi
mai Afon Clydach a lifai dani. Ni faliai am hynny. Onid

oedd afonydd Clydach rif y gwlith? Gadawodd y car mewn cilfach wrth ochr yr hewl a cherdded i fyny'r rhiw am ychydig lathenni. Gyferbyn ag ef, ar y llaw dde, safai bwthyn. Bwthyn deulawr, di-raen, di-nod. Di-nod, heblaw bod yr ardd fechan yn gyforiog o flodau'r enfys, a bod plac uwchben y drws yn dwyn y geiriau

Am fod fy Iesu'n fyw,
Byw hefyd fydd ei saint.

Yr oedd hefyd, wrth ochr y tŷ, hen laethdy pren, ac ar ei ddrws y llythrennau C O L. Byseddodd Steffan y llythrennau. Chwiliodd am y rhai a gollwyd, neu ryw olion neu gysgod ohonynt. Chwiliodd am y llythrennau newydd roedd Llio wedi'u hychwanegu wrth ail-fedyddio'r tŷ. Doedd dim golwg ohonynt. Caeodd ei lygaid a thynnu blaenau'i fysedd ar draws y pren gan geisio canfod, ym mhob rhic a phant, ryw awgrym o'r enw llawn, yr hen neu'r newydd. Ond yr hyn a deimlai, yn fwy na dim, oedd paent, hen baent crychlyd, yn plicio o dan ei ewinedd.

Agorodd ei lygaid.

– Oes pobl ma?

Doedd dim sŵn o'r tu mewn. Clywodd awyren uwchben. Edrychodd i fyny, gan gysgodi ei lygaid â'i law. Roedd yr awyren ei hun yn rhy fach ac yn rhy bell i'w gweld yn glir, ond gadawai lwybr gwyn ar ei hôl. Yn wir, yr oedd glesni'r awyr yn frodwaith o linellau gwyn, yn groesymgroes.

– Oes 'na bobl ma? Llio? Wyt ti yma, Llio…?

Ceisiodd edrych trwy'r ffenestri, ond roedd y llenni ynghau. Aeth at y drws. Curodd y pren di-raen â'i ddwrn, yn dawel i ddechrau – *tap-tap* – ac yna'n fwy penderfynol – *cnoc-cnoc!*

– Oes rhywun gartre'?

Ni ddaeth ateb. Cydiodd Steffan yn handlen y drws ffrynt a cheisio ei throi. Ac fe drodd. Fe drodd, yn anfoddog braidd, gan wichian ei henaint rhydlyd, ond fe drodd. Rhoes gic fach i waelod y drws. Agorodd led adain chwannen. Anelodd ei ysgwydd dde at y drws a rhoi hwp bach iddo. *Hwp!* Agorodd rhyw damaid o smic o fymryn yn fwy. A'i ysgwydd chwith. *Hwp!* Ond doedd dim yn tycio. Penderfynodd Steffan fod rhyw rwystr sylweddol yr ochr arall i'r drws Ni wyddai beth gallai'r rhwystr honno fod, ond cafodd ei ddychymyg ddigon o amser i ystyried tri esboniad annymunol, sef:

(1) corff ei gymar

HWP!!

(2) corff Mr Ebbway,

HWP!!!!

a

(3) thirlithriad

HWP!!!!!!

Ar ôl bustachu fel hyn am funud, gallai Steffan weld digon i sylweddoli nad un o'r pethau hyn oedd y broblem wedi'r cyfan. Nid corff oedd y rhwystr. Na ffawt daearegol. Na, yr hyn a rwystrai ei fynediad oedd, yn hytrach, gruglwyth anferth o bapur. Yn wir, y cruglwyth mwyaf o bapur a welodd Steffan erioed, yn fwy hyd yn oed na'r mynyddoedd papur a fu'n rhan o'i dirwedd feunyddiol yn y Weinyddiaeth. A sylweddolodd, yr un pryd, na allai neb fod wedi mynd trwy'r drws hwn ers amser maith. Gwasgodd ei hun trwy'r adwy. Dim ers misoedd. Edrychodd o'i gwmpas ar y pecynnau a'r amlenni a'r hysbysebion a'r papurau a'r taflenni a'r calendrau a'r gwahoddiadau a'r cylchgronau a'r cynigion arbennig, yn wal soled hyd ei fogail. Dim ers blynyddoedd. Ceisiodd gloddio llwybr trwyddynt, a methu, gan mor dynn yr oedd y

cyfan wedi'i saco i bob twll a chornel. Ystyriai nifer o ddulliau
posibl o hwyluso'r dasg, gan gynnwys

(1) taflu'r papurau allan trwy'r drws, fesul un a dau

(2) eu torri nhw'n fân fân fân gyda'r siswrn ar ei gyllell
    Byddin y Swistir, ac yna cerdded yn ddiffwdan trwy'r
    conffeti –

(3) gweiddi am help.

Yna sylweddolodd, gyda rhyddhad, fod ffordd haws o
gyrraedd yr ochr draw. Gyda chymorth y drws y tu ôl iddo,
dringodd i frig y domen. Yna, â chryn fedrusrwydd, fe
groesodd Steffan bob bryn a phant, weithiau'n sefyll ar y goes
chwith, weithiau ar y goes dde, ond gan gadw ei gydbwysedd
yn ddi-ffael hyd y diwedd, fel un o feistri Bondie Beach yn
syrffio'r don. Wedi goresgyn y llwybr papur anghyffredin
hwn, disgynnodd i lawr diogel y pasej, lle sgrialai'r corynnod
a'r moch coed a'r chwilod bach dienw yr oedd wedi tarfu
ar eu llonyddwch maith. Saethai pelydrau haul diwetydd
eu llafnau trwy wydr y drws ffrynt a goleuo dawns y myrdd
lychynau.

   Tynnodd Steffan ei ffôn symudol o'i boced. Doedd dim
signal. Pa syndod, meddyliai, a'r tŷ mewn pant, a'r bryniau a'r
coedwigoedd yn plygu drosto ar bob ochr? Rhyfedd, hefyd,
na fyddai Llio wedi gadael neges iddo, ar y drws y tu allan,
ar sil y ffenest, dan garreg yn rhywle. Roedd drws bob ochr
iddo. Cydiodd ym mwlyn pren y drws ar y llaw chwith a
dechrau ei droi. Ond wedyn, beth petai Llio neu Mr Ebbway
wedi dod i mewn trwy'r drws cefn?

– Llio… Wyt ti 'na?

   Llyncwyd ei eiriau gan y distawrwydd, y gwacter, a'r
llwch. Ond oedd 'na ddrws cefn i'w gael? Gollyngodd ei afael
yn y bwlyn a cherdded ar ei union i'r gegin fach yng nghefn y
tŷ gan obeithio, petai Llio wedi dod i mewn y ffordd honno,

y byddai wedi gadael neges yno. Roedd bwrdd yn y gegin, a phedair stôl o bob tu iddo. Ar y bwrdd roedd lliain wen, lân, ac ar y lliain wen roedd platiau a chwpanau a chyllyll a ffyrc i ddau.

Ac amlen wen, a'i enw ef arni.

33 Y Pedwerydd Pecyn

AGORODD STEFFAN yr amlen a thynnu allan chwe dalen o
bapur. Darllenodd.

Gadawodd Werner ei gar mewn cilfach wrth ochr yr
hewl a cherdded i fyny'r rhiw am ychydig lathenni.
Gyferbyn ag ef, ar y llaw dde, safai bwthyn. Bwthyn
deulawr, di-raen, di-nod. Di-nod, heblaw bod yr ardd
fechan yn gyforiog o flodau'r enfys.

Yr oedd hefyd, wrth ochr y tŷ, hen laethdy pren,
ac arno'r llythrennau C O L. Byseddodd Werner y
llythrennau. Dyma'r lle, meddyliai. Curodd y drws ac
aros. Curodd eto.

– *Ist jemand daheim?*

Doedd dim sŵn o'r tu mewn

– Oes 'na bobol ma…? Sabine…? *Sind Sie da?*

Ceisiodd edrych trwy'r ffenestri, ond roedd y llenni
ynghau. Cydiodd yn handlen y drws. Er syndod iddo,
doedd e ddim dan glo. Serch hynny, roedd angen hwb i'w
agor, nid am ei fod yn stiff, ond am fod ychydig o lythyrau
wedi mynd yn sownd oddi tano.

– Helo! Oes pobol ma?

Cododd y llythyrau o'r llawr. Ysgrifen ei dad oedd ar
bob amlen. Leipzig Mai 1947. Leipzig Tachwedd 1947.
Leipzig Ebrill 1948. A'r un diweddaraf. Leipzig Ionawr
1949. Y llynedd. Roedden nhw wedi cyrraedd, felly.

Gallai ddweud cymaint â hynny wrtho. Ond doedd yr un ohonynt wedi cael ei agor. Ac roeddent eisoes wedi dechrau pydru yn y lleithder. A dim parsel.

Saethai pelydrau haul diwetydd eu llafnau trwy wydr lliw'r drws ffrynt a goleuo dawns y myrdd lychynau. Rhyfedd, meddyliai Werner, os oedd Sabine wedi symud, na fyddai wedi gwneud trefniadau i anfon ei phost ymlaen. Ac os oedd hi wedi marw? At bwy, wedyn, yr âi â neges ei dad?

Roedd drws bob ochr iddo. Oedodd am eiliad. A oedd diben mynd ymhellach os nad oedd hi'n byw yma bellach? Os nad oedd y parsel...? Ond aeth ei chwilfrydedd yn drech nag ef. Cydiodd ym mwlyn pren y drws ar y chwith a'i droi. Roedd y stafell yn dywyll ac yn drewi o leithder. Daeth llygedyn o olau trwy'r ffenest lle'r oedd un o'r llenni wedi dod yn rhydd o fachyn. Yn y golau hwnnw gallai weld bod bwrdd hir yn llenwi canol y stafell. O'i gwmpas roedd rhyw naw neu ddeg o stolion. Y tu hwnt i'r rheiny, yn nwfn y cysgodion, roedd pentyrrau o betheuach na fedrai ddirnad eu natur yn iawn. Agorodd y llenni. Er ei ofal mawr, cododd cwmwl o lwch ohonynt, a disgynnodd cyrff corynnod ac adenydd gwyfynod a choesau chwilod bach dienw yn gawod amdano. Pesychodd. Rhwbiodd ei lygaid.

Llyfrau oedd y pentyrrau ar un ochr y stafell. Llyfrau mewn iaith nad oedd Werner yn gyfarwydd â hi ond, o fyseddu trwy rai ohonynt, a edrychai'n debyg i lyfrau emynau a Beiblau a llyfrau plant bach. Ac ar y wal uwchben y pentyrrau hyn, yr oedd ffotograff mewn ffrâm bren drwchus. Llun o blant ysgol, tybiai Werner, y rhai lleiaf yn gwenu, yr hynaf yn hunanfeddiannol o ddwys. Ynghyd â'u hathrawes a gweinidog. Safent y tu allan i'r bwthyn lle'r oedd Werner ei hun yn sefyll. Gwisgai'r athrawes yn ôl ffasiwn yr oes o'r blaen, ei gwallt wedi'i glymu'n dynn y tu ôl i'w phen. Roedd hefyd ryw dystysgrif, a honno wedi'i fframio yn yr un modd. Ond dim golwg ohoni *hi* yn unman. Nid hwn oedd ei byd *hi*, siawns?

Ar ochr arall y stafell, roedd rhagor o lyfrau, wedi'u sarnu ar hyd y llawr, ynghyd â rhyw gelfi, fel y tybiai Werner, a guddiwyd dan liain wen, er mwyn eu diogelu, siŵr o fod, rhag y llwch a'r haul. Cwpwrdd, efallai. Neu harmoniwm, o bosibl. Un mwy na'r cyffredin, hefyd: roedd y gwrthrych hwn, beth bynnag ydoedd, yn dalach na Werner ei hun.

Cododd gŵr y lliain. A gweld pren. Pren wedi'i baentio'n las a gwyn, a'i fowldio'n gain. Gwaelod cwpwrdd, meddyliai Werner. Wardrob, efallai. Cododd fwy. A gweld rhagor o bren. Ond heb yr un drws na drâr.

Cododd fwy. A gweld mwy o bren. A mwy. A sgidiau... Na, bŵts... A choesau... A defnydd... A breichiau...? A...

Yn araf, yn ofalus, yn betrus, tynnodd y lliain i ffwrdd.

Eisteddodd Werner i lawr wrth y bwrdd a dechrau sgrifennu.

Fy Nhad Annwyl

Ni ddeuthum o hyd i Sabine. Ni fedraf gadarnhau, ychwaith, er holi trwy'r gymdogaeth, a fu hi'n byw yn y tŷ hwn ai peidio. Cyrhaeddodd dy lythyrau, bob un, ond maent yn dal heb eu hagor. A gyrhaeddodd y parsel? A fu Sabine yma erioed? Ni fedraf ateb y cwestiynau hyn gyda sicrwydd.

Ni fedraf ond disgrifio'r hyn a welais yma.

Mae yma gerflun o natur anghyffredin. Cerflun o lanc ifanc ydyw, wedi'i wisgo mewn dillad o ryw oes hynafol, a het blu ar ei ben. Ni wn sut y daeth yma na pham, ond y mae'n sefyll ar sylfaen garreg, ac ar y garreg honno y mae plac pres neu efydd yn dwyn y geiriau canlynol:

Jacques de Vaucanson

Le Fluteur Automate

Dillad o frethyn sydd am y ffigwr hwn, nid dillad wedi'u cerfio. O bren y gwnaed y ffigwr ei hun, mae'n debyg, o'r hyn y gallaf ei weld – sef, yr wyneb a rhan o'r gwddf – a'r rheiny wedi'u paentio'n gelfydd iawn, er bod y paent yn plicio yma a thraw, gan ddangos y pren llwyd oddi tano. Ni fedraf weld y dwylo oherwydd bod menig amdanynt, menig sidan tynn, rwy'n credu, neu ryw ddefnydd cyffelyb sy'n ddigon tenau a meddal i blygu i siâp y bysedd. Y mae'r llanc yn dal ffliwt yn ei ddwylo, nid rhyw degan o beth, ychwaith, ond ffliwt bren gyfan, a'i osgo'n gywir fel petai ar fin canu'r offeryn. Mae ei wefusau, hyd yn oed, yn ffugio'r union embouchure *sydd ei hangen i wneud hynny. A'r bysedd, fel y crybwyllais, wedi'u hystumio hefyd i wneud yr un peth.*

Ond hyn sydd ryfeddaf, 'Nhad. Ar un ochr y plinth y mae dwy res o dyllau crynion, chwe thwll ym mhob rhes, ac ym mhob twll gallaf weld gwerthyd copr, a phob gwerthyd yn loyw lân, er bod ôl traul arnynt. Ac o dan bob twll a gwerthyd y mae enw cân neu alaw, megis 'Da Unten im Tale', neu 'O Mensch, bewein' dein' Sunde gross', neu, weithiau, dim ond y gair 'Gigue' neu 'Gavotte' neu 'Tanz'. Yn un o'r tyllau hyn y mae handlen haearn, a charn pren iddi. Ac o dan yr handlen, gwelaf y geiriau, 'Froher Tag, verlangte Stunden'. Alaw anghyfarwydd i mi, mae'n rhaid cyfaddef.

Mentraf droi'r handlen ac mae'r amhosibl yn digwydd. Daw'r cyfan yn fyw. Yn fyw, ac eto'n wyrthiol fecanyddol hefyd. Yn chwalu'r gwahaniaeth rhwng y ddau. Oherwydd nid blwch cerdd mo hwn, na rhyw gramoffon o'r oes a fu. Y mae e'n anadlu, 'Nhad, y mae e'n anadlu! Ni wn i ba beirianwaith cymhleth sydd ym mherfeddion y creadur hwn, pa feginau a phibau ac olwynion sy'n cuddio o'r golwg, ond gallaf weld y tafod yn symud. Gallaf weld y gwefusau hefyd yn tynhau ac yn ymlacio, yn cau ac yn agor, yn tynnu i mewn ac yn gwthio allan. A'r bysedd! Dylech weld y bysedd, 'Nhad, yn symud mor chwim ac mor gywir. Ac yn fwy na dim, pan roddaf fy llaw o flaen ei geg, gallaf deimlo ei anadl ar fy nghroen. Fel petai'n gerddor byw. A'r fath gerddoriaeth! Pob nodyn yn berffaith, yn ddigyfnewid berffaith, hyd dragywydd, ni waeth pa mor aml y bydd gofyn iddo chwarae'r darn.

Nid hyd dragywydd yn hollol, wrth gwrs, oherwydd gwaith blinderus, i un dyn beth bynnag, yw troi'r handlen a chadw'r cyfan i fynd am fwy na rhyw dair munud ar y tro. Wrth gymryd hoe fach, edrychaf eto ar wneuthuriad y Ffliwtydd Awtomatig hwn, a rhyfeddu at ddyfeisgarwch M. Vaucanson. Sylwaf yn enwedig ar flaenau'r bysedd, lle mae'r bonheddwr hwn, neu ryw olynydd iddo fwy na thebyg, wedi glynu darnau o ledr wrth y menig er mwyn sicrhau bod y bysedd yn cuddio'r tyllau'n llwyr. (Ffliwt baroque o'r iawn ryw, felly, 'Nhad, heb allweddau, neu unrhyw gyfleusterau cyfoes felly.) Ysywaeth, oherwydd gorddefnydd, neu freuder y sidan, neu ryw nam bach ar wneuthuriad y llaw, mae blaen bys bach y faneg dde wedi rhwygo fymryn bach. Er nad yw

*hynny i'w weld yn ymyrryd â safon y perfformiad, y mae'n
bendant yn amharu ar harddwch y ffigwr a hynny am un
rheswm syml. Y mae'r menig, fel y dywedais ynghynt, yn
dynn ond yn feddal. Y maent hefyd o liw nid annhebyg i liw
croen, er mai croen afiach o welw fyddai'r croen hwnnw, rhaid
cyfaddef. Bid a fo am hynny, ymddengys i'r llygad diniwed
fel petai asgwrn y bys bach, ie, yr asgwrn, wedi rhwygo'r
croen gwelw hwn, am mai dyna sydd i'w weld yn ymwthio
trwy'r rhwyg. Asgwrn blaen bys.*

*Rhyfeddaf eto at ymdrechion digymrodedd M. Vaucanson
i efelychu Natur, hyd yn oed yn y manylion lleiaf; hyd
yn oed pan fo'r manylion hynny wedi'u cuddio o'r golwg.
Nid oes pall ar ei gywreindeb. Ac er mwyn gwerthfawrogi'n
well y cywreindeb hwnnw, fe geisiaf dynnu'r menig. Rhaid
cyfaddef i mi deimlo braidd yn chwithig wrth wneud hynny.
Rhyfygus, dyna'r gair. Ond, heb fentro... fel y dywed yr
hen air. Dechreuaf gyda'r llaw chwith, gan fod bysedd y llaw
honno'n digwydd bod yn weddol syth ar y pryd. A'r siom
a gaf! Bysedd pren amrwd, a cholfachau bach pres rhwng y
cymalau. A'r darnau wedi'u clymu wrth ei gilydd gyda rhyw
wifrau hyll a thameidiau bach o ledr. Mae'r cyfan mor ddi-
urddas. Dyfeisgar. Ac eto mor ddienaid. Pwped, nid cerddor,
yw hwn, wedi'r cyfan.*

*Ond beth, wedyn, am y llaw dde? Mae'n rhaid i
mi droi'r handlen eto er mwyn symud yr alaw ymlaen
ryw ychydig. Yr wyf wrthi am funud gron cyn i'r
bysedd ymsythu'n ddigon i mi gael mynd at y faneg yn
ddidramgwydd. Fe'i tynnaf. Ac fe welaf yn union ble mae'r
pren yn peidio a ble mae'r esgyrn yn dechrau.*

Y bysedd, 'Nhad. Bysedd y llaw dde. Esgyrn ydyn nhw.

Y Weinyddiaeth er Hyrwyddo Buddiannau
Strategol y Celfyddydau
ADRAN DRYLOYWDER AC ATEBOLRWYDD

Trawsgrifiad o drafodaethau Is-banel Asesu 2/5b
(Gorllewin a Chanolbarth)
11–14 Mehefin 2005
Sesiwn 7

tt 407–418

Presennol: *Dr Antonia Gwynn (Cadeirydd), Bartholomew Osborne Joskin, Vavasor Iorwerth Powell, Lara Onllwyn Toozle, Llinos Rabaiotti (Swyddog Monitro)*

Ymddiheuriadau: *Ioan Thomas (Llywodraeth Cynulliad Cymru), Michelle Troika (Bwrdd Marchnata Cynnyrch Papur)*

Testun: **Cymorthdaliadau i Artistiaid**

/… gerflun Trubyetskpoi o Tsar Alecsandr III yn arloesol, mae'n dweud fan hyn

BOJ   … ac yn ôl adroddiad yr aseswr allanol…

LOT   … fe fyddai'r cyntaf i wneud gwaith ymchwil manwl ar y ffwrn frics draddodiadol yn Nwyrain Ewrop…

VIP   … yn ddamcaniaethol…

LOT   … ac yn ymarferol

Cad.   Ydyn ni'n gytûn, felly? Ydyn? Wyth deg y cant o'r costau? Naw mil…? A'r amodau…?

Ll R   Ar yr amod bod yr artist yn cynnal pum gweithdy cynhyrchu ffyrnau brics mewn ardaloedd a amddifadwyd o'r cyfryw weithgarwch.

Cad.   O'r gorau. Cais nesaf.

Cais 20049528

Cad.   Mae 'da ni wyth munud i ystyried hwn. Tua chant o eiriau'r un, ddywedwn i. A dyna chwarter fy ngeiriau i wedi mynd yn barod!

BOJ   (yn chwerthin)

VIP   Oes te ar y ffordd, Madam Cadeirydd?

Cad.   Llinos?

Ll R   Trefna' i rywbeth nawr.

BOJ   A bisgedi, os oes modd.

(Gohirir trafod pellach am dair munud)

Cad.   Iawn. Y cais nesaf. Steffan Muller. Yn gofyn am £5,321 er mwyn ymestyn a chwblhau nofel gofiannol. Llinos?

Ll R	Fel 'dych chi'n gwybod, rwy'n siŵr, Steffan oedd fy rhagflaenydd yn y swydd hon…
BOJ	Llaeth?
Ll R	Dim gormod, diolch. Ac mae e wedi anfon drafft…
LOT	Ei hunangofiant o 'di hwn, felly?
Ll R	Cofiant mae e'n ei alw fe, nid hunangofiant, ond Steffan yw enw'r prif gymeriad, os yw hynny…
VIP	Ond ei fywyd e sydd dan sylw yn y llyfr, ife? A'i waith yn y Weinyddiaeth?
Ll R	Ei waith, ie… Ond am ei fywyd… D'on i ddim yn nabod y dyn… 'Alla i ddim…
Cad.	A'r aseswyr allanol? Mae gyda ni dri adroddiad fan hyn.
Ll R	… Am fod elfennau arbenigol yn y…
Cad.	Arbenigol?
Ll R	… Yn ymwneud â cherddoriaeth a mathemateg…
BOJ	Bisgïen?
Ll R	Diolch.
Cad	… A'r trydydd sgrwnsh?
Ll R	Wel, barn sgrwnsh lenyddol, wedyn, sgrwnsh.
Cad.	Burion! Sgrwnsh. Ga'i gofnodi bod pawb wedi darllen yr adroddiadau…? A bod yr adroddiadau'n cael eu hymgorffori yn y cofnodion maes o law? Sgrwnsh. O'r gorau.

408

Cais 20049528

Adroddiad 1

Dr L Oborov, Coleg Cerdd a Drama Cymru

Diolch am y gwahoddiad i gyflwyno sylwadau ar gyfeiriadaeth gerddorol y deipysgrif, *Esgyrn Bach*. Deallaf mai nofel gofiannol yw'r deipysgrif hon, i fod, beth bynnag ydi'r creadur erthyl hwnnw. Os felly, *pwy ydy'r gwrthrych?* Dyna'r cwestiwn a ofynnwch. Ac fe gewch ateb, er bod yr ateb hwnnw'n ddigon amlwg, dybiwn i, ac fe'm synnwyd braidd na fyddai'r Weinyddiaeth wedi gweld ei ffordd yn glir i gael hyd iddo *heb* fynd ar ofyn arbenigwyr allanol. I beth mae dyn yn talu ei drethi, dywedwch?

Boed a fo am hynny, fe wyddoch, mae'n debyg, nad Bach yw'r testun, er gwaethaf y teitl. Yn wir, yn fy marn i, taflu llwch i'n llygaid *yw bwriad* yr holl ribidirês amdano ef a'r *ricecar* a'r organ ac yn y blaen, *er mwyn tynnu ein sylw oddi ar y gwir wrthrych*, i ba bwrpas does gen i'r un llefeleth. A phwy, felly, *ydi'r* gwrthrych hwnnw? Ystyriais y dystiolaeth i gyd, a gorchwyl digon diflas oedd hynny, credwch chi fi. Ydy, mae'r gwaith yn cyfeirio at amryw ddigwyddiadau, damcaniaethau a ffigurau ym myd cerdd, a'r rheiny, at ei gilydd, yn gymaint o lwch i'r llygaid â Herr Bach ei hun. Ond y mae un eithriad, ac un eithriad yn unig: a'r cyfansoddwr, Anton Bruckner, ydy hwnnw. Mae'n wir mai dim ond rhyw grybwyll ei enw wrth fynd heibio a wneir. Ond am y cyfeiriadau cudd a slei – y lled awgrymu, y sibrwd o'r tu ôl i bared sydd mor nodweddiadol o'r llysywen yma o awdur – mae'r rheiny'n rhemp. Yn wir, bron na ddywedwn i fod yna ymdrech ar bob tudalen – a honno'n ymdrech drwsgl,

hefyd, gan amlaf – i'w dynnu i sylw'r darllenydd druan.
Ei gyflwyno heb ei enwi, megis – peth *irritating* ar y naw.
Onid dyna lun ohono wrth yr organ? Onid dyna ddiben
yr holl wamalu am rifau ac esgyrn? Fel y gwyddys, yr
oedd gan Bruckner ei nodweddion hynod, a hynny'n
tystio, ddywedwn i, i'w athrylith unigryw. Lle buasai dyn
cyffredin, diddychymyg, priddlyd wedi *chwantu* menyw
siapus, cyfri'r blodau ar ei ffrog wnâi'r hen Anton, a thrwy
hynny brofi nad oedd yn gaeth i hualau'r cnawd. Tynnaf fy
het iddo! Ac onid yr un ymwybod dwfn â'i feidroldeb ei
hun a'i cymhellai hefyd i ymweld â'r *morgue*, yn enwedig
ar ôl rhyw alanas? Ac i fynd i weld esgyrn Beethoven,
wedyn, yn cael eu tynnu o'r ddaear? (Onid trosiad amrwd
o'r digwyddiad hwnnw yw'r disgrifiad o ddatgladdiad
Bach?)

Ac yn fwy na dim, yr enw. Anton. Oherwydd yn enw'r
cyfansoddwr y ceir yr ateb i'r cwestiwn ynglŷn â gwrthrych
y llyfr hwn. Canys nid un Anton sydd yma, ond tri, ac fel
y dywedodd Charles Dodgson mewn cyd-destun nid
annhebyg, *What I tell you three times is true.* Mae Anton
Bruckner. Ond y mae, yn ogystal, Anton yr hipo, a hwnnw
hefyd yn destun sen a dirmyg. Ac yn drydydd, ac yn bennaf
oll, y mae'r Anton arall hwnnw – Yr Athro Anton Carl
Schwartzenburg. Ie, fy *hen gyfaill a chydweithiwr, Anton
Schwartzenburg*, neb llai. Efe, yn ddiau, ydyw gwir destun
y gyfrol wenwynllyd yma. A gwawdlun plentynnaidd, onid
enllibus, yw'r portread ohono. Nid yw'n crybwyll dim
o'i gampau mawr. Onid yr Athro Schwartzenburg, trwy
drylwyredd ei hymchwiliadau i hanes cerddoriaeth yn Sir
Aberteifi, a ddatgelodd y gyfrinach ffiaidd am y mwncïod
ym Mhlas Nanteos? Ac i beth? I gael ei gyffelybu â hipo? Pa

wobr yw honno am oes o wasanaeth diflino? Wedi dweud hynny, credaf fod mwy o sylwedd yn perthyn i'r creadur bondigrybwyll hwnnw nag i'r un creadur arall yn y lol botes maip mwyaf di-faeth a lyncais i erioed.

Cais 20049528

Adroddiad 2

Dr S. Ramanujan, Adran Gyfrifiadureg, Prifysgol Morgannwg

Parthed y deipysgrif, *esgyrn bach.*

(1) Rhoddais y cyfan trwy'r cyfrifiadur a'i ddadansoddi er mwyn canfod patrymau iteraidd etc. Deuthum o hyd i gant tri deg tri o batrymau estynedig (sef, cyfresi'n cynnwys mwy na phum elfen), ond cyfyngaf fy sylwadau at un patrwm yn unig, sef y patrwm a amlygir yn y sgwrs rhwng Steffan a Llio ar tt 87–9. Fe welwch fod nifer y geiriau ym mrawddegau Steffan yn cynyddu o 1 i 2 i 3 i 8 i 13 i 21 i 35 a brawddegau Llio yn lleihau o 54 i 35 i 21 i 13 i 8 i 5 i 3 i 2 i 1. Rhifau Fibonacci yw'r rhain ac maent yn ymgorffori patrymau a welir yn adeiledd Natur (cregyn y môr, petalau, dail, afalau, etc., etc.). Ond ni farnaf fod hwn, na'r un arall o'r patrymau hyn yn arwyddocaol, yn rhannol am fod pob un ynghudd, y tu hwnt i gyrraedd y darllenydd; ond hefyd am fod unrhyw destun o'r maint hwn yn *sicr* o gynnwys cyfresi a phatrymau o'r fath. Ymhlith 43,496 o eiriau, 199,987 o lythrennau/ ffigurau, 115 o baragraffau, etc. etc., gellid disgwyl rhwng 125 a 140 o'r rhain. Y mae *esgyrn bach* yn ddigon agos at y rhif cyfartaledd, felly. Diffyg 'cyd-ddigwyddiadau' a fyddai'n syfrdanol, nid eu presenoldeb.

(2) Ystyriais, hefyd, a oedd arwyddocâd, nid yn unig mewn cyfresi ond mewn rhifau unigol sy'n cael sylw arbennig

yn y testun, megis nifer y ceisiadau (172) a'r Côd Post
SA20 y mae Prif Weithredwr y Weinyddiaeth yn pledio
ei achos â'r fath sêl, ond methais â dirnad dim o bwys.
Bwriadol ddiystyr yw'r pethau hyn, o bosibl, ond diystyr,
yn bendant.

(3) Pwysicach, efallai, yw'r llun o chwaraewyr snwcer y
mae Steffan yn ei weld yn y *Western Mail*. Petai ond wedi
dilyn y llwybr hwnnw, dichon y byddai wedi dysgu ei wers
yn llawer cynt. Mae camgymeriad tyngedfennol Steffan i'w
weld yn ddirdynnol o glir ar tt. 80 lle dywedir amdano:
'*petai ei frawd, brawd nad oedd wedi'i weld ers deng mlynedd, yn
taro'r bêl wen yn erbyn y bêl goch ar yr un bwrdd snwcer, fel yr
arferai ei wneud yn llawer rhy aml, a'i tharo gyda'r un nerth, o'r
un ongl, yna byddai'r bêl yn cwympo – ploc! – i'r boced bob tro.*'
Symbol o sicrwydd, o'r dibynadwy a'r rhagweladwy, yw'r
peli snwcer iddo ef. Ac ar un ystyr y mae yn llygad ei le.
Dyma lun o ddwy bêl snwcer, llun sy'n dangos y berthynas
dragwyddol ddigyfnewid rhwng ongl y trawiad ac ongl y
symudiad canlynol, rhwng achos ac effaith.

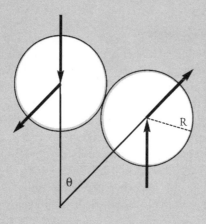

Ond y geometreg yn unig sy'n ddigyfnewid, wrth gwrs. Yn y byd go iawn, mae rhyw elfen o ansicrwydd, waeth pa nor fychan, yn rhwym o effeithio ar bob trawiad. Ac mae'r effaith honno yn cynyddu bob tro mae pêl yn taro pêl. Profwyd hyn i gyd gan Heisenberg. (Ac onid llun ohono sydd yma? Bu'n gerddor brwd ar hyd ei oes.)

Uchafswm o 12 trawiad sydd ei angen, meddai ef, i greu gwyriad o 90° oddi wrth yr ongl derfynol y byddai dyn yn ei ddisgwyl o ddibynnu ar geometreg. *A dyna'r gorau y gallai'r peiriant gorau ei gyflawni o dan yr amgylchiadau gorau.* Cymaint llai dibynadwy yw'r fraich a'r llygad a'r ymennydd meidrol.

Cymeraf mai dilyn hynt un ai Heisenberg neu'r brawd (yr un a fu'n chwarae snwcer) yw bwriad yr ymgeisydd, wrth ymestyn y stori; ac mai dyna yw'r elfen 'gofiannol' y sonnir amdano.

Cais 20049528
Adroddiad 3
Yr Athro B. J. Jones

Pleser yw cynnig sylwadau ar *esgyrn bach*. Diolch hefyd am anfon ataf gopïau o'r adroddiadau eraill. Gofynnais i mi fy hun a oedd y deipysgrif hon yn bodloni prif ofynion pob nofel dda. A yw ei hawdur yn meddu ar arddull groyw, naturiol? A all lunio ymddiddanion rhywiog? A all roi stori at ei gilydd wedi'i chynllunio a'i hamseru'n gymen? A all greu cymeriadau credadwy, byw? Ofnaf mai 'Na' oedd fy ateb i i bob un o'r cwestiynau hyn, ac yn enwedig yr olaf. Nid yw'r awdur yn ymdrafferthu i ddisgrifio ei arwr, hyd yn oed. Beth yw lliw ei wallt? Beth yw ei daldra? A yw'n dew neu'n denau? A oes ganddo

413

goes bren? Nis gwyddom. Ac mae ei gymar yn diflannu o fewn ychydig dudalennau. Byddai rhai yn dweud iddi gael dihangfa ffodus. Dyn a ŵyr ble y cafodd hi'r cyfle i feichiogi, heb sôn am yr awydd.

Ac eto, onid gwendid pennaf y deipysgrif hon yw'r ffaith bod unrhyw ystyr a allasai fod ynddi, er gwaethaf y brychau niferus hyn, yn cael eu lluchio y tu hwnt i gyrraedd y darllenydd mwyaf diwyd? I beth mae'r cyfeiriadau astrus yn dda ond i arddangos clyfrwch coeglyd yr awdur? Cwestiwn Dr Oborov. A chwestiwn da.

A chwestiwn, wedi i mi ail-ddarllen y testun, y tybiwn i y dylwn geisio ei ateb. Oherwydd, er llymed fy meirniadaeth o'r testun hwnnw ar sawl cyfrif, fe ddeuthum, ymhen y rhawg, i weld rhyw rinwedd ynddo, wedi'r cyfan, ac i feddwl amdano mewn ffordd dra gwahanol i'm cyd-ddarllenwyr. Y mae eu dehongliad hwy o'r gwaith, gyda phob parch iddynt, yn gyfyngedig i'w priod feysydd hwy eu hunain ac oherwydd hynny, rhan yn unig o'r darlun a welant, gan golli'r weledigaeth gyfan. Gwelant y dillad, ond nid y corff.

Yn *esgyrn bach*, gwelwn ddechrau tynnu'r dillad hynny. Ys dywedodd y Pêr Ganiedydd, 'Ffarwel weledig, groesaw anweledig bethau': geiriau a ddylai fod yn arwyddair i bawb sy'n ymhél â Natur a'r celfyddydau fel ei gilydd. Oblegid, nid oes yn y byd hwn yr un eiliad pryd nad effeithir ar y gweledig gan yr anweledig. Gollyngwn y diarhebol afal. Ac wele, am reswm cudd ac anwel, y mae'n cwympo. Disgyrchiant sy'n llechu dan ein traed ac yn tynnu'r tamaid gostyngedig hwnnw i'r llawr. Rhan annatod o'r *cyfan*, felly, yw'r anweledig, ac ni cheir dealltwriaeth *gyfansawdd* heb gofleidio'r egwyddor hwn.

Ystyrier Yr Eneiniog ym mhennod 5. Yn ôl yr awdur, y mae'r truan hwn yn melltithio nid dim ond *englynion* am yr hipo, yr anifail sydd mor wrthun ganddo, ond hefyd *limrigau*. Fe gofiwch i un o'r englynion gael ei ddyfynnu yn *esgyrn bach*; onid arwyddocaol, felly, *na* ddyfynnir limrig hefyd? Onid dyma arwydd *gweledig* yn dangos y ffordd tuag at yr *anweledig*? Bu hynny'n ddigon o sbardun i'r darllenydd hwn geisio llenwi'r bwlch. A dyma'r hyn a ganfûm o chwilio'r cyfuniad, *limrig* + *hipo*.

> Neithiwr mi welais, trwy'r dagrau,
> y lloer yn cusanu'r glannau,
> a draw yn y llaid,
> hen hipo fy nhaid
> yn chwilio ei gariad yntau.

Testunol, felly, o gofio cariadon coll Steffan a thad Werner. Ond fe â'r testunoldeb yn ddyfnach na hynny, a dim ond beirniadaeth gynhwysfawr, gyfansawdd fedr blymio'r dyfnderoedd hynny. Mae'r mathemategydd yn amau a yw awdur *esgyrn bach* yn ymwybodol o'r patrymau rhifol yn ei waith. Ac oherwydd ei sgeptigaeth, nid yw'n dilyn y llwybr i'r pen. Dyma fi'n dilyn yr un llwybr yn union, ac yn cyrchu'r nod!

Gadewch inni ystyried y limrig yn fwy manwl.

Y mae yma **un** gerdd.

Mae ynddi **ddwy** linell fer a **thair** llinell hir.

Mae'r llinellau byrion yn cynnwys **pum** sill yr un; y rhai hirion, **wyth**.

Ceir yn y gerdd **dair sill ar ddeg** sy'n acennog, **un ar hugain** sy'n ddiacen: cyfanswm, felly, **o bedair sill ar ddeg ar hugain**.

Sy'n rhoi inni'r gyfres: **1, 2, 3, 5, 8, 13, 21, 34.** Cyfres Fibonacci.

Y limrig absennol hwn, maentumiaf i, yw craidd y deipysgrif, ac yn ei swildod swynol yr ydys yn ymwybodol odiaeth o'r Tafod cudd sydd yn rhagflaenu Mynegiant, o'r cariad cynhaliol sy'n clymu'r ddaear a'r môr, y cariad y mae'r bardd yn ei ddagrau a hen hipo ei daid, ill dau, yn ei chwennych.

Ai awdur y limrig, felly, yw gwrthrych y cofiant? Ie, yn bendant, ddywedwn i: yr awdur absennol, anweledig. Yr awdur absennol sydd yn llechu y tu hwnt i'r testun. Oherwydd nid **un** yw man cychwyn cyfres Fibonacci. **DIM** yw'r gwraidd. **0. Sero.** Ac yn *esgyn bach* yr allwedd

Cad.   'Bennodd adroddiad Yr Athro Jones braidd yn ddisymwth, Llinos, naddo?

Ll R   Oherwydd y cyfyngiad…

Cad   …Y cyfyngiad?

Ll R   …Ar hyd adroddiadau… 735 o eiriau yw'r uchafswm ar hyn o bryd, er bod disgwyl i hynny gael ei gwtogi eto… Bu'r Athro Jones wastad yn dueddol o fynd dros y terfyn…

Cad.   O'r gorau…. Beth yw'r ffactorau asesu?

Ll R   Mae *esgyrn bach* yn sgorio yn uchel o ran portreadu Cymru mewn cyd-destun Ewropeaidd…

BOJ   Pwysig…

Ll R   …Ac o ran codi pontydd rhwng y celfyddydau

VIP   Hanfodol…

LL R	…Ond nid yw'n addas ar gyfer Cyfnod Allweddol 4… sydd yn flaenoriaeth gynnon ni…
Cad.	Na?
Ll R	…Ers i'r gweisg wahardd defnydd o'r ffurfiau amhersonol… a'r modd dibynnol… a geiriau astrus… fel…
VIP	… Fel 'astrus'!
BOJ	(yn chwerthin)
VIP	Na allen nhw… eu… eu…
BOJ	… eu carthu nhw o'r testun?
Ll R	Gallen. Ond mae 'na ystyriaeth arall sy'n bwysicach na hynny, mae arna i ofn… Dyw ein Graddfa Ddifreintiedigrwydd ddim yn rhoi blaenoriaeth i ardal bost yr ymgeisydd. A dweud y gwir, dyw e'n sgorio dim, hyd y gwela'i…
Cad.	Mm… A pha ardal yw honno, Llinos?
Ll R	SA dau ddeg, Cadeirydd
BOJ	Mm?
Ll R	Tyddyn ar ochr mynydd rywle ar bwys Llanymddyfri…
BOJ	A!
Ll R	Ardal denau iawn ei phoblogaeth… Rhy denau i fod yn rhan o'r Strategaeth Hygyrchedd… na'r Strategaeth Adfywio'r Gymuned Trwy Gydweithredu Traws-ddiwylliannol… na'r Strat
VIP	Rhy denau i besgi, felly!
BOJ	(yn chwerthin)
Cad.	Ond 'na ddywedodd rhywun yn rhywle bod SA

dau ddeg yn ardal a oedd i gael ei ffafrio gan y Weinyddiaeth...?

LI R   Do... Yn y nofel...

VIP   A!... Yn y nofel... Wel..

BOJ   Wel... 'Na fe, te.

Cad.   ... Ond bod y Prif Weithredwr ei hunan...

LIR   ... Ac efallai bod y Prif Weithredwr, ac yntau'n sigledig ei Gymraeg, wedi dweud 'dau ddeg' pan oedd e mewn gwirionedd yn golygu... 'deuddeg'.

Cad.   Deuddeg...?

LI R   SA deuddeg... Un deg dau... Ardal neilltuol o ddifreintiedig... Port Talbot, Castell Nedd... Ffor' 'na...

Cad.   Wela i... Wela i... Mae'n ymddangos, felly, bod Steffan druan wedi gosod ei nofel yn y lle anghywir...

LI R   ... Ac wedi symud i fyw i'r lle anghywir.

Cad.   Dim, felly, i gais 20049528... Ydyn ni'n gytûn?

BOJ   Dim.

VIP   Dim.

Am restr gyflawn o nofelau cyfoes Y Lolfa,
a'n holl lyfrau eraill, mynnwch gopi o'n
Catalog newydd, rhad – neu hwyliwch i
mewn i'n gwefan

www.ylolfa.com

i chwilio ac archebu ar-lein.

TALYBONT CEREDIGION CYMRU SY24 5AP
e-bost ylolfa@ylolfa.com
gwefan www.ylolfa.com
ffôn (01970) 832 304
ffacs 832 782